Prospects of Differential Geometry and its Related Fields

Prospects of Differential Geometry and its Related Fields

Proceedings of the 3rd International Colloquium on Differential Geometry and its Related Fields

Veliko Tarnovo, Bulgaria 3 – 7 September 2012

editors

Toshiaki Adachi

Nagoya Institute of Technology, Japan

Hideya Hashimoto

Meijo University, Japan

Milen J Hristov

St Cyril and St Methodius University of Veliko Tarnovo, Bulgaria

NEW JERSEY • LONDON • SINGAPORE • BEIJING • SHANGHAI • HONG KONG • TAIPEI • CHENNAI

Published by

World Scientific Publishing Co. Pte. Ltd.
5 Toh Tuck Link, Singapore 596224
USA office: 27 Warren Street, Suite 401-402, Hackensack, NJ 07601
UK office: 57 Shelton Street, Covent Garden, London WC2H 9HE

British Library Cataloguing-in-Publication Data
A catalogue record for this book is available from the British Library.

Cover image: Samovdoska Charshiya, Veliko Tarnovo. Photographed by T. Adachi.

PROSPECTS OF DIFFERENTIAL GEOMETRY AND ITS RELATED FIELDS
Proceedings of the 3rd International Colloquium on Differential Geometry and its Related Fields

ISBN 978-981-4541-80-0

Desk Editor: Lai Fun Kwong
Cover Designer: Hui Chee Lim

Printed in Singapore

PREFACE

The *3rd International Colloquium on Differential Geometry and its Related Fields* (ICDG2012) was held at St. Cyril and St. Methodius University of Veliko Tarnovo, Bulgaria, during the period 3–7 September, 2012. During this colloquium, participants gave surveys on recent developments in the fields of differential geometry, mathematical physics and coding theory and explain their results by describing the background why they came to study their problems.

This book contains selected contributions covering Riemannian metrics on Lie groups, harmonic maps, minimal surfaces, exceptional geometry, submanifold theory, geometric treatments of magnetic fields, information geometry, geometric designs, coding theory and so on. The editors expect these articles to provide significant information for researchers and to serve as a good guide to graduate students and provide insights on further development of studies in differential geometry and its related fields.

We gratefully acknowledge the financial support for the conference activities by Daiko Foundations and Nagoya Institute of Technology. We appreciate very much the scientific reviewers who read the articles carefully and gave many important suggestions to authors. Finally, we express our hearty thanks to the staff of St. Cyril and St. Methodius University for their hospitality. We note that St. Cyril and St. Methodius University had built a new faculty, the Faculty of Mathematics and Informatics, in 2011. We hope our series of colloquium will aid in the development of this faculty and promote friendship between Bulgarian scientists and scientists from all over the world.

The Editors
17 May, 2013

The 3rd International Colloquium on Differential Geometry and its Related Fields

3–7 September, 2012 – Veliko Tarnovo, Bulgaria

ORGANIZING COMMITTEE

T. Adachi – Nagoya Institute of Technology, Nagoya, Japan
S. H. Bouyuklieva – St. Cyril and St. Methodius University of Veliko Tarnovo, Veliko Tarnovo, Bulgaria
H. Hashimoto – Meijo University, Nagoya, Japan
M. J. Hristov – St. Cyril and St. Methodius University of Veliko Tarnovo, Veliko Tarnovo, Bulgaria

SCIENTIFIC ADVISORY COMMITTEE

G. Ganchev – Bulgarian Academy of Sciences, Sofia, Bulgaria
K. Sekigawa – Niigata University, Niigata, Japan
T. Sunada – Meiji University, Tokyo, Japan

PRESENTATIONS

1. **Stefka Bouyuklieva, Emil Arsov & Zlatko Varvanov** (Univ. Veliko Tarnovo),
   Applications of the Gaussian integers in coding theory
2. **Hajime Urakawa** (Tohoku Univ.),
   Geometry of biharmonic maps
   – existence, non-existence and classification problems of proper biharmonic maps –
3. **Milen Hristov & Slavi Hadjiivanov** (Univ. Veliko Tarnovo),
   Affine and conformal transformations of rational Bézier curves
4. **Nikolay Ivanov** (Univ. Veliko Tarnovo),
   String theory and noncommutative geometry
5. **Milen Hristov** (Univ. Veliko Tarnovo),
   Trajectories of rational Bézier curves and its convolutions
6. **Pengfei Bai** (Nagoya Inst. Tech.),
   Estimates of volumes of trajectory-balls for Kähler magnetic fields
7. **Misa Ohashi** (Osaka City Univ.),
   On the G_2-congruence classes of curves in the purely imaginary octonions
8. **Yusuke Sakane** (Osaka Univ.),
   Homogeneous Einstein metrics on generalized flag manifolds
   – cases with G_2-type t-roots –
9. **Peter Leifer** (Crimea State Univ.),
   The role of the $CP(N-1)$ geometry in the intrinsic unification of the general relativity and QFT
10. **Georgi Ganchev** (Bulgarian Acad. Sci.),
    Surfaces in the four-dimensional Euclidean or Minkowski space
11. **Midori Goto** (Fukuoka Inst. Tech.),
    On generalized Liouville manifolds

12. **Toshiaki Adachi** (Nagoya Inst. Tech.),
 Trajectories for Sasakian magnetic fields and circles on a nonflat complex space form
13. **Mancho Manev & Miroslava Ivanova** (Plovdiv Univ.),
 Natural connections with torsion expressed by the metric tensors on almost contact manifolds with B-metric and curvature tensors of φ-Kähler type
14. **Mancho Manev** (Plovdiv Univ.),
 Canonical type connections and natural connections with totally skew-symmetric torsion on almost contact manifolds with B-metric
15. **Galia Nakova** (Univ. Veliko Tarnovo),
 Radial transversal lightlike hypersurfaces of almost complex manifolds with Norden metric
16. **Keiko Uohashi** (Tohoku Gakuin Univ.),
 α–connections on level surfaces in a Hessian domain
17. **Hiroshi Matsuzoe** (Nagoya Inst. Tech.),
 Geometry of deformed exponential families
18. **Hideya Hashimoto** (Meijo Univ.),
 On decomposition of Stiefel manifolds by G_2 and $Spin(7)$
19. **Alexander Petkov & Stefan Ivanov** (Sofia Univ.),
 The sharp lower bound of the first eigenvalue of the sub-Laplacian on a quaternionic contact manifold
20. **Ivan Minchev, Dimiter Vassilev & Stefan Ivanov** (Sofia Univ.),
 Quaternionic hypersurfaces of hyper-Kähler manifolds

St. Cyril and St. Methodius University, September 2012

CONTENTS

Proceedings of the 3rd International
Colloquium on Differential Geometry
and its Related Fields
Veliko Tarnovo, September 3–7, 2012

GEOMETRY OF BIHARMONIC MAPS: L^2-RIGIDITY, BIHARMONIC LAGRANGIAN SUBMANIFOLDS OF KÄHLER MANIFOLDS, AND CONFORMAL CHANGE OF METRICS

Hajime URAKAWA

Institute for International Education, Tohoku University,
Sendai, 980-8576, Japan
E-mail: urakawa@math.is.tohoku.ac.jp

This article explains our recent contributions to studies of biharmonic maps concentrated to three topics, (1) our recent results on L^2- rigidity of biharmonic maps with S. Gudmundsson, N. Nakauchi, on the generalized Chen's conjecture (cf. [16]), (2) some classification of biharmonic Lagrangian submanifolds in complex space form generalizing Sasahara's results (cf. [21]) due to a joint work with Sh. Maeta (cf. [18]), and (3) conformal change of metrics and biharmonic maps due to a joint work with H. Naito (cf. [19]).

Keywords: Harmonic map; Biharmonic map; Generalized Chen's conjecture; Lagrangian submanifold; Conformal change.

1. Introduction and the generalized Chen's conjecture

Theory of harmonic maps was initiated by Eells and Sampson in 1964 (cf. [14]) and plays a central role in geometry. A mapping φ from a Riemannian manifold (M, g) into another Riemannian manifold (N, h) is *harmonic* if it is a critical point of the energy functional

$$E(\psi) = \frac{1}{2}\int_M |d\psi|^2 \, v_g$$

among the space $C^\infty(M, N)$ of C^∞-mappings from M into N. Namely, for every one parameter family $\varphi_t : M \to N$ $(\epsilon < t < \epsilon)$ with $\varphi_0 = \varphi$,

$$\left.\frac{d}{dt}\right|_{t=0} E(\varphi_t) = -\int_M \langle \tau(\varphi), V \rangle \, v_g = 0,$$

where V is a variation vector field along φ defined by

$$V(x) = \left.\frac{d}{dt}\right|_{t=0} \varphi_t(x) \in T_{\varphi(x)}N \quad (x \in M).$$

and, $\tau(\varphi)$ is the *tension field* defined by

$$\begin{aligned}\tau(\varphi) &= \mathrm{Trace}_g(\nabla d\varphi)\\ &= \sum_{i=1}^{m}(\widetilde{\nabla}_{e_i}d\varphi)(e_i)\\ &= \sum_{i=1}^{m}\{\overline{\nabla}_{e_i}(d\varphi(e_i)) - d\varphi(\nabla_{e_i}e_i)\}\\ &= -\delta(d\varphi), \qquad (1)\end{aligned}$$

where $\{e_i\}_{i=1}^m$ is a local orthonormal frame field on (M,g), ∇, ∇^N are Levi-Civita connections of (M,g), (N,h), and $\overline{\nabla}$, $\widetilde{\nabla}$ are the induced connections on $\varphi^{-1}TN$ from ∇^N and on $T^*M\otimes\varphi^{-1}TN$ from ∇ and $\overline{\nabla}$, respectively. Then, φ is harmonic if and only if $\tau(\varphi)=0$.

The second variational formula of the energy functional $E(\psi)$ was given by R.T. Smith in 1975 (cf. [25]), i.e., if $\varphi:(M,g)\to(N,h)$ is a harmonic map, then

$$\left.\frac{d^2}{dt^2}\right|_{t=0} E(\varphi_t) = \int_M \langle J(V), V\rangle\, v_g,$$

where J is the second order elliptic operator called the *Jacobi operator* defined by

$$J(V) = \overline{\Delta}V - \mathfrak{R}(V). \quad (V\in\Gamma(\varphi^{-1}TN)).$$

Here,

$$\overline{\Delta}V = -\sum_{i=1}^{m}(\overline{\nabla}_{e_i}\overline{\nabla}_{e_i}V - \overline{\nabla}_{\nabla_{e_i}e_i}V),$$

called the *rough Laplacian*, and

$$\mathfrak{R}(V) = \sum_{i=1}^{m} R^N(V, d\varphi(e_i))d\varphi(e_i).$$

A study of biharmonic maps was begun by J. Eells and L. Lemaire in 1983 (cf. [13]) and also G-Y. Jiang in 1986 (cf. [17]). A smooth map $\varphi:(M,g)\to(N,h)$ is called *biharmonic* if it is a critical point of the *2-energy* E_2 which is defined by

$$E_2(\psi) = \frac{1}{2}\int_M |\tau(\varphi)|^2\, v_g = \frac{1}{2}\int_M |\delta d\psi|^2\, v_g.$$

The first and second variational formulas were given by G-Y. Jiang (cf. [17]) as follows.

Theorem 1.1 (cf. [17]). *The first variational formula is given as*

$$\left.\frac{d}{dt}\right|_{t=0} E_2(\varphi_t) = -\int_M \langle \tau_2(\varphi), V\rangle\, v_g, \tag{2}$$

where $\tau_2(\varphi)$ is called the bi-tension field defined by $\tau_2(\varphi) = J(\tau(\varphi))$.

So, one can define biharmonic map as follows.

Definition 1.1. A smooth map $\varphi : (M, g) \to (N, h)$ is said to be *biharmonic* if

$$\tau_2(\varphi) = J(\tau(\varphi)) = 0. \tag{3}$$

Then, the second variational formula of the bienergy is given as follows.

Theorem 1.2 (cf. [17]). *Let $\varphi : (M, g) \to (N, h)$ be a biharmonic map. Then,*

$$\left.\frac{d^2}{dt^2}\right|_{t=0} E_2(f_t) = \int_M \langle J_2(V), V\rangle v_g, \tag{4}$$

where $J_2(V) = J(J(V)) - \mathfrak{R}_2(V)$ and

$$\begin{aligned}\mathfrak{R}_2(V) = R^N &\ (\tau(\varphi), V)\tau(\varphi)\\ &+\ 2\,\mathrm{tr}_g(R^N(d\varphi(\cdot), \tau(\varphi))\overline{\nabla}.V) + 2\,\mathrm{tr}_g(R^N(d\varphi(\cdot), V)\overline{\nabla}.\tau(\varphi))\\ &+\ \mathrm{tr}_g((\nabla^N_{d\varphi(\cdot)}R^N)(d\varphi(\cdot), \tau(\varphi))V)\\ &+\ \mathrm{tr}_g((\nabla_{\tau(\varphi)}R^N)(d\varphi(\cdot), V)d\varphi(\cdot)).\end{aligned}$$

By the definitions, every harmonic map is always biharmonic. It is a natural question to ask when the converse is true. B-Y. Chen proposed (cf. [5]) the following problem.

B-Y. Chen's conjecture: *Let $\varphi : (M, g) \to (\mathbb{R}^n, g_0)$ be a biharmonic isometric immersion into the standard Euclidean space $(\mathbb{R}^n, g_0)$. Then, it is harmonic, i.e., minimal.*

He also showed it is true if $m = \dim M = 2$ (cf. [5]).

The following generalized B-Y. Chen's conjecture was proposed by R. Caddeo, S. Montaldo and P. Piu (cf. [4]), and its negative answer was given by Y-L. Ou and L. Tang (cf. [20]) when (M, g) is an incomplete manifold.

The generalized B-Y. Chen's conjecture: *Let $\varphi : (M, g) \to (N, h)$ be a biharmonic isometric immersion whose sectional curvature of the target space (N, h) is non-positive. Then, it is harmonic, i.e., minimal.*

2. L^2-rigidity theorem of biharmonic maps

In this section, we show that the generalized B-Y. Chen's conjecture is true under the assumption with finite bienergy and finite energy (cf. [16]).

Theorem 2.1. *Let (M, g) be a complete Riemannian manifold, and (N, h) a Riemannian manifold whose sectional curvature is non-positive.*

(1) *Assume that $\varphi : (M, g) \to (N, h)$ is a biharmonic map with finite energy $(\varphi) < \infty$ and finite bienergy $E_2(\varphi) < \infty$. Then, it is harmonic.*

(2) *In the case* $\mathrm{Vol}(M, g) = \infty$, *the condition $E(\varphi) < \infty$ is needless. I.e., if $\varphi : (M, g) \to (N, h)$ is a biharmonic map with finite bienergy $E_2(\varphi) < \infty$, then it is harmonic.*

(*Sketch of Proof*) Take a fixed point $x_0 \in M$, any positive number $0 < r < \infty$ and a cutoff function η on M, i.e., $0 \leq \eta(x) \leq 1$ $(x \in M)$, $\eta(x) = 1$ $(x \in B_r(x_0))$, $\eta(x) = 0$ $(x \notin B_{2r}(x_0)$, and $|\nabla \eta| \leq \frac{2}{r}$ $(x \in M)$, where $B_r(x_0)$ is a metric ball of radius r centered at x_0. Let $\varphi : (M, g) \to (N, h)$ be a biharmonic map, i.e.,

$$\tau_2(\varphi) = \overline{\Delta}(\tau(\varphi)) - \sum_{i=1}^{m} R^N(\tau(\varphi), d\varphi(e_i))d\varphi(e_i) = 0.$$

Since the sectional curvature of (N, h) is non-positive, we have

$$\int_M \langle \overline{\Delta}(\tau(\varphi)), \eta^2\, \tau(\varphi)\rangle\, v_g = \int_M \eta^2 \sum_{i=1}^{m} \langle R^N(\tau(\varphi), d\varphi(e_i))d\varphi(e_i), \tau(\varphi)\rangle\, v_g \leq 0. \tag{5}$$

Since $\overline{\Delta} = \overline{\nabla}^*\, \overline{\nabla}$, we have

$$\begin{aligned}
0 &\geq \int_M \langle \overline{\Delta}(\tau(\varphi)), \eta^2\, \tau(\varphi)\rangle\, v_g \\
&= \int_M \langle \overline{\nabla}(\tau(\varphi), \overline{\nabla}(\eta^2\, \tau(\varphi))\rangle\, v_g \\
&= \int_M \sum_{i=1}^{m} \langle \overline{\nabla}_{e_i} \tau(\varphi), \overline{\nabla}_{e_i}(\eta^2\, \tau(\varphi))\rangle\, v_g \\
&= \int_M \sum_{i=1}^{m} \left\{ \eta^2\, \langle \overline{\nabla}_{e_i} \tau(\varphi), \overline{\nabla}_{e_i} \rangle + e_i(\eta^2)\, \langle \overline{\nabla}_{e_i} \tau(\varphi), \tau(\varphi)\rangle \right\} v_g \\
&= \int_M \eta^2 \sum_{i=1}^{m} |\overline{\nabla}_{e_i} \tau(\varphi)|^2\, v_g + 2 \int_M \sum_{i=1}^{m} \langle \eta\, \overline{\nabla}_{e_i} \tau(\varphi), e_i(\eta)\, \tau(\varphi)\rangle\, v_g.
\end{aligned}$$

Thus, we have

$$\int_M \eta^2 \sum_{i=1}^m |\overline{\nabla}_{e_i}\tau(\varphi)|^2 \, v_g \leq -2 \int_M \sum_{i=1}^m \langle \eta \, \overline{\nabla}_{e_i}\tau(\varphi), e_i(\eta) \, \tau(\varphi)\rangle \, v_g$$
$$\leq \epsilon \int_M \sum_{i=1}^m |V_i|^2 \, v_g + \frac{1}{\epsilon} \int_M \sum_{i=1}^m |W_i|^2 \, v_g. \quad (6)$$

Because, by letting $V_i = \eta \, \overline{\nabla}_{e_i}\tau(\varphi)$ and $W_i = e_i(\eta) \, \tau(\varphi)$, it holds that

$$\pm 2 \, \langle V_i, W_i\rangle \leq \epsilon \, |V_i|^2 + \frac{1}{\epsilon} \, |W_i|^2 \quad (7)$$

for all positive number $\epsilon > 0$. Then, we put $\epsilon = \frac{1}{2}$, and then we have

$$\int_M \eta^2 \sum_{i=1}^m |\overline{\nabla}_{e_i}\tau(\varphi)|^2 \, v_g \ \leq \ \frac{1}{2} \int_M \eta^2 \sum_{i=1}^m |\overline{\nabla}_{e_i}\tau(\varphi)|^2 \, v_g$$
$$+ 2 \int_M \sum_{i=1}^m e_i(\eta)^2 \, |\tau(\varphi)|^2 \, v_g, \quad (8)$$

which implies that

$$\int_M \eta^2 \sum_{i=1}^m |\overline{\nabla}_{e_i}\tau(\varphi)|^2 \, v_g \leq 4 \int_M |\nabla \eta|^2 \, |\tau(\varphi)|^2 \, v_g \leq \frac{16}{r^2} \int_M |\tau(\varphi)| \, v_g. \quad (9)$$

Here, since (M, g) is complete and non-compact, we can let r tend to infinity. Then the right hand side of (9) goes to zero since $E_2(\varphi) = \frac{1}{2} \int_M |\tau(\varphi)| \, v_g < \infty$. Since $\eta = 1$ on $B_r(x_0)$, the left hand side of (9) goes to $\int_M \sum_{i=1}^m |\overline{\nabla}_{e_i}\tau(\varphi)|^2 \, v_g$. Therefore, we obtain

$$\int_M \sum_{i=1}^m |\overline{\nabla}_{e_i}\tau(\varphi)|^2 \, v_g = 0. \quad (10)$$

Thus, we have

$$\overline{\nabla}_X \tau(\varphi) = 0, \quad (11)$$

in particular, we have $|\tau(\varphi)|$ is constant, say c, since

$$X \, |\tau(\varphi)|^2 = 2\langle \overline{\nabla}_X \tau(\varphi), \tau(\varphi)\rangle = 0.$$

Then, if $\mathrm{Vol}(M, g) = \infty$, and if $c \neq 0$,

$$\tau_2(\varphi) = \frac{1}{2} \int_M |\tau(\varphi)|^2 \, v_g = \frac{c^2}{2} \, \mathrm{Vol}(M, g) = \infty,$$

which contradicts to the finiteness of bienergy, so that c must be zero, i.e., φ is harmonic. We have (2).

For (1), we consider a 1-form α on M defined by

$$\alpha(X) = \langle d\varphi(X), \tau(\varphi)\rangle,$$

for all vector fields X on M. Then, one can prove that $\int_M |\alpha|\, v_g < \infty$ and $-\delta\alpha = |\tau(\varphi)|^2$. Therefore, by Gaffney's theorem, we have

$$0 = \int_M (-\delta\alpha)\, v_g = \int_M |\tau(\varphi)|^2\, v_g,$$

which implies $\tau(\varphi) = 0$. □

We have two applications of Theorem 2.1. We first consider an isometric immersion $\varphi : (M, g) \to (N, h)$. Then, $\tau(\varphi) = m\,\xi$ where $m = \dim(M)$ and ξ is the mean curvature normal vector field. Therefore, we obtain

Corollary 2.1. *Let (M, g) be a complete Riemannian manifold and (N, h) be a Riemannian manifold whose sectional curvature is non-positive. Assume that an isometric immesion $\varphi : (M, g) \to (N, h)$ is biharmonic, and $\int_M |\xi|^2\, v_g < \infty$. Then, $\xi = 0$, φ is minimal.*

This Corollary 2.1 shows that the generalized B-Y. Chen's conjecture is true under the condition that $\int_M |\xi|^2\, v_g < \infty$.

The next application is to horizontally conformal submersions. Z-P. Wang and Y-L. Ou showed ([26])

Theorem 2.2. *Let φ be a Riemannian submersion of a space form $(M^m(c), g_0)$ of constant sectional curvature c into a Riemannian surface (N^2, h). If φ is biharmonic, then it is harmonic.*

We want to treat with *horizontally conformal submersions* onto a higher dimensional Riemannian manifold (N, h). Let $\varphi : (M, g) \to (N, h)$ be a submersion such that the tangent space T_xM has the orthogonal direct decomposition:

$$T_xM = V_x \oplus H_x,$$

where $V_x = \mathrm{Ker}(d\varphi_x)$, $d\varphi_x$ is an isomorphism from H_x onto $T_{\varphi(x)}N$ and there exists a C^∞ function λ on M, called *dilation*, satisfying by the definition that

$$h(d\varphi_x(X), d\varphi_x(Y)) = \lambda^2(x)\, g(X, Y), \quad X, Y \in H_x.$$

In this case, it is known (cf. Baird and Eells [1]) that the tension field is given by

$$\tau(\varphi) = \frac{n-2}{2}\, \lambda^2\, d\varphi\left(\mathrm{grad}_H\left(\frac{1}{\lambda^2}\right)\right) - (m-n)\, d\varphi(\hat{\mathbf{H}}), \tag{12}$$

where grad_H is the H-component of grad and $\hat{\mathbf{H}}$ is the trace (with respect to g) of the second fundamental form of each fiber $\phi^{-1}(y)$ $(y \in N)$ of φ. The map φ is said to be *horizontally homothetic* if the dilation λ is constant along horizontally curves in M.

To see the difficulties of horizontally conformal submersions of the higher dimensional target space $n = \dim N \geq 3$, recall the following theorem due to Baird and Eells [1]:

Theorem 2.3. *Let* $\varphi : (M^m, g) \to (N^n, h)$ *be a horizontally conformal submersion* $(m > n \geq 2)$. *If*

(1) $n = 2$, *then* φ *is harmonic if and only if* φ *has minimal fibers,*

(2) $n \geq 3$, *two of the following conditions imply the other:*

 (a) φ *is a harmonic map,*

 (b) φ *has minimal fibers,*

 (c) φ *is horizontally homothetic.*

Then, our theorem is

Theorem 2.4. *Let* (M, g) *be a complete Riemannian manifold, and* (N, h), *a Riemannian manifold whose sectional curvature is non-positive, and* $\varphi : (M, g) \to (N, h)$, *a horizontally conformal submersion. Assume that*

$$\int_M \lambda^2 \left| \frac{n-2}{2} \lambda^2 \, \mathrm{grad}_H \left(\frac{1}{\lambda^2} \right) - (m-n) \, \hat{\mathbf{H}} \right|_g^2 v_g < \infty,$$

and either $\int_M \lambda^2 \, v_g < \infty$ *or* $\mathrm{Vol}(M, g) = \infty$. *If* φ *is biharmonic, then it is harmonic.*

Then, we have also

Corollary 2.2. *Let* (M, g) *be a complete Riemannian manifold,* (N, h), *a Riemannian surface whose sectional curvature is non-positive, and* $\varphi : (M, g) \to (N, h)$, *a horizontally conformal submersion. Assume that* $\int_M \lambda^2 \, |\hat{\mathbf{H}}|_g{}^2 \, v_g < \infty$ *and either* $\int_M \lambda^2 \, v_g < \infty$ *or* $\mathrm{Vol}(M, g) = \infty$. *If* φ *is biharmonic, then it is harmonic.*

3. Lagrangian submanifolds of Kähler manifolds

In this section, we state results of biharmonic Lagrangian submanifolds in Kähler manifolds due to a joint work with Sh. Maeta (cf. [18]), and also T. Sasahara's work on this topics (cf. [21–24]).

Let $\varphi : (M, g) \to (N, h)$ be an isometric immersion. Then, the Gauss formula is

$$\nabla^N_X Y = d\varphi(\nabla_X Y) + B(X, Y),$$

for all vector fields X and Y on M, and B denotes the second fundamental form. The Weingarten formula is

$$\nabla^N_X \xi = -A_\xi X + \nabla^\perp_X \xi, \quad X \in \mathfrak{X}(M),\ \xi \in \Gamma(TM^\perp),$$

where A_ξ is the shape operator for a normal vector field ξ on M and $\nabla^\perp$ stands for the normal connection of the normal bundle on M in N. The second fundamental form and the shape operator are related by

$$\langle B(X, Y), \xi\rangle = \langle A_\xi X, Y\rangle.$$

The equations of Gauss and Codazzi are given by

$$\begin{aligned} \langle R^N(X, Y)Z, W\rangle &= \langle R(X, Y)Z, W\rangle + \langle A_{B(X,Z)}Y, W\rangle, \\ (\nabla^\perp_X B)(Y, Z) &= (\nabla^\perp_Y B)(X, Z) + \langle A_{B(Y,Z)}X, W\rangle, \end{aligned}$$

where $\nabla^\perp B$ is given by

$$(\nabla^\perp_X B)(Y, Z) = \nabla^\perp_X(B(Y, Z)) - B(\nabla_X Y, Z) - B(Y, \nabla_X Z).$$

If $\varphi : (M, g) \to (N, h)$ is a biharmonic isometric immersion, then M is called a *biharmonic submanifold.* In this case, the tension field satisfies $\tau(\varphi) = m\,\mathbf{H}$, where $\mathbf{H}$ is the harmonic mean curvature vector field along φ (we wrote it by ξ in the previous section). The bitension field $\tau_2(\varphi)$ is rewritten as

$$\tau_2(\varphi) = m\left\{\Delta\,\mathbf{H} - \sum_{i=1}^m R^N(\mathbf{H}, d\varphi(e_i))d\varphi(e_i)\right\}, \tag{13}$$

and φ is biharmonic if and only if

$$\Delta\,\mathbf{H} - \sum_{i=1}^m R^N(\mathbf{H}, d\varphi(e_i))d\varphi(e_i) = 0. \tag{14}$$

Let us recall some fundamental facts on Lagrangian submanifolds in Kähler manifolds following Chen and Ogiue [12]. Let $(N^m, J, \langle\cdot,\cdot\rangle)$ be a Kähler manifold of complex dimension m, where J is the complex structure and $\langle\cdot,\cdot\rangle$ denotes the Kähler metric, which satisfies that $\langle JU, JV\rangle = \langle U, V\rangle$ and $d\Phi = 0$, where $\Phi(U, V) = \langle U, JV\rangle$, $(U, V \in \mathfrak{X}(N))$ is the fundamental 2-form. Let (M^m, g) be a *Lagrangian submanifold* in $(N^m, J, \langle\cdot,\cdot\rangle)$, that is, for all $x \in M$, $J(T_xM) \subset T_xM^\perp$, where $T_xM^\perp$ is the normal space at x. Then, it is well known that the following three equations hold:

$$\begin{aligned} \nabla^\perp_X JY &= J(\nabla_X Y), \\ R^N(JX, JY) &= R^N(X,Y), \\ R^N(U,V)\cdot J &= J \cdot R^N(U,V), \end{aligned}$$

for all X, Y on M, and all vector fields U,V on N. We obtain the biharmonic equations for a Lagrangian submanifold in a Kähler manifold as follows.

Lemma 3.1. *Let $\varphi : (M,g) \to (N, \langle\cdot,\cdot\rangle)$ be an isometric immersion of (M,g) into $(N,\langle\cdot,\cdot\rangle)$. Then it is biharmonic if and only if*

$$\mathrm{trace}_g\,(\nabla A_{\mathbf{H}}) + \mathrm{trace}_g\,\left(A_{\nabla^\perp_\bullet \mathbf{H}}(\bullet)\right) - \left(\sum_{i=1}^m R^N(\mathbf{H}, e_i)e_i\right)^T = 0, \tag{15}$$

$$\Delta^\perp \mathbf{H} + \mathrm{trace}_g B(A_{\mathbf{H}}(\bullet), \bullet) - \left(\sum_{i=1}^m R^N(\mathbf{H}, e_i)e_i\right)^\perp = 0, \tag{16}$$

where $(\cdot)^T$ is the tangential part and $(\cdot)^\perp$ is the normal part.

By using Lemma 3.1, we obtain the following theorem.

Theorem 3.1. *Let $(N^m, J, \langle\cdot,\cdot\rangle)$ be a Kähler manifold of complex dimension m. Assume that $\varphi : (M^m, g) \to (N^m, J, \langle\cdot,\cdot\rangle)$ is a Lagrangian submanifold. Then φ is biharmonic if and only if the following two equations hold:*

$$\begin{aligned} &\mathrm{trace}_g\,(\nabla A_{\mathbf{H}}) + \mathrm{trace}_g\,\left(A_{\nabla^\perp_\bullet \mathbf{H}}(\bullet)\right) \\ &\quad - \sum_{i=1}^m \left\langle \mathrm{trace}_g\,(\nabla^\perp_{e_i} B) - \mathrm{trace}_g\,(\nabla^\perp_\bullet B)\,(e_i, \bullet), \mathbf{H} \right\rangle e_i = 0, \end{aligned} \tag{17}$$

$$\begin{aligned} &\Delta^\perp \mathbf{H} + \mathrm{trace}_g B\left(A_{\mathbf{H}}(\bullet), \bullet\right) + \sum_{i=1}^m Ric^N(J\mathbf{H}, e_i) J e_i \\ &\quad - \sum_{i=1}^m Ric(J\mathbf{H}, e_i) J e_i - J\,\mathrm{trace}_g A_{B(J\mathbf{H},\bullet)}(\bullet) + m\, J A_{\mathbf{H}}(J\mathbf{H}) = 0, \end{aligned} \tag{18}$$

where Ric and Ric^N are the Ricci tensors of (M^m, g) and $(N^m, \langle\cdot,\cdot\rangle)$, respectively.

In the case that $(N, J, \langle\cdot,\cdot\rangle)$ is the simply connected complex m-dimensional complex space form $(N^m(4\epsilon), J, \langle\cdot,\cdot\rangle)$ of constant holomorphic sectional curvature 4ϵ, we have

Proposition 3.1. *Let $(N^m(4\epsilon), J, \langle\cdot,\cdot\rangle)$ be a complex space form of complex dimension m of constant holomorphic sectional curvature 4ϵ. Assume that $\varphi : (M^m, g) \to (N^m(4\epsilon), J, \langle\cdot,\cdot\rangle)$ is a Lagrangian submanifold. Then φ is biharmonic if and only if*

$$\mathrm{trace}_g\,(\nabla A_{\mathbf{H}}) + \mathrm{trace}_g\left(A_{\nabla^{\perp}_{\bullet}\mathbf{H}}(\bullet)\right) = 0, \tag{19}$$

$$\Delta^{\perp}\mathbf{H} + \mathrm{trace}_g B\,(A_{\mathbf{H}}(\bullet), \bullet) - (m+3)\epsilon\,\mathbf{H} = 0. \tag{20}$$

We want to classify biharmonic PNMC Lagrangian H-umbilical submanifolds in complex space forms. Let us recall the notion of Lagrangian H-umbilical submanifolds introduced by B-Y. Chen:

Definition 3.1 (cf. B-Y. Chen). If the second fundamental form of a Lagrangian submanifold M in a Kähler manifold takes the following form:

$$\begin{aligned} &B\,(e_1, e_1) = \lambda J e_1, \quad B(e_i, e_i) = \mu J e_1, \\ &B\,(e_1, e_i) = \mu J e_i, \quad B(e_i, e_j) = 0,\ (i \neq j), \quad (i, j = 2, \cdots, m), \end{aligned}$$

for some functions λ and μ with respect to a certain orthonormal frame field $\{e_1, \cdots, e_m\}$ on M, then M is called a *Lagrangian H-umbilical submanifold.*

Lagrangian H-umbilical submanifolds are the simplest examples of Lagrangian submanifolds next to totally geodesic submanifolds. Since it is known that there are no totally umbilical Lagrangian submanifolds in the complex space forms $N^m(4\varepsilon)$ with $m \geq 2$, we shall consider H-umbilical Lagrangian submanifolds.

In this case, the harmonic mean curvature vector $\mathbf{H}$ is

$$\mathbf{H} = \frac{\lambda + (m-1)\mu}{m} J e_1.$$

In the following, we shall denote by $a = \{\lambda + (m-1)\mu\}/m$.

B.Y. Chen also introduced the notion of PNMC submanifolds:

Definition 3.2. A submanifold M in a Riemannian manifold is said to have *parallel normalized mean curvature vector field* (PNMC) if it has nowhere zero mean curvature and the unit vector field in the direction of the mean curvature vector field is parallel in the normal bundle, i.e.,

$$\nabla^{\perp}\left(\frac{\mathbf{H}}{|\mathbf{H}|}\right) = 0. \tag{21}$$

We denote as $\nabla_{e_i} e_j = \sum_{l=1}^{m} \omega_j^l(e_i) e_l$ $(i, j = 1, \cdots, m)$. Then we obtain the following lemma.

Lemma 3.2. *Let M^m be an m-dimensional Lagrangian H-umbilical submanifold in a complex space form. For an orthonormal frame field $\{e_i\}_{i=1}^m$, we have the following relations:*

$$\begin{aligned}
e_j\lambda &= (2\mu - \lambda)\omega_j^1(e_1), \quad j > 1, \\
e_1\mu &= (\lambda - 2\mu)\omega_1^l(e_l), \quad \text{for all } l = 2, \cdots, m, \\
(\lambda - 2\mu)\omega_1^i(e_j) &= 0, \quad i \neq j > 1, \\
e_j\mu &= 0, \quad j > 1, \\
\mu\,\omega_1^j(e_1) &= 0 \\
\mu\,\omega_1^2(e_2) &= \cdots = \mu\,\omega_1^m(e_m), \\
\mu\,\omega_1^i(e_j) &= 0, \quad i \neq j > 1.
\end{aligned} \tag{22}$$

Due to this lemma, we can show the classification of all the biharmonic PNMC Lagrangian H-umbilical submanifolds of the complex space forms (cf. Maeta and Urakawa [18]).

Theorem 3.2. *Let $(N^m(4\epsilon), J, \langle\cdot,\cdot\rangle)$ be a complex space form of complex dimension m, where $\epsilon \in \{-1, 0, 1\}$. Assume that $\varphi : (M^m, g) \to (N^m(4\varepsilon), J, \langle\cdot,\cdot\rangle)$ is a Lagrangian H-umbilical submanifold which has PNMC. Then, φ is biharmonic if and only if $\epsilon = 1$ and $\varphi(M)$ is congruent to an m-dimensional submanifold of $\mathbb{CP}^m(4)$ given by*

$$\pi\left(\sqrt{\frac{\mu^2}{\mu^2+1}}e^{-\frac{i}{\mu}x}, \sqrt{\frac{1}{\mu^2+1}}e^{i\mu x}y_1, \cdots, \sqrt{\frac{1}{\mu^2+1}}e^{i\mu x}y_m\right) \subset \mathbb{C}P^m(4), \tag{23}$$

where $x, y_1, \cdots, y_m$ are real numbers satisfying ${y_1}^2 + \cdots + {y_m}^2 = 1$ and $\mu = \pm\sqrt{\{m + 5 \pm \sqrt{m^2 + 6m + 25}\}/(2m)}$.

Remark 3.1. Theorem 3.2 gives examples of biharmonic but not minimal Lagrangian submanifolds of the complex projective space $\mathbb{CP}^m(4)$.

4. Conformal change of metrics and biharmonic maps

In this section, we show recent results of a joint work with H. Naito (cf. [19]). First, let us recall the work of P. Baird and D. Kamissoko (cf. [2]) constructing biharmonic maps by conformal changing of a Riemannian metric of the domain manifold, but our setting is a little bit different from them. Let us consider a C^∞ mapping $\varphi:\ (M^m,\widetilde{g})\rightarrow(N^n,h)$ with $\widetilde{g}=f^{2/(m-2)}g$, $f\in C^\infty(M)$, $f>0$. ($m:=\dim M>2$) Then, for $\varphi\in C^\infty(M,N)$, the bitension field, denoted by $\tau_2(\varphi;\widetilde{g},h)$, is given by

$$\begin{aligned} f^{2m/(m-2)}\,\tau_2(\varphi;\widetilde{g},h) &= -\frac{m-6}{m-2}\,f\,\overline{\nabla}_X\tau_g(\varphi)+f^2\,J_g(\tau_g(\varphi)) \\ &\quad -\left\{\frac{4}{(m-2)^2}\,|X|_g{}^2+\frac{2}{m-2}\,f(\Delta_g f)\right\}\tau_g(\varphi) \\ &\quad -f^{-1}\left\{\frac{m^2}{(m-2)^2}\,|X|_g{}^2+\frac{m}{m-2}\,f\,(\Delta_g f)\right\}d\varphi(X) \\ &\quad +\frac{m+2}{m-2}\,\overline{\nabla}_X d\varphi(X)+f\,J_g(d\varphi(X)), \end{aligned} \tag{24}$$

where $X=\nabla^g f\in\mathfrak{X}(M)$. Then, let us consider the Euclidean space $(M,g)=(\mathbb{R}^m,g_0)$, $(m\geq 3)$, the standard Euclidean space, and $f\in C^\infty(\mathbb{R}^m)$ is given by

$$f(x_1,x_2,\cdots,x_m)=f(x_1)=f(x),$$

where we denote by $x_1=x$. Then, id : $(\mathbb{R}^m,f^{2/(m-2)}g_0)\rightarrow(\mathbb{R}^m,g_0)$ is biharmonic if and only if f satisfies the ordinary differential equation:

$$f^2f'''-2\,\frac{m+1}{m-2}\,ff'f''+\frac{m^2}{(m-2)^2}\,f'^3=0. \tag{25}$$

Then, we obtain

Theorem 4.1 (cf. Naito and Urakawa [19]). *Assume that* $m\geq 3$. *Then,*

(1) $(m\geq 5)$ *There exists no positive global* C^∞ *solution* f *on* $\mathbb{R}$ *of the ODE* (25);

(2) $(m=4)$ $f(x)=a/\cosh(b\,x+c)$ *is a global positive* C^∞ *solution of the ODE* (25) *for every* $a>0$, b *and* c;

(3) $(m=3)$ *There exists a positive bounded* C^∞ *solution* f *on* $\mathbb{R}$ *of the ODE* (25).

Remark 4.1. In the case that $m=3$, it is an interesting but difficult problem for us whether or not there exists a positive periodic solution f on $\mathbb{R}$ of the ODE (25) with $m=3$.

Then, our main theorem is the following:

Theorem 4.2 (cf. H. Naito and H. Urakawa [19]). *Let f be a positive bounded C^∞ solution on $\mathbb{R}$ of*

$$f^2 f''' - 8\, f f' f'' + 9\, f'^3 = 0 \quad (m = 3),$$

as in Theorem 4.1 (3), *and let* $\widetilde{f}(x, y) := f(x)$, $(x, y) \in \mathbb{R} \times \Sigma^2$.

Then, for any harmonic map $\varphi : (\Sigma^{m-1}, g) \to (P, h)$, *let us define* $\widetilde{\varphi} : \mathbb{R} \times \Sigma^{m-1} \ni (x, y) \mapsto \big(px + q, \varphi(y)\big) \in \mathbb{R} \times P$ $(m = 3, 4)$. *Then,*

(1) $\widetilde{\varphi} : \big(\mathbb{R} \times \Sigma^2, \widetilde{f}^{\,2}\,(dx^2 + g)\big) \to (\mathbb{R} \times P, dx^2 + h)$ *is biharmonic, but not harmonic if* $p \neq 0$;

(2) $\widetilde{\varphi} : \big(\mathbb{R} \times \Sigma^3, \frac{1}{\cosh x}\,(dx^2 + g)\big) \to (\mathbb{R} \times P, dx^2 + h)$ *is biharmonic, but not harmonic if* $p \neq 0$.

Bibliography

1. P. Baird and J. Eells, *A conservation law for harmonic maps*, Lecture Notes in Math., Springer, **894**, 1–25 (1981).
2. P. Baird and D. Kamissoko, *On constructing biharmonic maps and metrics, Ann. Global Anal. Geom.*, **23**, 65–75 (2003).
3. A. Balmus, S. Montaldo and C. Oniciuc, *Biharmonic PNMC submanifolds in spheres*, to appear in Ark. Mat.
4. R. Caddeo, S. Montaldo and P. Piu, *On biharmonic maps, Contemp. Math. Amer. Math. Soc.*, **288**, 286–290 (2001).
5. B-Y. Chen, *Som open problems and conjectures on submanifolds of finite type, Soochow J. Math.*, **17**, 169–188 (1991).
6. B. Y. Chen , *Total mean curvature and submanifolds of finite type*, Series in Pure Mathematics, **1**. World Scientific Publishing Co., Singapore (1984).
7. B-Y. Chen, *Complex extensors and Lagrangian submanifolds in complex Euclidean spaces*, Tohoku. Math. J. **49**, 277-297 (1997).
8. B-Y. Chen, *Interaction of Legendre curves and Lagrangian submanifolds*, Israel J. Math. **99**, 69-108 (1997).
9. B-Y. Chen, *Representation of flat Lagrangian H-umbilical submanifolds in complex Euclidean spaces*, Tohoku. Math J. **51**, 13-21 (1999).
10. B-Y. Chen, *Surfaces with parallel normalized mean curvature vector*, Monatsh. Math. **90**, 185-194 (1980).
11. B-Y. Chen and S. Ishikawa, *Biharmonic pseudo-Riemannian submanifolds in pseudo-Euclidean spaces*, Kyushu J. Math. **52**, no. 1-3, 101-108 (1998).
12. B-Y. Chen and K. Ogiue, *On totally real submanifolds*, Trans. Amer. Math. Soc. **193**, 257-266 (1974).
13. J. Eells and L. Lemaire, *Selected Topics in Harmonic Maps*, CBMS, **50**, Amer. Math. Soc. (1983).
14. J. Eells and J.H. Sampson, *Harmonic mappings of Riemannian manifolds, American J. Math.*, **86**, 109–160 (1964).

15. D. Fetcu, E. Loubeau, S. Montaldo and C. Oniciuc, *Biharmonic submanifolds of* $\mathbb{C}P^n$, Math. Z. **266**, no. 3, 505-531 (2010).
16. S. Gudmundsson, N. Nakauchi and H. Urakawa, *Biharmonic maps into a Riemannian manifold of non-positive curvature*, arXiv: 1201.6457v4 (2012).
17. G-Y. Jiang, *2-harmonic maps and their first and second variational formulas*, *Chinese Ann. Math.*, **7A**, 388–402 (1986); translated from Chinese to English by H. Urakawa, *Note di Matematica*, **28**, 209–232 (2009).
18. Sh. Maeta and H. Urakawa, *Biharmonic Lagrangian submanifolds in Kähler manifolds*, accepted in Glasgow Math. J. (2012).
19. H. Naito and H. Urakawa, *Conformal change of Riemannian metrics and biharmonic maps*, arXiv: 1301.7150v1.
20. Y-L. Ou and L. Tang, *The generalized Chen's conjecture on biharmonic submanifolds is false*, arXiv: 1006.1838v1.
21. T. Sasahara, *A classification result for biminimal Lagrangian surfaces in complex space form, J. Geometry Physics* **60**, 884–895 (2010).
22. T. Sasahara, *Biharmonic Lagrangian surfaces of constant mean curvature in complex space forms,* Glasgow Math. J. **49**, 497-507 (2007).
23. T. Sasahara, *Biminimal Lagrangian H-umbilical submanifolds in complex space forms*, Geom. Dedicata, **160**, 185–193 (2012).
24. T. Sasahara, *A classification result for biminimal Lagrangian surfaces in complex space forms*, J. Geom. Phys. **60**, 884-895 (2010).
25. R.T. Smith, *The second variation formula for harmonic mappings*, Proc. Amer. Math. Soc., **47**, 229–236 (1975).
26. Z-P. Wang and Y-L. Ou, *Biharmonic Riemannian submersions from 3-manifolds*, Math. Z., **269**, 917–925 (2011).

Received November 13, 2012
Revised December 14, 2012

Proceedings of the 3rd International
Colloquium on Differential Geometry
and its Related Fields
Veliko Tarnovo, September 3–7, 2012

HOMOGENEOUS EINSTEIN METRICS ON GENERALIZED FLAG MANIFOLDS WITH G_2-TYPE $\mathfrak{t}$-ROOTS

Andreas ARVANITOYEORGOS

Department of Mathematics, University of Patras,
GR-26500 Rion, Greece
E-mail: arvanito@math.upatras.gr

Ioannis CHRYSIKOS

Department of Mathematics and Statistics, Masaryk University,
Brno 611 37, Czech Republic
E-mail: chrysikosi@math.muni.cz

Yusuke SAKANE

Department of Pure and Applied Mathematics,
Graduate School of Information Science and Technology,
Osaka University, Toyonaka, Osaka 560-043, Japan
E-mail: sakane@math.sci.osaka-u.ac.jp

We construct the Einstein equation for an invariant Riemannian metric on generalized flag manifolds G/K with G_2-type $\mathfrak{t}$-roots. By computing a Gröbner basis for a system of polynomials on six variables, we prove that such a generalized flag manifold G/K, which is not the full flag manifold G_2/T, admits exactly one invariant Kähler Einstein metric and six non Kähler invariant Einstein metrics up to isometry and scalar.

Keywords: Homogeneous Einstein metric; Generalized flag manifold; Riemannian submersion; Gröbner basis.

1. Introduction

A Riemannian manifold (M, g) is called Einstein if the Ricci tensor r of the metric g satisfies $r = \lambda g$ for some $\lambda \in \mathbb{R}$. In this paper we discuss homogeneous Einstein metrics on generalized flag manifolds. A generalized flag manifold is an adjoint orbit of a compact semisimple Lie group G, or equivalently a compact homogeneous space of the form $M = G/K = G/C(S)$, where $C(S)$ is the centralizer of a torus S in G. Einstein metrics on generalized flag manifolds have been studied by several authors (Alekseevsky, Arvanitoyeorgos, Chrysikos, Graev, Kimura, Sakane).

Generalized flag manifolds G/K admit a finite number of G-invariant complex structures and invariant Kähler Einstein metrics on generalized flag manifolds were investigated by Alekseevsky and Perelomov [2]. The problem of finding non Kähler Einstein metrics on generalized flag manifolds was initially studied by Alekseevsky [1] and Arvanitoyeorgos [4]. Kimura [16] studied all invariant Einstein metrics for generalized flag manifolds with three isotropy summands. (See also [6] for two isotropy summands and [3] for three isotropy summands.)

In the recent works [5], [7] and [8] all invariant Einstein metrics were found for generalized flag manifolds with four isotropy summands. Moreover, we classified all invariant Einstein metrics on generalized flag manifolds with five isotropy summands in the works [13] and [11]. For six isotropy summands case, all invariant Einstein metrics were found for generalized flag manifolds $Sp(3)/(U(1) \times U(1) \times Sp(1))$, $Sp(4)/(U(1) \times U(1) \times Sp(2))$ and $Sp(4)/(U(2) \times U(1) \times Sp(1))$ in [10], and for the exceptional full flag manifold G_2/T in [9].

In this paper we study generalized flag manifolds with G_2-type $\mathfrak{t}$-roots which have six isotropy summands. These spaces are a generalization of the exceptional full flag manifold G_2/T. We give explicit expressions of Ricci tensor of the invariant metrics on these generalized flag manifolds. To give such expressions we use a Riemannian submersion from a generalized flag manifold to another generalized flag manifold with totally geodesic fibers and Kähler Einstein metrics on the generalized flag manifolds. Note that Graev [15] has estimated the number of complex Einstein metrics and studied Einstein metrics by a different method.

Our main results are the following:

Theorem 1.1. *A generalized flag manifold G/K with G_2-type $\mathfrak{t}$-roots, which is not the full flag manifold G_2/T, admits exactly one invariant Kähler Einstein metric and six non Kähler invariant Einstein metrics up to isometry and scalar. These are the spaces $F_4/U(3) \times U(1)$, $E_6/U(3) \times U(3)$, $E_7/U(6) \times U(1)$ and $E_8/E_6 \times U(1) \times U(1)$. The full flag manifold G_2/T admits exactly one invariant Kähler Einstein metric and two non Kähler invariant Einstein metrics up to isometry and scalar.*

2. Ricci tensor of a compact homogeneous space G/K

Let G be a compact semisimple Lie group, K a connected closed subgroup of G and let $\mathfrak{g}$ be the Lie algebra of G, $\mathfrak{k}$ the Lie algebra of K. Note that the Killing form of a compact Lie algebra $\mathfrak{g}$ is negative definite. We set $B = -$

Killing form. Then B is an $\mathrm{Ad}(G)$-invariant inner product on $\mathfrak{g}$.

Let $\mathfrak{m}$ be the orthogonal complement of $\mathfrak{k}$ in $\mathfrak{g}$ with respect to B. Then we have $\mathfrak{g} = \mathfrak{k} \oplus \mathfrak{m}$, $[\mathfrak{k}, \mathfrak{m}] \subset \mathfrak{m}$ and a decomposition of $\mathfrak{m}$ into irreducible $\mathrm{Ad}(K)$-modules:

$$\mathfrak{m} = \mathfrak{m}_1 \oplus \cdots \oplus \mathfrak{m}_q.$$

We assume that $\mathrm{Ad}(K)$-modules $\mathfrak{m}_j$ $(j = 1, \ldots, q)$ are mutually non-equivalent. Then a G-invariant metric on G/K can be written as

$$\langle\ ,\ \rangle = x_1 B|_{\mathfrak{m}_1} + \cdots + x_q B|_{\mathfrak{m}_q}, \tag{1}$$

for positive real numbers $x_1, \ldots, x_q$.

Note that G-invariant symmetric covariant 2-tensors on G/K are the same form as the metrics. In particular, the Ricci tensor r of a G-invariant Riemannian metric on G/K is of the same form as (1).

Let $\{e_\alpha\}$ be a B-orthonormal basis adapted to the decomposition of $\mathfrak{m}$, i.e., $e_\alpha \in \mathfrak{m}_i$ for some i, and $\alpha < \beta$ if $i < j$ (with $e_\alpha \in \mathfrak{m}_i$ and $e_\beta \in \mathfrak{m}_j$). We put $A^\gamma_{\alpha\beta} = B\left([e_\alpha, e_\beta], e_\gamma\right)$, so that $[e_\alpha, e_\beta] = \sum_\gamma A^\gamma_{\alpha\beta} e_\gamma$, and set $\begin{bmatrix} k \\ ij \end{bmatrix} = \sum (A^\gamma_{\alpha\beta})^2$, where the sum is taken over all indices α, β, γ with $e_\alpha \in \mathfrak{m}_i$, $e_\beta \in \mathfrak{m}_j$, $e_\gamma \in \mathfrak{m}_k$ (cf. [21]). Then, the positive number $\begin{bmatrix} k \\ ij \end{bmatrix}$ is independent of the B-orthonormal bases chosen for $\mathfrak{m}_i, \mathfrak{m}_j, \mathfrak{m}_k$, and

$$\begin{bmatrix} k \\ ij \end{bmatrix} = \begin{bmatrix} k \\ ji \end{bmatrix} = \begin{bmatrix} j \\ ki \end{bmatrix}, \tag{2}$$

that is, $\begin{bmatrix} k \\ ij \end{bmatrix}$ is symmetric with respect to indices i, j, k.

Let $d_k = \dim \mathfrak{m}_k$. Then we have

Lemma 2.1 ([19]). *The components $r_1, \ldots, r_q$ of Ricci tensor r of the metric $\langle\ ,\ \rangle$ of the form (1) on G/K are given by*

$$r_k = \frac{1}{2x_k} + \frac{1}{4d_k} \sum_{j,i} \frac{x_k}{x_j x_i} \begin{bmatrix} k \\ ji \end{bmatrix} - \frac{1}{2d_k} \sum_{j,i} \frac{x_j}{x_k x_i} \begin{bmatrix} j \\ ki \end{bmatrix} \quad (k = 1,\ \ldots, q) \tag{3}$$

where the sum is taken over $i, j = 1, \ldots, q$.

3. Riemannian submersion

Let G be a compact semisimple Lie group and K, L two closed subgroups of G with $K \subset L$. Then we have a natural fibration $\pi : G/K \to G/L$ with fiber L/K.

Let $\mathfrak{p}$ be the orthogonal complement of $\mathfrak{l}$ in $\mathfrak{g}$ with respect to B, and $\mathfrak{n}$ be the orthogonal complement of $\mathfrak{k}$ in $\mathfrak{l}$. Then we have $\mathfrak{g} = \mathfrak{l} \oplus \mathfrak{p} = \mathfrak{k} \oplus \mathfrak{n} \oplus \mathfrak{p}$. An $\mathrm{Ad}_G(L)$-invariant scalar product on $\mathfrak{p}$ defines a G-invariant metric $\check{g}$ on G/L, and an $\mathrm{Ad}_L(K)$-invariant scalar product on $\mathfrak{n}$ defines an L-invariant metric $\hat{g}$ on L/K. The orthogonal direct sum for these scalar products on $\mathfrak{n} \oplus \mathfrak{p}$ defines a G-invariant metric g on G/K.

Theorem 3.1. *The map π is a Riemannian submersion from $(G/K, g)$ to $(G/L, \check{g})$ with totally geodesic fibers isometric to $(L/K, \hat{g})$.*

Note that $\mathfrak{n}$ is the vertical subspace of the submersion and $\mathfrak{p}$ is the horizontal subspace.

For a Riemannian submersion, O'Neill [18] has introduced two tensors A and T. Since the fibers are totally geodesic in our case, we have $T = 0$. We also have

$$A_X Y = \frac{1}{2}[X,\ Y]_{\mathfrak{n}} \quad \text{for } \ X, Y \in \mathfrak{p}.$$

Let $\{X_i\}$ be an orthonormal basis of $\mathfrak{p}$ and $\{U_j\}$ an orthonormal basis of $\mathfrak{n}$. We put for $X, Y \in \mathfrak{p}$, $g(A_X,\ A_Y) = \sum_i g(A_X X_i, A_Y X_i)$.

Let r, $\check{r}$ be the Ricci tensor of the metric g, $\check{g}$ respectively. Then we have

$$r(X,Y) = \check{r}(X,Y) - 2g(A_X, A_Y) \quad \text{for } \ X, Y \in \mathfrak{p}.$$

Let

$$\mathfrak{p} = \mathfrak{p}_1 \oplus \cdots \oplus \mathfrak{p}_\ell, \quad \mathfrak{n} = \mathfrak{n}_1 \oplus \cdots \oplus \mathfrak{n}_s$$

be a decomposition of $\mathfrak{p}$ into irreducible $\mathrm{Ad}(L)$-modules and a decomposition of $\mathfrak{n}$ into irreducible $\mathrm{Ad}(K)$-modules respectively. We assume that the $\mathrm{Ad}(L)$-modules $\mathfrak{p}_j$ $(j = 1, \ldots, \ell)$ are mutually non-equivalent. Note that each irreducible component $\mathfrak{p}_j$ as $\mathrm{Ad}(L)$-module can be decomposed into irreducible $\mathrm{Ad}(K)$-modules. To compute the values $\begin{bmatrix} k \\ ij \end{bmatrix}$ for G/K, we use information from the Riemannian submersion $\pi : (G/K, g) \to (G/L, \check{g})$ with totally geodesic fibers isometric to $(L/K, \hat{g})$. We consider a G-invariant metric on G/K defined by a Riemannian submersion $\pi : (G/K, g) \to (G/L, \check{g})$ of the form

$$g = y_1 B|_{\mathfrak{p}_1} + \cdots + y_\ell B|_{\mathfrak{p}_\ell} + z_1 B|_{\mathfrak{n}_1} + \cdots + z_s B|_{\mathfrak{n}_s} \tag{4}$$

for positive real numbers $y_1, \cdots, y_\ell, z_1, \cdots, z_s$.

We decompose each irreducible component $\mathfrak{p}_j$ into irreducible $\mathrm{Ad}(K)$-modules:

$$\mathfrak{p}_j = \mathfrak{m}_{j,1} \oplus \cdots \oplus \mathfrak{m}_{j,k_j}.$$

As before we assume that the $\mathrm{Ad}(K)$-modules $\mathfrak{m}_{j,t}$ ($j = 1, \ldots, \ell$, $t = 1, \ldots, k_j$) are mutually non-equivalent. Note that the metric of the form (4) can be written as

$$g = y_1 \sum_{t=1}^{k_1} B|_{\mathfrak{m}_{1,t}} + \cdots + y_\ell \sum_{t=1}^{k_\ell} B|_{\mathfrak{m}_{\ell,t}} + z_1 B|_{\mathfrak{n}_1} + \cdots + z_s B|_{\mathfrak{n}_s} \tag{5}$$

and this is a special case of the metric of the form (1).

Lemma 3.1 ([10, 11]). *Let $d_{j,t} = \dim \mathfrak{m}_{j,t}$. The components $r_{(j,t)}$ ($j = 1, \ldots, \ell$, $t = 1, \ldots, k_j$) of the Ricci tensor r for the metric* (5) *on G/K are given by*

$$r_{(j,t)} = \check{r}_j - \frac{1}{2d_{j,t}} \sum_{i=1}^{s} \sum_{j',t'} \frac{z_i}{y_j y_{j'}} \begin{bmatrix} & i & \\ (j,t) & & (j',t') \end{bmatrix}, \tag{6}$$

where $\check{r}_j$ are the components of Ricci tensor $\check{r}$ for the metric $\check{g}$ on G/L.

4. Decomposition associated to generalized flag manifolds

Let G be a compact semi-simple Lie group, $\mathfrak{g}$ the Lie algebra of G and $\mathfrak{h}$ a maximal abelian subalgebra of $\mathfrak{g}$. We denote by $\mathfrak{g}^{\mathbb{C}}$ and $\mathfrak{h}^{\mathbb{C}}$ the complexification of $\mathfrak{g}$ and $\mathfrak{h}$ respectively. We identify an element of the root system Δ of $\mathfrak{g}^{\mathbb{C}}$ relative to the Cartan subalgebra $\mathfrak{h}^{\mathbb{C}}$ with an element of $\mathfrak{h}_0 = \sqrt{-1}\mathfrak{h}$ by the duality defined by the Killing form of $\mathfrak{g}^{\mathbb{C}}$. Let $\Pi = \{\alpha_1, \ldots, \alpha_l\}$ be a fundamental system of Δ and $\{\Lambda_1, \ldots, \Lambda_l\}$ the fundamental weights of $\mathfrak{g}^{\mathbb{C}}$ corresponding to Π, that is

$$\frac{2(\Lambda_i, \alpha_j)}{(\alpha_j, \alpha_j)} = \delta_{ij} \qquad (1 \leq i, j \leq l).$$

Let Π_0 be a subset of Π and $\Pi \setminus \Pi_0 = \{\alpha_{i_1}, \ldots, \alpha_{i_r}\}$, where $1 \leq i_1 < \cdots < i_r \leq l$. We put $[\Pi_0] = \Delta \cap \{\Pi_0\}_{\mathbb{Z}}$, where $\{\Pi_0\}_{\mathbb{Z}}$ denotes the subspace of $\mathfrak{h}_0$ generated by Π_0.

Consider the root space decomposition of $\mathfrak{g}^{\mathbb{C}}$ relative to $\mathfrak{h}^{\mathbb{C}}$:

$$\mathfrak{g}^{\mathbb{C}} = \mathfrak{h}^{\mathbb{C}} + \sum_{\alpha \in \Delta} \mathfrak{g}^{\mathbb{C}}_{\alpha}.$$

Let Δ^+ be the set of all positive roots relative to Π. We put $\Delta^+_{\mathfrak{m}} = \Delta^+ \setminus [\Pi_0]$. We define a parabolic subalgebra $\mathfrak{u}$ of $\mathfrak{g}^{\mathbb{C}}$ by

$$\mathfrak{u} = \mathfrak{h}^{\mathbb{C}} + \sum_{\alpha \in [\Pi_0] \cup \Delta^+} \mathfrak{g}^{\mathbb{C}}_{\alpha}.$$

Then the nilradical $\mathfrak{n}$ of $\mathfrak{u}$ is given by

$$\mathfrak{n} = \sum_{\alpha \in \Delta^+_{\mathfrak{m}}} \mathfrak{g}^{\mathbb{C}}_{\alpha}.$$

Let $G^{\mathbb{C}}$ be a simply connected complex semisimple Lie group whose Lie algebra is $\mathfrak{g}^{\mathbb{C}}$ and U the parabolic subgroup of $G^{\mathbb{C}}$ generated by $\mathfrak{u}$. Then the complex homogeneous manifold $G^{\mathbb{C}}/U$ is compact simply connected and G acts transitively on $G^{\mathbb{C}}/U$. Note also that $K = G \cap U$ is a connected closed subgroup of G, $G^{\mathbb{C}}/U = G/K$ as C^{∞}-manifolds, and $G^{\mathbb{C}}/U$ admits a G-invariant Kähler metric.

Let $\mathfrak{k}$ be the Lie algebra of K and $\mathfrak{k}^{\mathbb{C}}$ the complexification of $\mathfrak{k}$. Then we have a direct decomposition

$$\mathfrak{u} = \mathfrak{k}^{\mathbb{C}} \oplus \mathfrak{n}, \qquad \mathfrak{k}^{\mathbb{C}} = \mathfrak{h}^{\mathbb{C}} + \sum_{\alpha \in [\Pi_0]} \mathfrak{g}^{\mathbb{C}}_{\alpha}.$$

Take a Weyl basis $E_{-\alpha} \in \mathfrak{g}^{\mathbb{C}}_{\alpha}$ $(\alpha \in \Delta)$ with

$$\begin{aligned} [E_\alpha, E_{-\alpha}] &= -\alpha \, (\alpha \in \Delta) \\ [E_\alpha, E_\beta] &= \begin{cases} N_{\alpha,\beta} E_{\alpha+\beta} & \text{if} \quad \alpha+\beta \in \Delta \\ 0 & \text{if} \quad \alpha+\beta \notin \Delta, \end{cases} \end{aligned}$$

where $N_{\alpha,\beta} = N_{-\alpha,-\beta} \in \mathbb{R}$. Then we have

$$\mathfrak{g} = \mathfrak{h} + \sum_{\alpha \in \Delta} \{\mathbb{R}(E_\alpha + E_{-\alpha}) + \mathbb{R}\sqrt{-1}(E_\alpha - E_{-\alpha})\}$$

and the Lie subalgebra $\mathfrak{k}$ is given by

$$\mathfrak{k} = \mathfrak{h} + \sum_{\alpha \in [\Pi_0]} \{\mathbb{R}(E_\alpha + E_{-\alpha}) + \mathbb{R}\sqrt{-1}(E_\alpha - E_{-\alpha})\}.$$

For integers $j_1, \ldots, j_r$ with $(j_1, \ldots, j_r) \neq (0, \ldots, 0)$, we put

$$\Delta(j_1, \ldots, j_r) = \left\{ \sum_{j=1}^{l} m_j \alpha_j \in \Delta^+ \;\middle|\; m_{i_1} = j_1, \ldots, m_{i_r} = j_r \right\}.$$

Note that $\Delta^+_{\mathfrak{m}} = \Delta^+ \setminus [\Pi_0] = \bigcup_{j_1,\ldots,j_r} \Delta(j_1, \ldots, j_r)$.

For $\Delta(j_1,\ldots,j_r) \neq \emptyset$, we define an $\mathrm{Ad}_G(K)$-invariant subspace $\mathfrak{m}(j_1,\ldots,j_r)$ of $\mathfrak{g}$ by

$$\mathfrak{m}(j_1,\ldots,j_r) = \sum_{\alpha\in\Delta(j_1,\ldots,j_r)} \left\{\mathbb{R}(E_\alpha + E_{-\alpha}) + \mathbb{R}\sqrt{-1}(E_\alpha - E_{-\alpha})\right\}.$$

Then we have a decomposition of $\mathfrak{m}$ into mutually non-equivalent irreducible $\mathrm{Ad}_G(K)$-modules $\mathfrak{m}(j_1,\ldots,j_r)$:

$$\mathfrak{m} = \sum_{j_1,\ldots,j_r} \mathfrak{m}(j_1,\ldots,j_r).$$

We put $\mathfrak{t} = \left\{H \in \mathfrak{h}_0 \mid (H,\ \Pi_0) = 0\right\}$. Then $\{\Lambda_{i_1},\ldots,\Lambda_{i_r}\}$ is a basis of $\mathfrak{t}$. Put $\mathfrak{s} = \sqrt{-1}\mathfrak{t}$. Then the Lie algebra $\mathfrak{k}$ is given by $\mathfrak{k} = \mathfrak{z}(\mathfrak{s})$ (the Lie algebra of centralizer of a torus S in G).

We consider the restriction map

$$\kappa : \mathfrak{h}_0^* \to \mathfrak{t}^* \quad \alpha \mapsto \alpha|_\mathfrak{t}$$

and set $\Delta_\mathfrak{t} = \kappa(\Delta)$. The elements of $\Delta_\mathfrak{t}$ are called $\mathfrak{t}$-roots.

There exists a 1-1 correspondence between $\mathfrak{t}$-roots ξ and irreducible submodules $\mathfrak{m}_\xi$ of the $\mathrm{Ad}_G(K)$-module $\mathfrak{m}^\mathbb{C}$ given by

$$\Delta_\mathfrak{t} \ni \xi \mapsto \mathfrak{m}_\xi = \sum_{\kappa(\alpha)=\xi} \mathfrak{g}_\alpha^\mathbb{C}.$$

Thus we have a decomposition of the $\mathrm{Ad}_G(K)$-module $\mathfrak{m}^\mathbb{C}$:

$$\mathfrak{m}^\mathbb{C} = \sum_{\xi\in\Delta_\mathfrak{t}} \mathfrak{m}_\xi.$$

Denote by $\Delta_\mathfrak{t}^+$ the set of all positive $\mathfrak{t}$-roots, that is, the restricton of the system Δ^+. Then we have $\mathfrak{n} = \sum_{\xi\in\Delta_\mathfrak{t}^+} \mathfrak{m}_\xi$. Denote by τ the complex conjugation of $\mathfrak{g}^\mathbb{C}$ with respect to $\mathfrak{g}$ (note that τ interchanges $\mathfrak{g}_\alpha^\mathbb{C}$ and $\mathfrak{g}_{-\alpha}^\mathbb{C}$) and by $\mathfrak{v}^\tau$ the set of fixed points of τ in a complex vector subspace $\mathfrak{v}$ of $\mathfrak{g}^\mathbb{C}$. Thus we have a decomposition of $\mathrm{Ad}_G(K)$-module $\mathfrak{m}$ into irreducible submodules:

$$\mathfrak{m} = \sum_{\xi\in\Delta_\mathfrak{t}^+} (\mathfrak{m}_\xi + \mathfrak{m}_{-\xi})^\tau.$$

There exists a natural 1-1 correspondence between $\Delta_\mathfrak{t}^+$ and the set $\{\Delta(j_1,\ldots,j_r) \neq \emptyset\}$. For a generalized flag manifold G/K, we have a decomposition of $\mathfrak{m}$ into mutually non-equivalent irreducible $\mathrm{Ad}_G(H)$-modules:

$$\mathfrak{m} = \sum_{\xi\in\Delta_\mathfrak{t}^+} (\mathfrak{m}_\xi + \mathfrak{m}_{-\xi})^\tau = \sum_{j_1,\ldots,j_r} \mathfrak{m}(j_1,\ldots,j_r).$$

Thus a G-invariant metric g on G/K can be written as

$$g = \sum_{\xi \in \Delta_{\mathfrak{t}}^{+}} x_\xi B|_{(\mathfrak{m}_\xi + \mathfrak{m}_{-\xi})^\tau} = \sum_{j_1,\dots,j_r} x_{j_1\cdots j_r} B|_{\mathfrak{m}(j_1,\dots,j_r)} \tag{7}$$

for positive real numbers x_ξ, $x_{j_1\cdots j_r}$.

5. The classification of generalized flag manifolds with G_2-type $\mathfrak{t}$-roots

The Dynkin mark of a simple root $\alpha_i \in \Pi$ $(i = 1, \dots, \ell)$ is the positive integer m_i in the expression of the highest root $\widetilde{\alpha} = \sum_{k=1}^{\ell} m_k \alpha_k$ in terms of simple roots. We denote by Mrk the function Mrk : $\Pi \to \mathbb{Z}^+$, $\alpha_i \mapsto m_i$. Recall that a system of positive roots of the Lie group G_2 is given by $\{\alpha_1, \alpha_2, \alpha_1 + \alpha_2, \alpha_1 + 2\alpha_2, \alpha_1 + 3\alpha_2, 2\alpha_1 + 3\alpha_2\}$, with heighest root $\widetilde{\alpha} = 2\alpha_1 + 3\alpha_2$. The corresponding painted Dynkin diagram of the full flag manifold G_2/T is

α_1 α_2

Now we list the painted Dynkin diagrams with two black nodes whose Dynkin marks are 2 and 3, that is,

$$\Pi \setminus \Pi_0 = \{\alpha_i, \alpha_j : \mathrm{Mrk}(\alpha_i) = 2, \ \mathrm{Mrk}(\alpha_j) = 3\}.$$

The highest root $\tilde{\alpha}$ of F_4 is given by $\tilde{\alpha} = 2\alpha_1 + 3\alpha_2 + 4\alpha_3 + 2\alpha_4$. Thus, for F_4 we find the following painted Dynkin diagrams:

type $F_4(1)$: α_1 α_2 α_3 α_4 — 2 3 4 2

type $F_4(2)$: α_1 α_2 α_3 α_4 — 2 3 4 2

The highest root $\tilde{\alpha}$ of E_6 is given by $\tilde{\alpha} = \alpha_1 + 2\alpha_2 + 3\alpha_3 + 2\alpha_4 + \alpha_5 + 2\alpha_6$. Thus, for E_6 we find the following painted Dynkin diagrams:

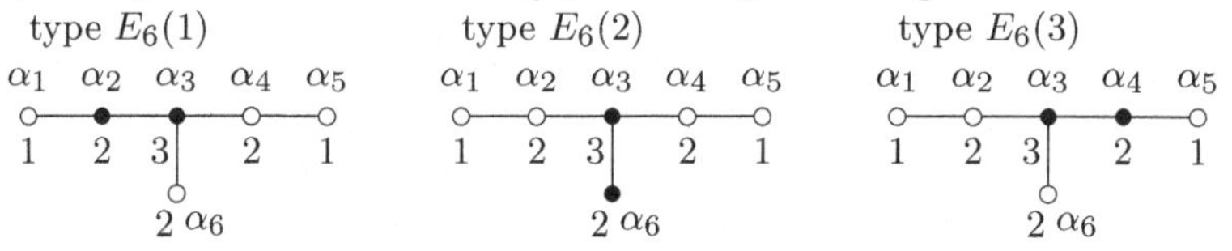

The highest root $\tilde{\alpha}$ of E_7 is given by $\tilde{\alpha} = \alpha_1 + 2\alpha_2 + 3\alpha_3 + 4\alpha_4 + 3\alpha_5 + 2\alpha_6 + 2\alpha_7$. Thus, for E_7 we find the following painted Dynkin diagrams:

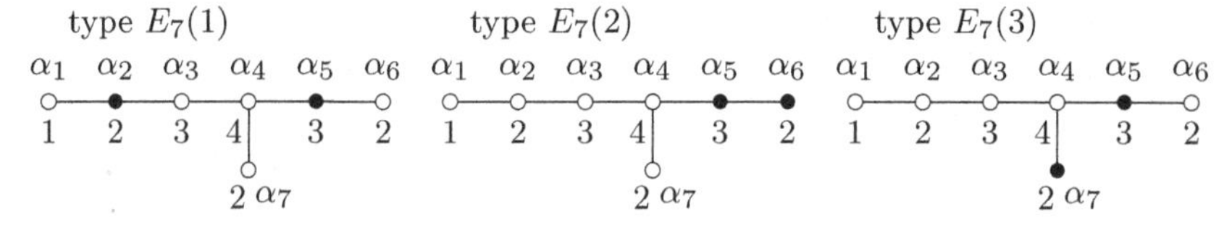

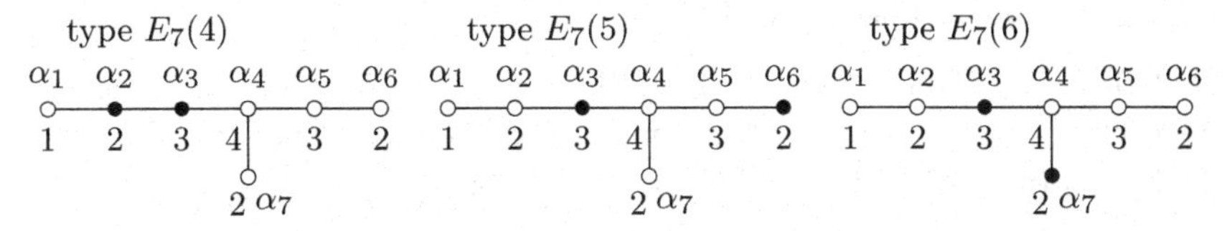

The highest root $\tilde{\alpha}$ of E_8 is given by $\tilde{\alpha} = 2\alpha_1+3\alpha_2+4\alpha_3+5\alpha_4+6\alpha_5+4\alpha_6+2\alpha_7+3\alpha_8$. Thus, for E_8 we find the following painted Dynkin diagrams:

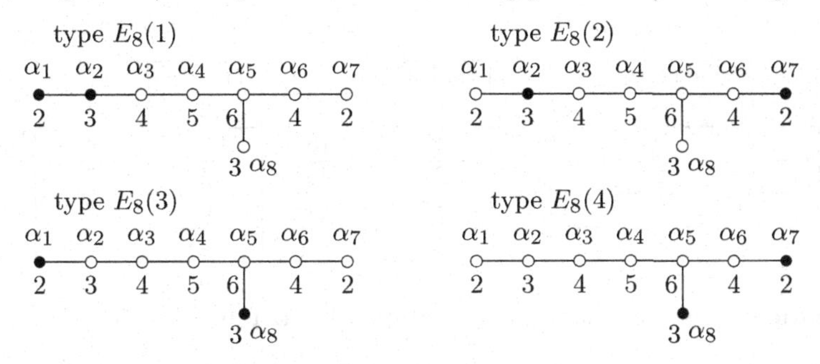

For $\Pi \setminus \Pi_0 = \{\alpha_i, \alpha_j\}$, we put $\xi_i = \kappa(\alpha_i)$ and $\xi_j = \kappa(\alpha_j)$. We give the set of all positive t-roots $\Delta_{\mathfrak{t}}^+$ in Table 1.

Table 1. Positive t-roots $\Delta_{\mathfrak{t}}^+$ for pairs (Π, Π_0)

Type	the set of all positive t-roots $\Delta_{\mathfrak{t}}^+$
$F_4(1)$	$\{\xi_1,\ \xi_2,\ \xi_1+\xi_2,\ \xi_1+2\xi_2,\ \xi_1+3\xi_2,\ 2\xi_1+3\xi_2\}$
$F_4(2)$	$\{\xi_2,\ \xi_4,\ \xi_2+\xi_4,\ \xi_2+2\xi_4,\ 2\xi_2,\ 2\xi_2+\xi_4,\ 2\xi_2+2\xi_4,\ 3\xi_2+2\xi_4\}$
$E_6(1)$	$\{\xi_2,\ \xi_3,\ \xi_2+\xi_3,\ \xi_2+2\xi_3,\ 2\xi_2+2\xi_3,\ 2\xi_2+3\xi_3\}$
$E_6(2)$	$\{\xi_3,\ \xi_6,\ \xi_3+\xi_6,\ 2\xi_3+\xi_6,\ 3\xi_3+\xi_6,\ 3\xi_3+2\xi_6\}$
$E_6(3)$	$\{\xi_3,\ \xi_4,\ \xi_3+\xi_4,\ 2\xi_3+\xi_4,\ 2\xi_3+2\xi_4,\ 3\xi_3+2\xi_4\}$
$E_7(1)$	$\{\xi_2,\ \xi_5,\ \xi_2+\xi_5,\ 2\xi_5,\ \xi_3+2\xi_5,\ 2\xi_3+\xi_5,\ 2\xi_2+2\xi_5,\ 2\xi_2+3\xi_5\}$
$E_7(2)$	$\{\xi_5,\ \xi_6,\ \xi_5+\xi_6,\ 2\xi_5+\xi_6,\ 3\xi_5+\xi_6,\ 3\xi_5+2\xi_6\}$
$E_7(3)$	$\{\xi_5,\ \xi_7,\ \xi_5+\xi_7,\ 2\xi_5+\xi_7,\ 2\xi_5+2\xi_7,\ 3\xi_5+2\xi_7\}$
$E_7(4)$	$\{\xi_2,\ \xi_3,\ \xi_2+\xi_3,\ \xi_2+2\xi_3,\ 2\xi_2+2\xi_3,\ 2\xi_2+3\xi_3\}$
$E_7(5)$	$\{\xi_3,\ \xi_6,\ \xi_3+\xi_6,\ 2\xi_3+\xi_6,\ 2\xi_3+2\xi_6,\ 3\xi_3+2\xi_6\}$
$E_7(6)$	$\{\xi_3,\ \xi_7,\ \xi_3+\xi_7,\ 2\xi_3+\xi_7,\ 2\xi_3+2\xi_7,\ 3\xi_3+\xi_7,\ 3\xi_3+2\xi_7\}$
$E_8(1)$	$\{\xi_1,\ \xi_2,\ \xi_1+\xi_2,\ \xi_1+2\xi_2,\ \xi_1+3\xi_2,\ 2\xi_1+3\xi_2\}$
$E_8(2)$	$\{\xi_2,\ \xi_7,\ \xi_2+\xi_7,\ 2\xi_2+\xi_7,\ \xi_2+2\xi_7,\ 2\xi_2+2\xi_7,\ 3\xi_2+2\xi_7\}$
$E_8(3)$	$\{\xi_1,\ \xi_8,\ \xi_1+\xi_8,\ 2\xi_8,\ \xi_1+2\xi_8,\ \xi_1+3\xi_8,\ 2\xi_1+3\xi_8\}$
$E_8(4)$	$\{\xi_7,\ \xi_8,\ \xi_7+\xi_8,\ \xi_7+2\xi_8,\ \xi_7+3\xi_8,\ 2\xi_7+2\xi_8,\ 2\xi_7+3\xi_8\}$
G_2	$\{\xi_1,\ \xi_2,\ \xi_1+\xi_2,\ \xi_1+2\xi_2,\ \xi_1+3\xi_2,\ 2\xi_1+3\xi_2\}$

From Table 1 we see that the painted Dynkin diagrams of types $F_4(1)$, $E_6(2)$, $E_7(2)$, $E_8(1)$ and G_2 correspond to generalized flag manifolds with

G_2-type $\mathfrak{t}$-roots. We put

$$\begin{aligned} &\mathfrak{m}(1,0) = (\mathfrak{m}_{\xi_i} + \mathfrak{m}_{-\xi_i})^{\tau}, && \mathfrak{m}(0,1) = (\mathfrak{m}_{\xi_j} + \mathfrak{m}_{-\xi_j})^{\tau}, \\ &\mathfrak{m}(1,1) = (\mathfrak{m}_{\xi_i+\xi_j} + \mathfrak{m}_{-\xi_i-\xi_j})^{\tau}, && \mathfrak{m}(1,2) = (\mathfrak{m}_{\xi_i+2\xi_j} + \mathfrak{m}_{-\xi_i-2\xi_j})^{\tau}, \\ &\mathfrak{m}(1,3) = (\mathfrak{m}_{\xi_i+3\xi_j} + \mathfrak{m}_{-\xi_i-3\xi_j})^{\tau}, && \mathfrak{m}(2,3) = (\mathfrak{m}_{2\xi_i+3\xi_j} + \mathfrak{m}_{-2\xi_i-3\xi_j})^{\tau}. \end{aligned} \tag{8}$$

By using tables of positive roots (eg. Table B in Appendix of [14, pp. 528–531]), we obtain the dimensions of these spaces as in the Table 2.

Table 2. Dimensions of irreducible components with G_2-type $\mathfrak{t}$-roots

Type	$\mathfrak{m}(1,0)$	$\mathfrak{m}(0,1)$	$\mathfrak{m}(1,1)$	$\mathfrak{m}(1,2)$	$\mathfrak{m}(1,3)$	$\mathfrak{m}(2,3)$
$F_4(1)$	2	12	12	12	2	2
$E_6(2)$	2	18	18	18	2	2
$E_7(2)$	2	30	30	30	2	2
$E_8(1)$	2	54	54	54	2	2
G_2	2	2	2	2	2	2

6. Kähler Einstein metrics of a generalized flag manifold

We put $Z_{\mathfrak{t}} = \{\Lambda \in \mathfrak{t} \mid 2(\Lambda, \alpha)/(\alpha, \alpha) \in \mathbb{Z} \text{ for each } \alpha \in \Delta\}$. Then $Z_{\mathfrak{t}}$ is a lattice of $\mathfrak{t}$ generated by $\{\Lambda_{i_1}, \ldots, \Lambda_{i_r}\}$. For each $\Lambda \in Z_{\mathfrak{t}}$ there exists a unique holomorphic character χ_Λ of U such that $\chi_\Lambda(\exp H) = \exp \Lambda(H)$ for each $H \in \mathfrak{h}^{\mathbb{C}}$. Then the correspondence $\Lambda \to \chi_\Lambda$ gives an isomorphism of $Z_{\mathfrak{t}}$ to the group of holomorphic characters of U.

Let F_Λ denote the holomorphic line bundle on $G^{\mathbb{C}}/U$ associated to the principal bundle $U \to G^{\mathbb{C}} \to G^{\mathbb{C}}/U$ by the holomorphic character χ_Λ, and $H(G^{\mathbb{C}}/U, \underline{\mathbb{C}}^*)$ the group of isomorphism classes of holomorphic line bundles on $G^{\mathbb{C}}/U$.

The correspondence $\Lambda \mapsto F_\Lambda : Z_{\mathfrak{t}} \to H(G^{\mathbb{C}}/U, \underline{\mathbb{C}}^*)$ induces a homomorphism. Also the correspondence $F \mapsto c_1(F)$ defines a homomorphism of $H(G^{\mathbb{C}}/U, \underline{\mathbb{C}}^*)$ to $H^2(M, \mathbb{Z})$.

Then it is known that the homomorphisms

$$Z_{\mathfrak{t}} \xrightarrow{F} H(G^{\mathbb{C}}/U, \underline{\mathbb{C}}^*) \xrightarrow{c_1} H^2(M, \mathbb{Z})$$

are in fact isomorphisms. In particular, we have that second Betti number $b_2(M)$ of M is given by

$$b_2(M) = \dim \mathfrak{t} = \text{the cardinality of } \Pi \setminus \Pi_0$$

We put $Z_{\mathfrak{t}}^+ = \{\lambda \in Z_{\mathfrak{t}} \mid (\lambda, \alpha) > 0 \text{ for } \alpha \in \Pi - \Pi_0\}$. Then we see that $Z_{\mathfrak{t}}^+ = \sum_{\alpha \in \Pi \setminus \Pi_0} \mathbb{Z}^+ \Lambda_\alpha$. We define elements $\delta_{\mathfrak{m}}$ and δ of $\sqrt{-1}\mathfrak{h}$ by

$$\delta_{\mathfrak{m}} = \frac{1}{2} \sum_{\alpha \in \Delta_{\mathfrak{m}}^+} \alpha \quad \text{and} \quad \delta = \frac{1}{2} \sum_{\alpha \in \Delta^+} \alpha.$$

Let $c_1(M)$ be the first Chern class of M. Then we have

$$2\delta_{\mathfrak{m}} \in Z_{\mathfrak{t}}^{+}, \quad c_1(M) = c_1(F_{2\delta_{\mathfrak{m}}}).$$

Put $k_\alpha = 2(2\delta_{\mathfrak{m}}, \alpha)/(\alpha, \alpha)$ for $\alpha \in \Pi \setminus \Pi_0$. Then $2\delta_{\mathfrak{m}} = \sum_{\alpha \in \Pi \setminus \Pi_0} k_\alpha \Lambda_\alpha = k_{\alpha_{i_1}} \Lambda_{\alpha_{i_1}} + \cdots + k_{\alpha_{i_r}} \Lambda_{\alpha_{i_r}}$ and each $k_{\alpha_{i_s}}$ is a positive integer.

The G-invariant metric $g_{2\delta_{\mathfrak{m}}}$ on G/K corresponding to $2\delta_{\mathfrak{m}}$, which is a Kähler Einstein metric, is given by

$$g_{2\delta_{\mathfrak{m}}} = \sum_{j_1, \cdots, j_r} \left(\sum_{\ell=1}^{r} k_{\alpha_{i_\ell}} j_\ell \, (\alpha_{i_\ell}, \alpha_{i_\ell})/2 \right) B|_{\mathfrak{m}(j_1, \cdots, j_r)}.$$

Example 6.1. For generalized flag manifolds G/K with G_2-type $\mathfrak{t}$-roots, we put $\Pi \setminus \Pi_0 = \{\alpha_i, \alpha_j : \mathrm{Mrk}(\alpha_i) = 2, \ \mathrm{Mrk}(\alpha_j) = 3\}$. Then we see that

$$2\delta_{\mathfrak{m}} = 2\Lambda_{\alpha_i} + 2t\Lambda_{\alpha_j}$$

for types $F_4(1)$, $E_6(2)$, $E_7(2)$ and $E_8(1)$. The number t is given in Table 3.

Table 3.

Type	$F_4(1)$	$E_6(2)$	$E_7(2)$	$E_8(1)$
t	2	3	5	9

In fact, we see that, for the type $F_4(1)$,

$$\begin{aligned} 2\delta_{\mathfrak{m}} &= 2\delta - 2(\alpha_3 + \alpha_4) = 2(\Lambda_1 + \Lambda_2 + \Lambda_3 + \Lambda_4) \\ &\quad -2((-\Lambda_2 + 2\Lambda_3 - \Lambda_4) + (-\Lambda_3 + 2\Lambda_4)) = 2(\Lambda_1 + 2\Lambda_2), \end{aligned}$$

for the type $E_6(2)$,

$$\begin{aligned} 2\delta_{\mathfrak{m}} &= 2\delta - 2(\alpha_1 + \alpha_2) - 2(\alpha_4 + \alpha_5) = 2(\Lambda_1 + \Lambda_2 + \Lambda_3 + \Lambda_4 + \Lambda_5 + \Lambda_6) \\ &\quad - 2((2\Lambda_1 - \Lambda_2) + (-\Lambda_1 + 2\Lambda_2 - \Lambda_3) + (-\Lambda_3 + 2\Lambda_4 - \Lambda_5) \\ &\quad + (-\Lambda_4 + 2\Lambda_5)) = 2(\Lambda_6 + 3\Lambda_3). \end{aligned}$$

For the type $E_7(2)$, we see that Π_0 is of type A_5 and thus

$$\begin{aligned} 2\delta_{\mathfrak{m}} &= 2\delta - (5\alpha_1 + 8\alpha_2 + 9\alpha_3 + 8\alpha_4 + 5\alpha_7) \\ &= 2(\Lambda_1 + \Lambda_2 + \Lambda_3 + \Lambda_4 + \Lambda_5 + \Lambda_6 + \Lambda_7) - (5(2\Lambda_1 - \Lambda_2) \\ &\quad + 8(-\Lambda_1 + 2\Lambda_2 - \Lambda_3) + 9(-\Lambda_2 + 2\Lambda_3 - \Lambda_4) \\ &\quad + 8(-\Lambda_3 + 2\Lambda_4 - \Lambda_5 - \Lambda_7) + 5(-\Lambda_4 + 2\Lambda_7)) = 2(\Lambda_6 + 5\Lambda_5). \end{aligned}$$

For the type $E_8(1)$, we see that Π_0 is of type E_6 and thus

$$\begin{aligned} 2\delta_{\mathfrak{m}} &= 2\delta - 2(8\alpha_3 + 15\alpha_4 + 21\alpha_5 + 15\alpha_6 + 8\alpha_7 + 11\alpha_8) \\ &= 2(\Lambda_1 + \Lambda_2 + \Lambda_3 + \Lambda_4 + \Lambda_5 + \Lambda_6 + \Lambda_7 + \Lambda_8) - 2(8(-\Lambda_2 + 2\Lambda_3 - \Lambda_4) \\ &\quad + 15(-\Lambda_3 + 2\Lambda_4 - \Lambda_5) + 21(-\Lambda_4 + 2\Lambda_5 - \Lambda_6 - \Lambda_8) \end{aligned}$$

$$+ 15(-\Lambda_5 + 2\Lambda_6 - \Lambda_7) + 8(-\Lambda_6 + 2\Lambda_7) + 11(-\Lambda_5 + 2\Lambda_8)) = 2(\Lambda_1 + 9\Lambda_2).$$

Thus the Kähler Einstein metric $g_{2\delta_{\mathfrak{m}}}$ on G/K is given by

$$\begin{aligned} g_{2\delta_{\mathfrak{m}}} &= 2B|_{\mathfrak{m}(1,0)} + 2tB|_{\mathfrak{m}(0,1)} + 2(1+t)B|_{\mathfrak{m}(1,1)} + 2(1+2t)B|_{\mathfrak{m}(1,2)} \\ &\quad + 2(1+3t)B|_{\mathfrak{m}(1,3)} + 2(2+3t)B|_{\mathfrak{m}(2,3)}. \end{aligned} \tag{9}$$

For the flag manifold G_2/T, we have $2\delta_{\mathfrak{m}} = 2\delta = 2\Lambda_{\alpha_1} + 2\Lambda_{\alpha_2}$. Thus

$$\begin{aligned} g_{2\delta_{\mathfrak{m}}} &= 6B|_{\mathfrak{m}(1,0)} + 2B|_{\mathfrak{m}(0,1)} + 8B|_{\mathfrak{m}(1,1)} + 10B|_{\mathfrak{m}(1,2)} \\ &\quad + 12B|_{\mathfrak{m}(1,3)} + 18B|_{\mathfrak{m}(2,3)}. \end{aligned} \tag{10}$$

Note that (10) is obtained from (9) by taking $t = 1/3$ and then multiplying by the scalar 3.

7. Generalized flag manifolds with two or three isotropy summands

From now on we assume that the Lie group G is simple. For a generalized flag manifold G/L, we denote by q the number of mutually non equivalent irreducible $\mathrm{Ad}_G(L)$-modules $\mathfrak{m}(j_1, \ldots, j_r)$ with $\mathfrak{m} = \sum_{j_1,\cdots,j_r} \mathfrak{m}(j_1, \ldots, j_r)$.

We consider the components r_i of Ricci tensor r of the metric g given by (7) for $q = 2, 3$.

In the case when $q = 2$, we see that only the case $r = b_2(G/L) = 1$ ocurrs. Note that the only non zero structure constant is $\left[\!\left[\begin{smallmatrix} 2 \\ 11 \end{smallmatrix} \right]\!\right]$. Put $\check{d}_1 = \dim \mathfrak{m}(1)$ and $\check{d}_2 = \dim \mathfrak{m}(2)$. For a G-invariant metric $\check{g} = x_1 \cdot B|_{\mathfrak{m}(1)} + x_2 \cdot B|_{\mathfrak{m}(2)}$, the components $\check{r}_1, \check{r}_2$ of Ricci tensor $\mathrm{ric}(\check{g})$ of the metric $\check{g}$ are given by

$$\begin{cases} \check{r}_1 = \dfrac{1}{2x_1} - \dfrac{x_2}{2\,\check{d}_1\,{x_1}^2} \left[\!\left[\begin{matrix} 2 \\ 11 \end{matrix} \right]\!\right] \\[2ex] \check{r}_2 = \dfrac{1}{2x_2} - \dfrac{1}{2\,\check{d}_2\,x_2} \left[\!\left[\begin{matrix} 1 \\ 21 \end{matrix} \right]\!\right] + \dfrac{x_2}{4\,\check{d}_2\,{x_1}^2} \left[\!\left[\begin{matrix} 2 \\ 11 \end{matrix} \right]\!\right] . \end{cases} \tag{11}$$

Since the metric $\check{g} = 1 \cdot B|_{\mathfrak{m}(1)} + 2 \cdot B|_{\mathfrak{m}(2)}$ is Kähler Einstein, we see that $\left[\!\left[\begin{smallmatrix} 2 \\ 11 \end{smallmatrix} \right]\!\right] = \dfrac{\check{d}_1 \check{d}_2}{\check{d}_1 + 4\check{d}_2}$.

Example 7.1. For a generalized flag manifold G/K with G_2-type $\mathfrak{t}$-roots we use the notations (8), we put $\mathfrak{l}_1 = \mathfrak{k} + \mathfrak{m}(0,1)$. Then $\mathfrak{l}_1$ is a Lie subalgebra of $\mathfrak{g}$ and we denote by L_1 the corresponding Lie subgroup of G. Note that

G/L_1 is a generalized flag manifold with two irreducible isotropy summands $\mathfrak{p}_1 = \mathfrak{m}(1)$ and $\mathfrak{p}_2 = \mathfrak{m}(2)$, where we see that

$$\mathfrak{p}_1 = \mathfrak{m}(1,0) \oplus \mathfrak{m}(1,1) \oplus \mathfrak{m}(1,2) \oplus \mathfrak{m}(1,3), \quad \mathfrak{p}_2 = \mathfrak{m}(2,3).$$

In the case when $q = 3$, we see that either $r = b_2(G/L) = 1$ or $r = b_2(G/L) = 2$ ocurrs. Here we consider the case of $r = b_2(G/L) = 1$. Note that the non zero structure constants are $\left[\!\left[\begin{smallmatrix} 2 \\ 11 \end{smallmatrix}\right]\!\right]$ and $\left[\!\left[\begin{smallmatrix} 3 \\ 12 \end{smallmatrix}\right]\!\right]$. Put $\check{d}_1 = \dim \mathfrak{m}(1)$, $\check{d}_2 = \dim \mathfrak{m}(2)$ and $\check{d}_3 = \dim \mathfrak{m}(3)$. For a G-invariant metric $\check{g} = v_1 \cdot B|_{\mathfrak{m}(1)} + v_2 \cdot B|_{\mathfrak{m}(2)} + v_3 \cdot B|_{\mathfrak{m}(3)}$, the components $\check{r}_1, \check{r}_2, \check{r}_3$ of Ricci tensor $\mathrm{ric}(\check{g})$ of the metric $\check{g}$ are given by

$$\begin{cases} \check{r}_1 = \dfrac{1}{2v_1} - \dfrac{v_2}{2\check{d}_1\, {v_1}^2}\left[\!\left[\begin{matrix} 2 \\ 11 \end{matrix}\right]\!\right] + \dfrac{1}{2\check{d}_1}\left[\!\left[\begin{matrix} 3 \\ 12 \end{matrix}\right]\!\right]\left(\dfrac{v_1}{v_2 v_3} - \dfrac{v_2}{v_1 v_3} - \dfrac{v_3}{v_1 v_2}\right) \\[2ex] \check{r}_2 = \dfrac{1}{2v_2} + \dfrac{1}{4\check{d}_2}\left[\!\left[\begin{matrix} 2 \\ 11 \end{matrix}\right]\!\right]\left(\dfrac{v_2}{{v_1}^2} - \dfrac{2}{v_2}\right) + \dfrac{1}{2\check{d}_2}\left[\!\left[\begin{matrix} 3 \\ 12 \end{matrix}\right]\!\right]\left(\dfrac{v_2}{v_1 v_3} - \dfrac{v_1}{v_2 v_3} - \dfrac{v_3}{v_1 v_2}\right) \\[2ex] \check{r}_3 = \dfrac{1}{2v_3} + \dfrac{1}{2\check{d}_3}\left[\!\left[\begin{matrix} 3 \\ 12 \end{matrix}\right]\!\right]\left(\dfrac{v_3}{v_1 v_2} - \dfrac{v_1}{v_2 v_3} - \dfrac{v_2}{v_1 v_3}\right). \end{cases} \tag{12}$$

Since the metric $\check{g} = 1 \cdot B|_{\mathfrak{m}(1)} + 2 \cdot B|_{\mathfrak{m}(2)} + 3 \cdot B|_{\mathfrak{m}(3)}$ is Kähler Einstein, we see that

$$\left[\!\left[\begin{matrix} 2 \\ 11 \end{matrix}\right]\!\right] = \frac{\check{d}_1\check{d}_2 + 2\check{d}_1\check{d}_3 - \check{d}_2\check{d}_3}{\check{d}_1 + 4\check{d}_2 + 9\check{d}_3}, \quad \left[\!\left[\begin{matrix} 3 \\ 12 \end{matrix}\right]\!\right] = \frac{(\check{d}_1 + \check{d}_2)\check{d}_3}{\check{d}_1 + 4\check{d}_2 + 9\check{d}_3}. \tag{13}$$

Example 7.2. For a generalized flag manifold G/K with G_2-type $\mathfrak{t}$-roots we use the notations (8), we put $\mathfrak{l}_2 = \mathfrak{k} + \mathfrak{m}(1,0)$. Then $\mathfrak{l}_2$ is a Lie subalgebra of $\mathfrak{g}$ and we denote by L_2 the corresponding Lie subgroup of G. Note that G/L_2 is a generalized flag manifold with three irreducible isotropy summands $\mathfrak{q}_1 = \mathfrak{m}(1)$, $\mathfrak{q}_2 = \mathfrak{m}(2)$ and $\mathfrak{q}_3 = \mathfrak{m}(3)$, where we see that

$$\mathfrak{q}_1 = \mathfrak{m}(0,1) \oplus \mathfrak{m}(1,1), \quad \mathfrak{q}_2 = \mathfrak{m}(1,2), \quad \mathfrak{q}_3 = \mathfrak{m}(1,3) \oplus \mathfrak{m}(2,3).$$

8. Generalized flag manifolds with G_2-type $\mathfrak{t}$-roots

We now consider the generalized flag manifolds G/K with G_2-type $\mathfrak{t}$-roots. We have a decomposition of $\mathfrak{m}$ into six mutually non equivalent irreducible $\mathrm{Ad}_G(K)$-modules

$$\mathfrak{m} = \mathfrak{m}(1,0) \oplus \mathfrak{m}(0,1) \oplus \mathfrak{m}(1,1) \oplus \mathfrak{m}(1,2) \oplus \mathfrak{m}(1,3) \oplus \mathfrak{m}(2,3).$$

We put $\mathfrak{m}_1 = \mathfrak{m}(1,0)$, $\mathfrak{m}_2 = \mathfrak{m}(0,1)$, $\mathfrak{m}_3 = \mathfrak{m}(1,1)$, $\mathfrak{m}_4 = \mathfrak{m}(1,2)$, $\mathfrak{m}_5 = \mathfrak{m}(1,3)$ and $\mathfrak{m}_6 = \mathfrak{m}(2,3)$. Then, from Table 2 we see that, for types $F_4(1)$,

$E_6(2)$, $E_7(2)$ and $E_8(1)$, $\dim \mathfrak{m}_1 = 2$, $\dim \mathfrak{m}_2 = 6t$, $\dim \mathfrak{m}_3 = 6t$, $\dim \mathfrak{m}_4 = 6t$, $\dim \mathfrak{m}_5 = 2$ and $\dim \mathfrak{m}_6 = 2$, where t is given in the Table 3, and, for type G_2, $\dim \mathfrak{m}_1 = 2$, $\dim \mathfrak{m}_2 = 2$, $\dim \mathfrak{m}_3 = 2$, $\dim \mathfrak{m}_4 = 2$, $\dim \mathfrak{m}_5 = 2$. We also see that

$$\begin{gathered}
[\mathfrak{m}_1, \mathfrak{m}_2] = \mathfrak{m}_3, \quad [\mathfrak{m}_1, \mathfrak{m}_3] = \mathfrak{m}_2, \quad [\mathfrak{m}_1, \mathfrak{m}_5] = \mathfrak{m}_6, \quad [\mathfrak{m}_1, \mathfrak{m}_6] = \mathfrak{m}_5, \\
[\mathfrak{m}_2, \mathfrak{m}_3] \subset \mathfrak{m}_1 \oplus \mathfrak{m}_4, \quad [\mathfrak{m}_2, \mathfrak{m}_4] \subset \mathfrak{m}_3 \oplus \mathfrak{m}_5, \quad [\mathfrak{m}_2, \mathfrak{m}_5] = \mathfrak{m}_4, \\
[\mathfrak{m}_3, \mathfrak{m}_4] \subset \mathfrak{m}_2 \oplus \mathfrak{m}_6, \quad [\mathfrak{m}_3, \mathfrak{m}_6] = \mathfrak{m}_4, \quad [\mathfrak{m}_4, \mathfrak{m}_5] = \mathfrak{m}_2, \\
[\mathfrak{m}_4, \mathfrak{m}_6] = \mathfrak{m}_3, \quad [\mathfrak{m}_5, \mathfrak{m}_6] = \mathfrak{m}_1.
\end{gathered}$$

Hence, we conclude that only the structure constants $\begin{bmatrix} 3 \\ 12 \end{bmatrix}$, $\begin{bmatrix} 6 \\ 15 \end{bmatrix}$, $\begin{bmatrix} 4 \\ 23 \end{bmatrix}$, $\begin{bmatrix} 5 \\ 24 \end{bmatrix}$, $\begin{bmatrix} 6 \\ 34 \end{bmatrix}$ and their symmetries are non-zero. We write G-invariant metrics g on G/K as

$$g = x_1 B|_{\mathfrak{m}_1} + x_2 B|_{\mathfrak{m}_2} + x_3 B|_{\mathfrak{m}_3} + x_4 B|_{\mathfrak{m}_4} + x_5 B|_{\mathfrak{m}_5} + x_6 B|_{\mathfrak{m}_6} \tag{14}$$

where x_j $(j = 1, \ldots, 6)$ are positive numbers.

Now from Lemma 2.1, we obtain the following proposition.

Proposition 8.1. *The components r_i $(i = 1, \ldots, 6)$ of the Ricci tensor for a G-invariant Riemannian metric on G/K determined by* (14) *are given as follows:*

$$\begin{aligned}
r_1 &= \frac{1}{2x_1} + \frac{1}{2d_1}\begin{bmatrix} 3 \\ 12 \end{bmatrix}\Big(\frac{x_1}{x_2 x_3} - \frac{x_2}{x_1 x_3} - \frac{x_3}{x_1 x_2}\Big) + \frac{1}{2d_1}\begin{bmatrix} 6 \\ 15 \end{bmatrix}\Big(\frac{x_1}{x_5 x_6} - \frac{x_6}{x_1 x_5} - \frac{x_5}{x_1 x_6}\Big), \\
r_2 &= \frac{1}{2x_2} + \frac{1}{2d_2}\begin{bmatrix} 3 \\ 12 \end{bmatrix}\Big(\frac{x_2}{x_1 x_3} - \frac{x_1}{x_2 x_3} - \frac{x_3}{x_1 x_2}\Big) \\
&\quad + \frac{1}{2d_2}\begin{bmatrix} 4 \\ 23 \end{bmatrix}\Big(\frac{x_2}{x_3 x_4} - \frac{x_3}{x_2 x_4} - \frac{x_4}{x_2 x_3}\Big) + \frac{1}{2d_2}\begin{bmatrix} 5 \\ 24 \end{bmatrix}\Big(\frac{x_2}{x_4 x_5} - \frac{x_4}{x_2 x_5} - \frac{x_5}{x_2 x_4}\Big), \\
r_3 &= \frac{1}{2x_3} + \frac{1}{2d_3}\begin{bmatrix} 3 \\ 12 \end{bmatrix}\Big(\frac{x_3}{x_1 x_2} - \frac{x_2}{x_1 x_3} - \frac{x_1}{x_2 x_3}\Big) \\
&\quad + \frac{1}{2d_3}\begin{bmatrix} 4 \\ 23 \end{bmatrix}\Big(\frac{x_3}{x_2 x_4} - \frac{x_4}{x_2 x_3} - \frac{x_2}{x_3 x_4}\Big) + \frac{1}{2d_3}\begin{bmatrix} 6 \\ 34 \end{bmatrix}\Big(\frac{x_3}{x_4 x_6} - \frac{x_4}{x_3 x_6} - \frac{x_6}{x_3 x_4}\Big), \\
r_4 &= \frac{1}{2x_4} + \frac{1}{2d_4}\begin{bmatrix} 4 \\ 23 \end{bmatrix}\Big(\frac{x_4}{x_2 x_3} - \frac{x_2}{x_3 x_4} - \frac{x_3}{x_2 x_4}\Big) \\
&\quad + \frac{1}{2d_4}\begin{bmatrix} 5 \\ 24 \end{bmatrix}\Big(\frac{x_4}{x_2 x_5} - \frac{x_5}{x_2 x_4} - \frac{x_2}{x_4 x_5}\Big) + \frac{1}{2d_4}\begin{bmatrix} 6 \\ 34 \end{bmatrix}\Big(\frac{x_4}{x_3 x_6} - \frac{x_3}{x_4 x_6} - \frac{x_6}{x_3 x_4}\Big),
\end{aligned}$$

$$r_5 = \frac{1}{2x_5} + \frac{1}{2d_5}\begin{bmatrix}5\\24\end{bmatrix}\left(\frac{x_5}{x_2x_4} - \frac{x_2}{x_4x_5} - \frac{x_4}{x_2x_5}\right) + \frac{1}{2d_5}\begin{bmatrix}6\\15\end{bmatrix}\left(\frac{x_5}{x_1x_6} - \frac{x_6}{x_1x_5} - \frac{x_1}{x_5x_6}\right),$$

$$r_6 = \frac{1}{2x_6} + \frac{1}{2d_6}\begin{bmatrix}6\\15\end{bmatrix}\left(\frac{x_6}{x_1x_5} - \frac{x_5}{x_1x_6} - \frac{x_1}{x_6x_5}\right) + \frac{1}{2d_6}\begin{bmatrix}6\\34\end{bmatrix}\left(\frac{x_6}{x_3x_4} - \frac{x_3}{x_4x_6} - \frac{x_4}{x_3x_6}\right).$$

Now we compute the non-negative numbers $\begin{bmatrix}k\\ij\end{bmatrix}$.

Proposition 8.2. *For the types $F_4(1)$, $E_6(2)$, $E_7(2)$ and $E_8(1)$, we have*

$$\begin{bmatrix}3\\12\end{bmatrix} = \frac{t}{t+1}, \begin{bmatrix}6\\15\end{bmatrix} = \frac{1}{3(t+1)}, \begin{bmatrix}4\\23\end{bmatrix} = t, \begin{bmatrix}5\\24\end{bmatrix} = \frac{t}{t+1}, \begin{bmatrix}6\\34\end{bmatrix} = \frac{t}{t+1}.$$

For the type G_2, we have

$$\begin{bmatrix}3\\12\end{bmatrix} = \frac{1}{4}, \begin{bmatrix}6\\15\end{bmatrix} = \frac{1}{4}, \begin{bmatrix}4\\23\end{bmatrix} = \frac{1}{3}, \begin{bmatrix}5\\24\end{bmatrix} = \frac{1}{4}, \begin{bmatrix}6\\34\end{bmatrix} = \frac{1}{4}.$$

Proof. We consider the subgroups L_1 and L_2 of G as in Examples 7.1 and 7.2. Then we have Riemannian submersions $G/K \to G/L_1$ and $G/K \to G/L_2$. Note that G/L_1 and G/L_2 are generalized flag manifolds with two and three isotropy summands respectively.

We consider a G-invariant metric g_1 on G/K defined by a Riemannian submersion $\pi : (G/K, g_1) \to (G/L_1, \breve{g}_1)$ with totally geodesic fibers isometric to $(L_1/K, \hat{g}_1)$. Then we see that the metric g_1 on G/K and the metric $\breve{g}_1$ on G/L_1 are of the forms

$$g_1 = y_1 B|_{\mathfrak{p}_1} + y_2 B|_{\mathfrak{p}_2} + z_1 B|_{\mathfrak{n}_1} \quad \text{and} \quad \breve{g}_1 = y_1 B|_{\mathfrak{p}_1} + y_2 B|_{\mathfrak{p}_2}.$$

We decompose irreducible components $\mathfrak{p}_1$ and $\mathfrak{p}_2$ into irreducible $\mathrm{Ad}(K)$-modules. Then we obtain that

$$\mathfrak{p}_1 = \mathfrak{m}_1 \oplus \mathfrak{m}_3 \oplus \mathfrak{m}_4 \oplus \mathfrak{m}_5, \quad \mathfrak{p}_2 = \mathfrak{m}_6, \quad \mathfrak{n}_1 = \mathfrak{m}_2.$$

Thus the metric g_1 can be written as

$$g_1 = y_1 B|_{\mathfrak{m}_1} + z_1 B|_{\mathfrak{m}_2} + y_1 B|_{\mathfrak{m}_3} + y_1 B|_{\mathfrak{m}_4} + y_1 B|_{\mathfrak{m}_5} + y_2 B|_{\mathfrak{m}_6}. \quad (15)$$

Since the metric (15) is a special case of the metric (14), from Proposition 8.1 we see that the components r_1, r_3, r_4, r_5 of the Ricci tensor for the metric g_1 on G/K are given by

$$r_1 = \frac{1}{2y_1} - \frac{1}{2d_1}\begin{bmatrix}3\\12\end{bmatrix}\frac{z_1}{{y_1}^2} - \frac{1}{2d_1}\begin{bmatrix}6\\15\end{bmatrix}\frac{y_2}{{y_1}^2}, \tag{16}$$

$$r_3 = \frac{1}{2y_1} - \frac{1}{2d_3}\begin{bmatrix}3\\12\end{bmatrix}\frac{z_1}{{y_1}^2} - \frac{1}{2d_3}\begin{bmatrix}4\\23\end{bmatrix}\frac{z_1}{{y_1}^2} - \frac{1}{2d_3}\begin{bmatrix}6\\34\end{bmatrix}\frac{y_2}{{y_1}^2} \tag{17}$$

$$r_4 = \frac{1}{2y_1} - \frac{1}{2d_4}\begin{bmatrix}4\\23\end{bmatrix}\frac{z_1}{{y_1}^2} - \frac{1}{2d_4}\begin{bmatrix}5\\24\end{bmatrix}\frac{z_1}{{y_1}^2} - \frac{1}{2d_4}\begin{bmatrix}6\\34\end{bmatrix}\frac{y_2}{{y_1}^2} \tag{18}$$

$$r_5 = \frac{1}{2y_1} - \frac{1}{2d_5}\begin{bmatrix}5\\24\end{bmatrix}\frac{z_1}{{y_1}^2} - \frac{1}{2d_5}\begin{bmatrix}6\\15\end{bmatrix}\frac{y_2}{{y_1}^2}. \tag{19}$$

From (11), we see that the component $\check{r}_1$ of the Ricci tensor for the metric $\check{g}_1$ on G/L_1 is given by

$$\check{r}_1 = \frac{1}{2y_1} - \frac{y_2}{2\,{y_1}^2}\frac{\check{d}_2}{\check{d}_1 + 4\check{d}_2}. \tag{20}$$

Note that, for the types $F_4(1)$, $E_6(2)$, $E_7(2)$ and $E_8(1)$, we have $d_1 = 2$, $d_2 = 6t$, $d = 6t$, $d_4 = 6t$, $d_5 = 2$ and $d_6 = 2$ and hence $\check{d}_1 = \dim \mathfrak{p}_1 = 12t + 4$, $\check{d}_2 = \dim \mathfrak{p}_2 = 2$. For the type G_2, we have $\check{d}_1 = \dim \mathfrak{p}_1 = 8$, $\check{d}_2 = \dim \mathfrak{p}_2 = 2$.

For the types $F_4(1)$, $E_6(2)$, $E_7(2)$ and $E_8(1)$, by comparing the component $\check{r}_1$ of the Ricci tensor for the metric $\check{g}_1$ on G/L_1 with the horizontal part of (16), (17), from Lemma 3.1, we obtain that $\begin{bmatrix}6\\15\end{bmatrix} = \frac{1}{3(t+1)}$ and $\begin{bmatrix}6\\34\end{bmatrix} = \frac{t}{t+1}$. For the type G_2, we obtain that $\begin{bmatrix}6\\15\end{bmatrix} = \frac{1}{4}$ and $\begin{bmatrix}6\\34\end{bmatrix} = \frac{1}{4}$ by a similar method.

We also consider a G-invariant metric g_2 on G/K defined by a Riemannian submersion $\pi : (G/K,\, g_2) \to (G/L_2,\, \check{g}_2)$ with totally geodesic fibers isometric to $(L_2/K,\, \hat{g}_2)$. Then we see that the metric g_2 on G/K and the metric $\check{g}_2$ on G/L_2 are of the forms

$$g_2 = v_1 B|_{\mathfrak{q}_1} + v_2 B|_{\mathfrak{q}_2} + v_3 B|_{\mathfrak{q}_3} + w_1 B|_{\mathfrak{s}_1} \text{ and } \check{g}_2 = v_1 B|_{\mathfrak{q}_1} + v_2 B|_{\mathfrak{q}_2} + v_3 B|_{\mathfrak{q}_3}.$$

We decompose irreducible components $\mathfrak{q}_1$, $\mathfrak{q}_2$ and $\mathfrak{q}_3$ into irreducible $\mathrm{Ad}(K)$-modules. Then we obtain that

$$\mathfrak{q}_1 = \mathfrak{m}_2 \oplus \mathfrak{m}_3, \quad \mathfrak{q}_2 = \mathfrak{m}_4, \quad \mathfrak{q}_3 = \mathfrak{m}_5 \oplus \mathfrak{m}_6, \quad \mathfrak{s}_1 = \mathfrak{m}_1.$$

Thus the metric g_2 can be written as

$$g_2 = w_1 B|_{\mathfrak{m}_1} + v_1 B|_{\mathfrak{m}_2} + v_1 B|_{\mathfrak{m}_3} + v_2 B|_{\mathfrak{m}_4} + v_3 B|_{\mathfrak{m}_5} + v_3 B|_{\mathfrak{m}_6}. \tag{21}$$

Since the metric (21) is a special case of the metric (14), from Proposition 8.1 we see that the components r_2, r_3 of the Ricci tensor for the G-invariant Riemannian metric g_2 on G/K are given by

$$r_2 = \frac{1}{2v_1} - \frac{1}{2d_2}\begin{bmatrix}3\\12\end{bmatrix}\frac{w_1}{v_1{}^2} - \frac{1}{2d_2}\begin{bmatrix}4\\23\end{bmatrix}\frac{v_2}{v_1{}^2} + \frac{1}{2d_2}\begin{bmatrix}5\\24\end{bmatrix}\left(\frac{v_1}{v_2v_3} - \frac{v_2}{v_1v_3} - \frac{v_3}{v_1v_2}\right),$$

$$r_3 = \frac{1}{2v_1} - \frac{1}{2d_3}\begin{bmatrix}3\\12\end{bmatrix}\frac{w_1}{v_1{}^2} - \frac{1}{2d_3}\begin{bmatrix}4\\23\end{bmatrix}\frac{v_2}{v_1{}^2} + \frac{1}{2d_3}\begin{bmatrix}6\\34\end{bmatrix}\left(\frac{v_1}{v_2v_3} - \frac{v_2}{v_1v_3} - \frac{v_3}{v_1v_2}\right).$$

From Lemma 3.1, by comparing the component $\check{r}_1$ of the Ricci tensor for the metric $\check{g}_2$ on G/L_2 given by (12) with the horizontal part of r_2 and r_3, we obtain that

$$\frac{1}{\check{d}_1}\left[\begin{bmatrix}2\\11\end{bmatrix}\right] = \frac{1}{d_2}\begin{bmatrix}4\\23\end{bmatrix}, \quad \frac{1}{\check{d}_1}\left[\begin{bmatrix}3\\12\end{bmatrix}\right] = \frac{1}{d_2}\begin{bmatrix}5\\24\end{bmatrix} = \frac{1}{d_3}\begin{bmatrix}6\\34\end{bmatrix}. \tag{22}$$

Note that for the types $F_4(1)$, $E_6(2)$, $E_7(2)$ and $E_8(1)$, we see that $\check{d}_1 = 12t$, $\check{d}_2 = 6t$ and $\check{d}_3 = 4$. By substituting these values in the equations (13), we obtain that $\begin{bmatrix}4\\23\end{bmatrix} = t$ and $\begin{bmatrix}5\\24\end{bmatrix} = \begin{bmatrix}6\\34\end{bmatrix} = \dfrac{t}{t+1}$. Now taking into account the explicit form (9) of the Kähler Einstein metric $g_{2\delta_{\mathfrak{m}}}$ on G/K in Example 6.1, and substituting the values $x_1 = 2$, $x_2 = 2t$, $x_3 = 2(1+t)$, $x_4 = 2(1+2t)$, $x_5 = 2(1+3t)$, $x_6 = 2(2+3t)$ for $r_1 = r_3$ in Proposition 8.1, we obtain that $\begin{bmatrix}3\\12\end{bmatrix} = \dfrac{t}{t+1}$.

For the type G_2, we have $\check{d}_1 = 4$, $\check{d}_2 = 2$ and $\check{d}_3 = 4$, and, by substituting these values in the equations (13), we obtain that $\begin{bmatrix}4\\23\end{bmatrix} = \dfrac{1}{3}$ and $\begin{bmatrix}5\\24\end{bmatrix} = \begin{bmatrix}6\\34\end{bmatrix} = \dfrac{1}{4}$. Note that the Kähler Einstein metric $g_{2\delta_{\mathfrak{m}}}$ on G/K is given by (10). By substituting the values $x_1 = 6$, $x_2 = 2$, $x_3 = 8$, $x_4 = 10$, $x_5 = 12$, $x_6 = 18$ for $r_1 = r_3$ in Proposition 8.1, we obtain that $\begin{bmatrix}3\\12\end{bmatrix} = \dfrac{1}{4}$. □

9. Proof of the theorems

Let $1 \leq p \leq 6$ and W_1 be a subgroup of the Weyl group of the Lie algebra $\mathfrak{g}$ on its root system. Then if each element $w \in W_1$ satisfies $w(\mathfrak{m}_p) = \mathfrak{m}_q$ for some q, then the action of W_1 induces an action on the components of the G-invariant metric (14). Thus each element of the group W_1 induces an isometry on the generalized flag manifold G/K. For generalized flag manifolds G/K with G_2-type $\mathfrak{t}$-roots, we put $\Pi \setminus \Pi_0 = \{\alpha_i, \alpha_j : \mathrm{Mrk}(\alpha_i) = 2,\ \mathrm{Mrk}(\alpha_j) = 3\}$ as in Section 5. We consider reflections s_{α_i} about the root

α_i, $s_{\widetilde{\alpha}}$ about the highest root $\widetilde{\alpha}$, $s_{\widetilde{\alpha}-\alpha_i}$ about the root $\widetilde{\alpha}-\alpha_i$, and let W_1 be the subgroup generated by these reflections.

Lemma 9.1. *The reflections s_{α_i}, $s_{\widetilde{\alpha}}$ and $s_{\widetilde{\alpha}-\alpha_i}$ satisfy the following:*

$$s_{\widetilde{\alpha}}(\mathfrak{m}_1)=\mathfrak{m}_5,\ s_{\widetilde{\alpha}}(\mathfrak{m}_2)=\mathfrak{m}_2,\ s_{\widetilde{\alpha}}(\mathfrak{m}_3)=\mathfrak{m}_4,\ s_{\widetilde{\alpha}}(\mathfrak{m}_6)=\mathfrak{m}_6,$$
$$s_{\alpha_i}(\mathfrak{m}_1)=\mathfrak{m}_1,\ s_{\alpha_i}(\mathfrak{m}_2)=\mathfrak{m}_3,\ s_{\alpha_i}(\mathfrak{m}_4)=\mathfrak{m}_4,\ s_{\alpha_i}(\mathfrak{m}_5)=\mathfrak{m}_6,$$
$$s_{\widetilde{\alpha}-\alpha_i}(\mathfrak{m}_1)=\mathfrak{m}_6,\ s_{\widetilde{\alpha}-\alpha_i}(\mathfrak{m}_2)=\mathfrak{m}_4,\ s_{\widetilde{\alpha}-\alpha_i}(\mathfrak{m}_3)=\mathfrak{m}_3,\ s_{\widetilde{\alpha}-\alpha_i}(\mathfrak{m}_5)=\mathfrak{m}_5.$$

Proof. Note that $\widetilde{\alpha}=2\alpha_i+3\alpha_j+\beta$ where $\beta\in\{\Pi_0\}_{\mathbb{Z}}$. Then we see that

$$s_{\widetilde{\alpha}}(\alpha_i)=\alpha_i-2\frac{(\alpha_i,\widetilde{\alpha})}{(\widetilde{\alpha},\widetilde{\alpha})}\widetilde{\alpha}=-(\widetilde{\alpha}-\alpha_i)=-(\alpha_i+3\alpha_j+\beta).$$

Thus we have $s_{\widetilde{\alpha}}(\mathfrak{m}_1)=\mathfrak{m}_5$. Since $(\widetilde{\alpha},\alpha_j)=0$, we have $s_{\widetilde{\alpha}}(\alpha_j)=\alpha_j$ and thus $s_{\widetilde{\alpha}}(\mathfrak{m}_2)=\mathfrak{m}_2$. We also have that $s_{\widetilde{\alpha}}(\alpha_i+\alpha_j)=-(\alpha_i+2\alpha_j+\beta)$ and thus $s_{\widetilde{\alpha}}(\mathfrak{m}_3)=\mathfrak{m}_4$.

Now we consider a positive root γ such that $\kappa(\gamma)=\xi_j$. Then we can write $\gamma=\alpha_j+\beta_0$ where $\beta_0\in\{\Pi_0\}_{\mathbb{Z}}$ and we have $s_{\alpha_i}(\gamma)=\gamma-2\frac{(\gamma,\alpha_i)}{(\alpha_i,\alpha_i)}\alpha_i=\alpha_i+\alpha_j+\beta_0$, and thus we see that $s_{\alpha_i}(\mathfrak{m}_2)=\mathfrak{m}_3$. We consider a positive root γ such that $\kappa(\gamma)=\xi_i+2\xi_j$. Then we can write $\gamma=\alpha_i+2\alpha_j+\beta_1$ where $\beta_1\in\{\Pi_0\}_{\mathbb{Z}}$ and we see that $s_{\alpha_i}(\gamma)=\gamma-2\frac{(\gamma,\alpha_i)}{(\alpha_i,\alpha_i)}\alpha_i=\gamma$ and hence $s_{\alpha_i}(\mathfrak{m}_4)=\mathfrak{m}_4$. We also see that

$$s_{\alpha_i}(\widetilde{\alpha})=\widetilde{\alpha}-2\frac{(\widetilde{\alpha},\alpha_i)}{(\alpha_i,\alpha_i)}\alpha_i=\widetilde{\alpha}-\alpha_i=\alpha_i+3\alpha_j+\beta$$

and hence $s_{\alpha_i}(\mathfrak{m}_6)=\mathfrak{m}_5$.

Now we have

$$s_{\widetilde{\alpha}-\alpha_i}(\alpha_i)=\alpha_i-2\frac{(\alpha_i,\widetilde{\alpha}-\alpha_i)}{(\widetilde{\alpha}-\alpha_i,\widetilde{\alpha}-\alpha_i)}(\widetilde{\alpha}-\alpha_i)=\alpha_i+(\widetilde{\alpha}-\alpha_i)=\widetilde{\alpha}$$

and hence $s_{\widetilde{\alpha}-\alpha_i}(\mathfrak{m}_1)=\mathfrak{m}_6$. We consider a positive root γ such that $\kappa(\gamma)=\xi_j$. Then we can write $\gamma=\alpha_j+\beta_0$ where $\beta_0\in\{\Pi_0\}_{\mathbb{Z}}$ and we have $s_{\widetilde{\alpha}-\alpha_i}(\gamma)=\gamma-2\frac{(\gamma,\widetilde{\alpha}-\alpha_i)}{(\widetilde{\alpha}-\alpha_i,\widetilde{\alpha}-\alpha_i)}(\widetilde{\alpha}-\alpha_i)$. Noting that $(\alpha_j+\beta_0,\widetilde{\alpha})=0$ and $\frac{2(\alpha_j+\beta_0,\alpha_i)}{(\widetilde{\alpha}-\alpha_i,\widetilde{\alpha}-\alpha_i)}=-1$, we see that

$$s_{\widetilde{\alpha}-\alpha_i}(\gamma)=\gamma-(\widetilde{\alpha}-\alpha_i)=-(\alpha_i+2\alpha_j+\beta-\beta_0)$$

and hence $s_{\widetilde{\alpha}-\alpha_i}(\mathfrak{m}_2)=\mathfrak{m}_4$. Now we consider a positive root γ such that $\kappa(\gamma)=\xi_i+\xi_j$. Then we can write $\gamma=\alpha_i+\alpha_j+\beta_1$ where $\beta_1\in\{\Pi_0\}_{\mathbb{Z}}$. Note that $(\alpha_i+\alpha_j+\beta_1,\widetilde{\alpha}-\alpha_i)=(\alpha_i,\widetilde{\alpha})-(\alpha_i+\alpha_j,\alpha_i)=(\alpha_i,\alpha_i+2\alpha_j)=0$. Thus we have $s_{\widetilde{\alpha}-\alpha_i}(\mathfrak{m}_3)=\mathfrak{m}_3$. □

We normalize the system of equations as

$$r_1=1,\quad r_1=r_5,\quad r_2=r_3,\quad r_3=r_4,\quad r_4=r_5,\quad r_5=r_6. \tag{23}$$

From Lemma 9.1, we see that if $(x_1,x_2,x_3,x_4,x_5,x_6)=(a_1,\,a_2,\,a_3,\,a_4,\,a_5,\,a_6)$ is a solution for the system of equations (23), then .

$$\begin{aligned}
(x_1,x_2,x_3,x_4,x_5,x_6)&=(a_5,a_2,a_4,a_3,a_1,a_6),\\
(x_1,x_2,x_3,x_4,x_5,x_6)&=(a_6,a_3,a_4,a_2,a_1,a_5),\\
(x_1,x_2,x_3,x_4,x_5,x_6)&=(a_1,a_3,a_2,a_4,a_6,a_5),\\
(x_1,x_2,x_3,x_4,x_5,x_6)&=(a_5,a_4,a_2,a_3,a_6,a_1),\\
(x_1,x_2,x_3,x_4,x_5,x_6)&=(a_6,a_4,a_3,a_2,a_5,a_1)
\end{aligned}$$

are also solutions of the system (23). These metrics are all isometric to each other.

Due to the above observations we can divide our study into two cases:
(1) Case of $(x_1-x_5)(x_1-x_6)(x_5-x_6)=0$
(2) Case of $(x_1-x_5)(x_1-x_6)(x_5-x_6)\neq 0$.

We see that solutions of the system of equations (23) are the solutions of the following system of equations with $x_1x_2x_3x_4x_5x_6\neq 0$:

$$\left\{\begin{aligned}
&3tx_1{}^2x_5x_6-12(t+1)x_1x_2x_3x_5x_6+6(t+1)x_2x_3x_5x_6\\
&-3tx_2{}^2x_5x_6-3tx_3{}^2x_5x_6+x_1{}^2x_2x_3-x_2x_3x_5{}^2-x_2x_3x_6{}^2=0,\\
&(t+1)x_1x_2{}^2x_5-12(t+1)x_1x_2x_3x_4x_5-(t+1)x_1x_3{}^2x_5\\
&+6(t+1)x_1x_3x_4x_5-(t+1)x_1x_4{}^2x_5-x_1{}^2x_4x_5+x_1x_2{}^2x_3\\
&-x_1x_3x_4{}^2-x_1x_3x_5{}^2+x_2{}^2x_4x_5-x_3{}^2x_4x_5=0,\\
&-(t+1)x_1x_2{}^2x_6-12(t+1)x_1x_2x_3x_4x_6+6(t+1)x_1x_2x_4x_6\\
&+(t+1)x_1x_3{}^2x_6-(t+1)x_1x_4{}^2x_6-x_1{}^2x_4x_6+x_1x_2x_3{}^2\\
&-x_1x_2x_4{}^2-x_1x_2x_6{}^2-x_2{}^2x_4x_6+x_3{}^2x_4x_6=0,\\
&-(t+1)x_2{}^2x_5x_6-12(t+1)x_2x_3x_4x_5x_6+6(t+1)x_2x_3x_5x_6\\
&-(t+1)x_3{}^2x_5x_6+(t+1)x_4{}^2x_5x_6-x_2{}^2x_3x_6-x_2x_3{}^2x_5\\
&+x_2x_4{}^2x_5-x_2x_5x_6{}^2+x_3x_4{}^2x_6-x_3x_5{}^2x_6=0,\\
&-3tx_1x_2{}^2x_6-12(t+1)x_1x_2x_4x_5x_6+6(t+1)x_1x_2x_4x_6\\
&-3tx_1x_4{}^2x_6+3tx_1x_5{}^2x_6x_1{}^2x_2x_4+x_2x_4x_5{}^2-x_2x_4x_6{}^2=0,\\
&-3tx_1x_3{}^2x_5-12(t+1)x_1x_3x_4x_5x_6+6(t+1)x_1x_3x_4x_5\\
&-3tx_1x_4{}^2x_5+3tx_1x_5x_6{}^2-x_1{}^2x_3x_4-x_3x_4x_5{}^2+x_3x_4x_6{}^2=0,
\end{aligned}\right. \tag{24}$$

where $t=2,3,5,9$ for types of $F_4(1)$, $E_6(2)$, $E_7(2)$, $E_8(1)$ respectively. Note that, for type G_2, the system of equations (23) is equivalent to the system of equations (24) with $x_1x_2x_3x_4x_5x_6\neq 0$ for $t=1/3$.

For an G-invariant Einstein metric g of the form (14) on $M = G/K$, we define a scale invariant given by $H_g = V_g^{1/d} S_g$, where S_g is the scalar curvature of the given metric g, $V_g = \prod_{i=1}^{6} x_i^{d_i}$ is the volume of the metric g and $d = \sum_{i=1}^{6} d_i = \dim G/K$. Then $H_g = V_g^{1/d} S_g$ is invariant under a common scaling of the variables x_i, and if two metrics are isometric, then they have the same scale invariant (cf. [5]).

(1) $(x_1 - x_5)(x_1 - x_6)(x_5 - x_6) = 0$.

Consider the case of $x_6 = x_5$ and denote by I the ideal of the polynomial ring $\mathbb{Q}[y, x_1, x_2, x_3, x_4, x_5]$ defined by the system of equations (24) (substituted $x_6 = x_5$) and the equation $y x_1 x_2 x_3 x_4 x_5 - 1 = 0$. Taking a lexicographic order $>$ with $y > x_5 > x_1 > x_4 > x_3 > x_2$ and computing a Gröbner basis of I, we see that the Gröbner basis contains the polynomial $x_2 - x_3$. Thus we see that if $x_6 = x_5$, then $x_2 = x_3$. Note that, from Lemma 9.1, by changing the role of variables, we see that if $x_1 = x_5$ then $x_3 = x_4$ and if $x_1 = x_6$ then $x_2 = x_4$

Case of $x_1 = x_5 = x_6$.

In this case we see that $x_2 = x_3 = x_4$ and the system of equations (24) with $x_1 x_2 \neq 0$ reduces to

$$3t{x_1}^2 - 12(t+1)x_1{x_2}^2 + 5{x_2}^2 = 0, \ \ 12(t+1){x_2}^2 - 5(t+1)x_2 + 2x_1 = 0.$$

For the types $F_4(1)$, $E_6(2)$, $E_7(2)$ and $E_8(1)$, solving the above system of equations for $t = 2, 3, 5, 9$, we get the following positive solutions and the approximate values of the scale invariant H_g:

Type	x_1	x_2	H_g
$F_4(1)$	$5\ (151 + 9\sqrt{65})/2304$	$(105 - \sqrt{65})/288$	2.12782
	$5(151 - 9\sqrt{65})/2304$	$(105 + \sqrt{65})/288$	2.09042
$E_6(2)$	$5(191 + 54\sqrt{5})/2904$	$25/66 - \sqrt{5}/88$	2.21077
	$5(191 - 54\sqrt{5})/2904$	$25/66 + \sqrt{5}/88$	2.14956
$E_7(2)$	$5(346 + 45\sqrt{35})/5202$	$(120 - \sqrt{35})/306$	2.30167
	$5(346 - 45\sqrt{35})/5202$	$(120 + \sqrt{35})/306$	2.22283
$E_8(1)$	$(851 + 135\sqrt{30})/2523$	$(175 - \sqrt{30})/435$	2.37875
	$(851 - 135\sqrt{30})/2523$	$(175 + \sqrt{30})/435$	2.29805

For the type G_2 the solutions are not real, that is, we obtain $\{x_1 = 5(47 + 2i\sqrt{35})/648,\ x_2 = (20 - i\sqrt{35})/72\}$, $\{x_1 = 5(47 - 2i\sqrt{35})/648,\ x_2 = (20 + i\sqrt{35})/72\}$.

Remark 9.1. These solutions have been obtained by Graev ([15], p. 15) by a different method up to scale.

Case of $x_1 \neq x_5 = x_6$.

For this case, we have that $x_2 = x_3$ and we denote by I the ideal of the polynomial ring $\mathbb{Q}[y, x_1, x_2, x_4, x_5]$ defined by the system of equations (24) (substituted $x_6 = x_5$ and $x_3 = x_2$) and the equation $yx_1x_2x_4x_5 - 1 = 0$. We take a lexicographic order $>$ with $y > x_5 > x_1 > x_4 > x_2$. By computing a Gröbner basis of I, we see that the Gröbner basis contains a polynomial $f_1(x_2)$ of x_2 with integer coefficients of degree 14 for $t = 2, 3, 5, 9$ and $t = 1/3$, and the other variables x_1, x_4, x_5 can be expressed by using polynomials $h_1(x_2)$, $h_2(x_2)$, $h_3(x_2)$ of x_2 with rational coefficients of degree 13 as

$$x_1 = h_1(x_2), \ x_4 = h_2(x_2), \ x_5 = h_3(x_2).$$

For $t = 2, 3, 5, 9$, by solving the equations $f_1(x_2) = 0$ for x_2 and substituting these values of x_2 into $h_1(x_2)$, $h_2(x_2)$, $h_3(x_2)$, we obtain the following approximate values of positive solutions and the scale invariant H_g:

Type	x_1	x_2	x_4	x_5	H_g
$F_4(1)$	0.213634	0.394142	0.387978	0.153853	2.09061
	0.131089	0.399189	0.286152	0.487934	2.10521
	0.183911	0.281824	0.496740	0.717569	2.12955
	0.308056	0.288143	0.453834	0.639469	2.13050
$E_6(2)$	0.162608	0.405047	0.40131	0.105968	2.14956
	0.0943681	0.400714	0.318384	0.540087	2.18137
	0.110463	0.292421	0.556791	0.824414	2.20791
	0.353910	0.297449	0.49441	0.712573	2.21603
$E_7(2)$	0.109810	0.411835	0.410122	0.0651614	2.22323
	0.0603151	0.405232	0.35016	0.591972	2.26806
	0.0650699	0.305663	0.600598	0.897282	2.29379
	0.385762	0.30791	0.546317	0.801150	2.31070
$E_8(1)$	0.0663727	0.415002	0.414383	0.0366812	2.29842
	0.0350012	0.409675	0.376131	0.633035	2.34664
	0.0363025	0.316678	0.629779	0.94361	2.37044
	0.409463	0.317459	0.592135	0.877534	2.39160

For the case G_2, that is $t = 1/3$, we obtain the following approximate values of positive solutions and the scale invariant H_g:

Type	x_1	x_2	x_4	x_5	H_g
G_2	0.637124	0.368377	0.0983455	0.356015	1.92118
	0.317633	0.436907	0.0927966	0.426906	1.90532

These Einstein metrics of the case G_2/T have been studied in [9].

Remark 9.2. This case has been studied by Graev ([15]). Note also that, together with the cases $x_1 = x_5 \neq x_6$ and $x_1 = x_6 \neq x_5$, the number of complex solutions of equations (24) for this type is $14 \times 3 = 42$.

(2) $(x_1 - x_5)(x_1 - x_6)(x_5 - x_6) \neq 0$.
We renormalize the system of equations (23) as

$$x_1 = 1, \quad r_1 = r_5, \quad r_2 = r_3, \quad r_3 = r_4, \quad r_4 = r_5, \quad r_5 = r_6. \tag{25}$$

We see that, up to scale, solutions of the system of equations (23) are the solutions of the following system of equations with $x_2x_3x_4x_5x_6 \neq 0$:

$$\left\{\begin{array}{l}
t{x_2}^2x_3x_6 - 3t{x_2}^2x_4x_5x_6 + 6(t+1)x_2x_3x_4x_5x_6 \\
- 6(t+1)x_2x_3x_4x_6 - 3t{x_3}^2x_4x_5x_6 + 3tx_3{x_4}^2x_6 - 3tx_3{x_5}^2x_6 \\
+ 3tx_4x_5x_6 - 2x_2x_3x_4{x_5}^2 + 2x_2x_3x_4 = 0, \\
(t+1){x_2}^2x_5x_6 - 6(t+1)x_2x_4x_5x_6 - 2(t+1){x_3}^2x_5x_6 \\
+ 6(t+1)x_3x_4x_5x_6 + {x_2}^2x_3x_6 + 2{x_2}^2x_4x_5x_6 - x_2{x_3}^2x_5 + x_2{x_4}^2x_5 \\
+ x_2x_5{x_6}^2 - 2{x_3}^2x_4x_5x_6 - x_3{x_4}^2x_6 - x_3{x_5}^2x_6 = 0, \\
-6(t+1)x_2x_3x_5x_6 + 6(t+1)x_2x_4x_5x_6 + 2(t+1){x_3}^2x_5x_6 \\
- 2(t+1){x_4}^2x_5x_6 + {x_2}^2x_3x_6 - {x_2}^2x_4x_5x_6 + 2x_2{x_3}^2x_5 - 2x_2{x_4}^2x_5 \\
+ {x_3}^2x_4x_5x_6 - x_3{x_4}^2x_6 + x_3{x_5}^2x_6 - x_4x_5x_6 = 0, \\
3t{x_2}^2x_3x_6 - (t+1){x_2}^2x_5x_6 - 6(t+1)x_2x_3x_4x_6 - (t+1){x_3}^2x_5x_6 \\
+ 6(t+1)x_2x_3x_5x_6 + 3tx_3{x_4}^2x_6 - 3tx_3{x_5}^2x_6 + (t+1){x_4}^2x_5x_6 \\
- {x_2}^2x_3x_6 - x_2{x_3}^2x_5 - x_2x_3x_4{x_5}^2 + x_2x_3x_4{x_6}^2 + x_2x_3x_4 \\
+ x_2{x_4}^2x_5 - x_2x_5{x_6}^2 + x_3{x_4}^2x_6 - x_3{x_5}^2x_6 = 0, \\
-3t{x_2}^2x_3x_6 + 3tx_2{x_3}^2x_5 - 6(t+1)x_2x_3x_4x_5 \\
+ 6(t+1)x_2x_3x_4x_6 + 3tx_2{x_4}^2x_5 - 3tx_2x_5{x_6}^2 - 3tx_3{x_4}^2x_6 \\
+ 3tx_3{x_5}^2x_6 + 2x_2x_3x_4{x_5}^2 - 2x_2x_3x_4{x_6}^2 = 0.
\end{array}\right. \tag{26}$$

We denote by I the ideal of $\mathbb{Q}[y, x_2, x_3, x_4, x_5, x_6]$ defined by the system of equations (26) and the equation $yx_2x_3x_4x_5x_6 - 1 = 0$. We take a lexicographic order $>$ with $y > x_6 > x_5 > x_4 > x_3 > x_2$. By computing a Gröbner basis of I for $t = 2, 3, 5, 9$, we see that the Gröbner basis contains a polynomial of the form

$$(x_2 - t)(x_2 - (t+1))\left((3t+1)x_2 - (2t+1)\right)\left((3t+1)x_2 - t\right) \times \\
\left((3t+2)x_2 - (2t+1)\right)\left((3t+2)x_2 - (t+1)\right) g_1(x_2, t),$$

where $g_1(x_2, t)$ is a polynomial of x_1 with integer coefficients of degree 102. For the case $(x_2 - t)(x_2 - (t+1))\left((3t+1)x_2 - (2t+1)\right)\left((3t+1)x_2 - t\right) \times \left((3t+2)x_2 - (2t+1)\right)\left((3t+2)x_2 - (t+1)\right) = 0$, we obtain 6 Kähler Einstein metrics which are isometric each other, up to scale. For the case $g_1(x_2, t) = 0$ where $t = 2, 3, 5, 9$, we see that the other variables x_3, x_4, x_5, x_6 can be expressed by polynomials of x_2. By solving the equations $g_1(x_2, t) = 0$ of x_2 and substituting positive values of x_2 into thepolynomials of x_2, we see that at least one of values x_3, x_4, x_5, x_6 is negative. Thus we have no Einstein metrics for these cases.

For the case G_2, that is $t = 1/3$, we see that the Gröbner basis contains a polynomial of the form

$$(3x_2 - 1)(3x_2 - 4)(6x_2 - 1)(6x_2 - 5)(9x_2 - 4)(9x_2 - 5)g_2(x_1),$$

where $g_2(x_2)$ is a polynomial of x_1 with integer coefficients of degree 84. For the case $(3x_2 - 1)(3x_2 - 4)(6x_2 - 1)(6x_2 - 5)(9x_2 - 4)(9x_2 - 5) = 0$, we obtain 6 Kähler Einstein metrics which are isometric to each other, up to scale. By solving the equations $g_2(x_2) = 0$ of x_2 and by substituting the positive values of x_2, we see that at least one of values x_3, x_4, x_5, x_6 is negative. Thus we have no Einstein metrics for these cases (cf. [9]).

Remark 9.3. Note that the number of complex solutions of the system of equations (23) are $2+42+6+102 = 152$ for the cases of type $F_4(1)$, $E_6(2)$, $E_7(2)$, $E_8(1)$ and $2 + 42 + 6 + 84 = 134$ for the type G_2. These numbers have been obtained by Graev ([15], p. 15, Propositions 3.3 and 3.4).

Acknowledgments

The second author was full-supported by Masaryk University under the Grant Agency of Czech Republic, project no. P 201/12/G028.

Bibliography

1. D. V. Alekseevsky: *Homogeneous Einstein metrics*, Differential Geometry and its Applications (Proceeding of the Conference), 1–21. Univ. of. J. E. Purkyne, Czechoslovakia (1987).
2. D. V. Alekseevsky and A. M. Perelomov: *Invariant Kähler-Einstein metrics on compact homogeneous spaces*, Funct. Anal. Appl. 20 (3) (1986) 171–182.
3. S. Anastassiou and I. Chrysikos: *The Ricci flow approach to homogeneous Einstein metrics on flag manifolds*, J. Geom. Phys. 61 (2011) 1587–1600.
4. A. Arvanitoyeorgos: *New invariant Einstein metrics on generalized flag manifolds*, Trans. Am. Math. Soc. 337(2) (1993) 981–995.

5. A. Arvanitoyeorgos and I. Chrysikos: *Invariant Einstein metrics on flag manifolds with four isotropy summands*, Ann. Glob. Anal. Geom. 37 (2) (2010) 185–219.
6. A. Arvanitoyeorgos and I. Chrysikos: *Invariant Einstein metrics on generalized flag manifolds with two isotropy summands*, J. Aust. Math. Soc. 90 (2011) 237–251.
7. A. Arvanitoyeorgos, I. Chrysikos and Y. Sakane: *Complete description of invariant Einstein metrics on the generalized flag manifold* $SO(2n)/U(p) \times U(n-p)$, Ann. Glob. Anal. Geom. 38 (4) (2010) 413–438.
8. A. Arvanitoyeorgos, I. Chrysikos and Y. Sakane: *Homogeneous Einstein metrics on the generalized flag manifold* $Sp(n)/(U(p) \times U(n-p))$, Differential Geom. Appl. 29 (2011) S16–S27.
9. A. Arvanitoyeorgos, I. Chrysikos and Y. Sakane: *Homogeneous Einstein metrics on* G_2/T, Proc. Amer. Math. Soc. 141 (7) (2013) 2485–2499.
10. A. Arvanitoyeorgos, I. Chrysikos and Y. Sakane: *Homogeneous Einstein metrics on generalized flag manifolds* $Sp(n)/(U(p) \times U(q) \times Sp(n-p-q))$, Proc. 2nd Intern. Coll. Differential Geom. Related Fields, Veliko Tarnovo 2010, World Sci. (2011) 1–24.
11. A. Arvanitoyeorgos, I. Chrysikos and Y. Sakane: *Homogeneous Einstein metrics on generalized flag manifolds with five isotropy summands*, arXiv:1207.2897 [math. DG] 13 Dec 2012.
12. A. L. Besse: *Einstein manifolds*, Springer-Verlag, Berlin Heidelberg 1987.
13. I. Chrysikos: *Flag manifolds, symmetric* $\mathfrak{t}$*-triples and Einstein metrics*, Differential Geom. Appl. 30 (2012) 642–659.
14. H. Freudenthal and H. de Vries: *Linear Lie groups*, Academic Press, New York 1969.
15. M. M. Graev: *On invariant Einstein metrics on Kähler homogeneous spaces* SU_4/T^3, G_2/T^2, $E_6/T^2 \cdot (A_2)^2$, $E_7/T^2 \cdot A_5$, $E_8/T^2 \cdot E_6$, $F_4/T^2 \cdot A_2$, arXiv:1111.0639v1 [math. DG] 2 Nov 2011.
16. M. Kimura: *Homogeneous Einstein metrics on certain Kähler C-spaces*, Adv. Stud. Pure Math. 18 (1) (1990) 303–320.
17. M. Nishiyama: *Classification of invariant complex structures on irreducible compact simply connected coset spaces*, Osaka J. Math. 21 (1984) 39–58.
18. B. O'Neill: *The fundamental equation of a submersion*, Michigan Math. J. 13 (1966) 459–469.
19. J-S. Park and Y. Sakane: *Invariant Einstein metrics on certain homogeneous spaces*, Tokyo J. Math. 20 (1) (1997) 51–61.
20. Y. Sakane: *Homogeneous Einstein metrics on flag manifolds*, Towards 100 years after Sophus Lie (Kazan, 1998), Lobachevskii J. Math. 4 (1999), 71–87.
21. M. Wang and W. Ziller: *Existence and non-existence of homogeneous Einstein metrics*, Invent. Math. 84 (1986) 177–194.

Received January 22, 2013
Revised April 18, 2013

Proceedings of the 3rd International
Colloquium on Differential Geometry
and its Related Fields
Veliko Tarnovo, September 3–7, 2012

APPLICATIONS OF THE GAUSSIAN INTEGERS IN CODING THEORY

Stefka BOUYUKLIEVA

Faculty of Mathematics and Informatics,
Veliko Tarnovo University, Veliko Tarnovo, Bulgaria
E-mail: stefka@uni-vt.bg

Codes over Gaussian integers are suitable for coding over two-dimensional signal space. We present a metric over QAM-like constellations modeled by quotient rings of Gaussian integers. Moreover, we give some simple examples for codes correcting errors of Mannheim weights one and two.

Keywords: Gaussian integers; Signal constellations; Mannheim distance; Block codes.

1. Introduction

A complex number whose real and imaginary parts are both integers is called a Gaussian integer. The set of all Gaussian integers equipped with addition and multiplication forms an Euclidean ring which we denote by

$$\mathcal{G} = \mathbb{Z}[i] = \{a + bi \mid a, b \in \mathbb{Z},\ i^2 = -1\}.$$

The norm of a Gaussian integer γ is the natural number defined as

$$N(\gamma) = a^2 + b^2 = \gamma \cdot \gamma^* = |\gamma|^2,$$

where γ^* is the complex conjugate of γ. The norm is multiplicative, which means that $N(\gamma_1\gamma_2) = N(\gamma_1)N(\gamma_2)$ for all $\gamma_1, \gamma_2 \in \mathcal{G}$. The reason we prefer to deal with norms on $\mathcal{G}$ instead of absolute values is that norms are integers and the divisibility properties of norms in $\mathbb{Z}$ will provide important information about divisibility properties in $\mathcal{G}$. The units of $\mathcal{G}$ are those elements with norm 1, i.e. the elements $1, -1, i$ and $-i$, and these are all invertible elements of $\mathcal{G}$. If γ is a Gaussian integer then $\pm\gamma$ and $\pm i\gamma$ are the *associates* of γ. Knowing a Gaussian integer up to multiplication by a unit is analogous to knowing an integer up to its sign. We present some properties of the Gaussian integers, which are important for the considered applications, in Section 2. Section 3 is devoted to the considered metric over fields represented by quotient rings of Gaussian integers.

The main aim of this paper is to present an interesting application of this ring in coding theory. The design of error-correcting codes for two-dimensional signal spaces has been widely considered in the technical literature. In Sections 4 and 5 we define a few classes of codes over the Gaussian integers and give some examples for codes correcting errors of Mannheim weights one and two.

2. Some properties of the Gaussian integers

In this section we introduce some properties connected with the divisibility and a modulo operation in the ring $\mathcal{G}$.

Divisibility in $\mathcal{G}$ is defied in the natural way: We say that β divides α ($\beta \mid \alpha$) if $\alpha = \beta\gamma$ for some $\gamma \in \mathcal{G}$. If β divides α then $N(\beta) \mid N(\alpha)$. Any Gaussian integer γ with $N(\gamma) > 1$ has eight obvious factors: ± 1, $\pm i$, $\pm\gamma$ and $\pm i\gamma$ which we call trivial. If γ has only trivial factors, it is called prime. All Gaussian primes are:

(1) $1 + i$ and its associates;
(2) the rational primes p with $p \equiv 3 \pmod 4$ and their associates;
(3) the factors $a + bi$ of the rational primes p with $p \equiv 1 \pmod 4$.

The expression of a Gaussian integer as a product of Gaussian primes is unique, apart from the order of the primes, the presence of unities, and ambiguities between associated primes.

One reason we are able to transfer a lot of results from the ring of integers $\mathbb{Z}$ to the ring of Gaussian integers $\mathcal{G}$ is the analogue of division-with-remainder in $\mathbb{Z}$.

Theorem 2.1 ([1]). *For any two Gaussian integers $\alpha, \beta \in \mathcal{G}$ with $\beta \neq 0$, there are $\gamma, \rho \in \mathcal{G}$ such that $\alpha = \beta\gamma + \rho$ and $N(\rho) < N(\beta)$. In fact, we can choose ρ so that $N(\rho) \leq \frac{1}{2}N(\beta)$.*

The numbers γ and ρ are the quotient and remainder, and the remainder is bounded in size (according to its norm) by the size of the divisor β. To calculate the quotient and remainder, we define a rounding of a complex number as $[a + bi] = [a] + [b]i$ where $[a]$ is the nearest integer to a, $a \in \mathbb{R}$ which means that $[a] = \lfloor a \rfloor$ if the fractional part $\{a\} \leq 1/2$ and $[a] = \lceil a \rceil$ if $\{a\} > 1/2$. Furthermore, for $\alpha, \beta \in \mathbb{C}$ we define $[\alpha/\beta] = [\alpha\beta^*/\beta\beta^*]$. If β

is a Gaussian integer then $[\alpha/\beta] = [\alpha\beta^*/N(\beta)]$. If we take $\gamma = [\alpha/\beta]$ and $\rho = \alpha - \beta\gamma$ then $N(\rho) \leq (1/2)N(\beta)$ (see the proof of Theorem 2.1 in [1]).

Example 2.1. $\alpha = 41 + 24i, \beta = 11 - 2i$, $N(\beta) = 125$. Then

$$\frac{\alpha}{\beta} = \frac{41+24i}{11-2i} = \frac{(41+24i)(11+2i)}{125} = \frac{403}{125} + \frac{346}{125}i \implies \left[\frac{\alpha}{\beta}\right] = 3+3i.$$

If we take $\gamma = 3+3i$ then $\alpha - \beta\gamma = 2 - 3i$ and since $N(2-3i) = 13 < N(\beta)$, $\rho = 2 - 3i$.

There is one interesting difference between the division Theorem 2.1 and the (usual) division theorem in $\mathbb{Z}$: The quotient and remainder are not unique in $\mathcal{G}$.

Example 2.2. Let $\alpha = 37 + 2i$ and $\beta = 11 + 2i$, so $N(\beta) = 125$. Now

$$\frac{\alpha}{\beta} = \frac{37+2i}{11+2i} = \frac{(37+2i)(11-2i)}{125} = \frac{411}{125} - \frac{52}{125}i \implies \left[\frac{\alpha}{\beta}\right] = 3.$$

So we can take $\gamma = 3$ and $\rho = \alpha - \beta\gamma = 4 - 4i$, $N(\rho) = 32 < 125$. However, it is also true that $\alpha = \beta(3 - i) + (2 + 7i)$, $N(2 + 7i) = 53 < 125$.

There is a way to picture what modular arithmetic in $\mathcal{G}$ means, by plotting the multiples of a Gaussian integer in $\mathcal{G}$. All multiples of $1 + 2i$ in $\mathcal{G}$ correspond to the vertices of the squares in Figure 1. We can use the same figure to make a list of representatives for $\mathcal{G}/(1 + 2i)$ taking the Gaussian integers inside a square and one of its vertices. Choosing the square with edges $1+2i$ and $-2+i$, we get a list of five Gaussian integers: 0, i, $2i$, $-1+i$, $-1 + 2i$. Every Gaussian integer is congruent modulo $1 + 2i$ to exactly one of these so they form a complete residue system modulo $1 + 2i$.

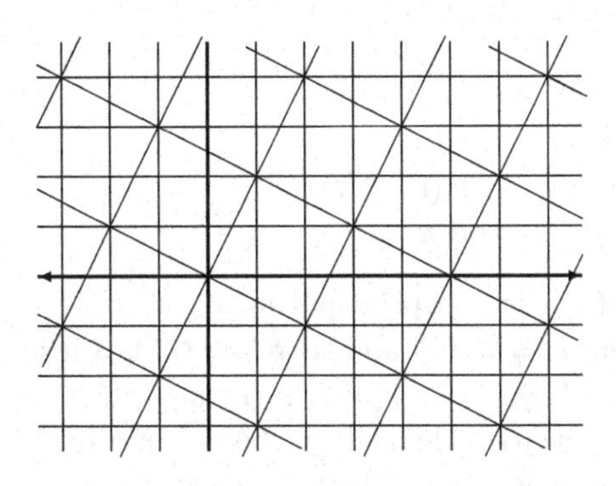

Figure 1. Multiples of $1 + 2i$.

We define a modulo operation in $\mathcal{G}$ in the usual way:

$$\alpha \equiv \beta \pmod{\gamma} \iff \gamma \mid (\alpha - \beta).$$

The congruences in $\mathcal{G}$ have similar properties as the congruences in $\mathbb{Z}$, for example if $\alpha_i \equiv \beta_i \pmod{\gamma}$, $i = 1, 2$, then $\alpha_1 + \alpha_2 \equiv \beta_1 + \beta_2 \pmod{\gamma}$, and $\alpha_1\alpha_2 \equiv \beta_1\beta_2 \pmod{\gamma}$. Moreover, for $\alpha, \beta \in \mathcal{G}$ with $\beta \neq 0$, the congruence $\alpha x \equiv 1 \pmod{\beta}$ is solvable if and only if α and β are relatively prime in $\mathcal{G}$.

Consider now the quotient ring $\mathcal{G}/(\beta)$ where (β) is the principal ideal of $\mathcal{G}$ generated by $\beta \in \mathcal{G}$. Recall that $\mathcal{G}$ is an Euclidean ring and therefore it is also a principal ideal domain. If $\beta \neq 0$ then $\mathcal{G}/(\beta)$ is finite, and $\#\mathcal{G}/(\beta) = N(\beta)$ (see [1]). It is easy to prove that (β) is a maximal ideal in $\mathcal{G}$ if and only if β is a Gaussian prime. Hence in such a case $\mathcal{G}/(\beta)$ is a field with $N(\beta)$ elements. For the different primes we have

(1) $\mathcal{G}/(1+i) \cong GF(2)$ (the same for $\beta = 1 - i$, $-1 + i$, or $-1 - i$);
(2) $\mathcal{G}/(p) \cong GF(p^2)$ for the rational primes p with $p \equiv 3 \pmod{4}$ (and their associates);
(3) $\mathcal{G}/(\pi) \cong GF(p)$ for the nontrivial factors $\pi = a + bi$ of the rational primes p with $p \equiv 1 \pmod{4}$.

To the end of this paper, we focus on Gaussian primes π which are factors of the rational primes p with $p \equiv 1 \pmod{4}$. The well known Fermat's theorem states that an odd prime p can be represented as a sum of squares if and only if $p \equiv 1 \pmod{4}$ (see [3]). Hence such primes p are the product of two conjugate Gaussian integers:

$$p = a^2 + b^2 = \pi \cdot \pi^*, \quad \pi = a + bi \in \mathcal{G}, \quad \pi^* = a - bi.$$

Let $\mu : \mathcal{G} \to \mathcal{G}$ be a modulo function according to π with the following properties:

(1) $\mu(\alpha) = \mu(\beta) \iff \alpha \equiv \beta \pmod{\pi}$;
(2) $\mu(0) = 0$, $\mu(1) = 1$;
(3) $\mu(\alpha + \beta) \equiv \mu(\alpha) + \mu(\beta) \pmod{\pi}$;
(4) $\mu(\alpha\beta) \equiv \mu(\alpha)\mu(\beta) \pmod{\pi}$.

Obviously, $\ker(\mu_\pi)$ is the principal ideal $(\pi) \lhd \mathcal{G}$. Denote by $\mathcal{G}_\pi$ the image $\Im(\mu_\pi)$. Then $\mathcal{G}_\pi \cong \mathcal{G}/(\pi)$ and therefore $\mathcal{G}_\pi$ is a field with p elements. Moreover, $\mathcal{G}_\pi$ is a complete residue system modulo π equipped with the addition and multiplication defined by the modulo function μ. If we take another modulo function, we will have a different complete residue system but the corresponding fields are isomorphic. In this section we consider $\mathcal{G}_\pi$

only as a complete residue system but we need the field structure in Sections 4 and 5.

Let $\mathcal{G}'_\pi$ be the complete residue system which corresponds to the modulo function μ' defined in the following way

$$\mu'(\alpha) = \alpha \bmod \pi = \alpha - \left[\frac{\alpha}{\pi}\right]\pi = \alpha - \left[\frac{\alpha\pi^*}{N(\pi)}\right]\pi. \tag{1}$$

So

$$\mathcal{G}'_{1+2i} = \{0, 1, -1, i, -i\}.$$

We can visualize the set $\mathcal{G}'_\pi$ as it is shown in Figure 2 where the points represent the elements of $\mathcal{G}'_\pi$ (in this case $\pi = 5 + 2i$). Since we have coding for communication channels in mind, we call these two-dimensional visualizations of $\mathcal{G}_\pi$ by the communication term signal constellation.

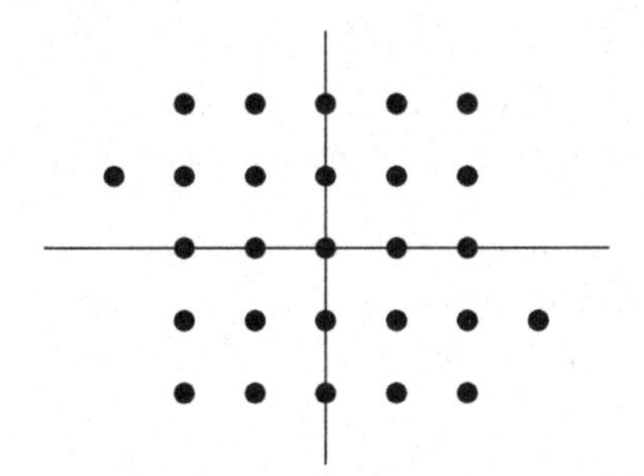

Figure 2. $\mathcal{G}'_{5+2i}$.

3. Constellations and distances

Codes over Gaussian integers are suitable for coding over two-dimensional signal space. Therefore it is important to use the two-dimensional signal constellations. The constellation diagram is a useful representation of QAM (Quadrature Amplitude Modulation) which is both an analog and a digital modulation scheme. The data are usually binary, the number of points in the grid is usually a power of 2. For instance 64-QAM and 256-QAM are often used in digital cable television and cable modem applications. In QAM, the constellation points are usually arranged in a square grid with equal vertical and horizontal spacing, although other configurations are possible like in the case which we consider. If the constellation points lie on a circle, they only affect the phase of the carrier: Such signaling schemes are termed Phase Shift Keying (PSK). The choice of constellation for a particular application depends on considerations such as power-bandwidth tradeoffs and implementation complexity. The constellation itself is typically chosen such that the center of mass of the points is at the origin (the zero point). Constellations associated with the Gaussian prime π the same as the constellations associated with a single error-correcting integer code (see [7]) have a variety of shapes, in the integer case depending on the value of t, and in the Gaussian case depending on the complete residue system. For example, the constellation in Figure 3 is associated with the same π as the constellation in Figure 2.

Hamming and Lee distances have been revealed to be inappropriate metrics to deal with QAM signal sets and other related constellations. In [4] Huber introduces a Mannheim distance in the following way: If $\alpha, \beta \in \mathcal{G}_\pi$ and $\gamma = \beta - \alpha \bmod \pi = \mu'(\beta - \alpha)$, then the Mannheim weight of γ is $w'_M(\gamma) = |Re(\gamma)| + |Im(\gamma)|$, and the Mannheim distance between α and β is $d'_M(\alpha, \beta) = w'_M(\gamma)$. Unfortunately, so defined distance is not a true metric since it does not fulfil the triangular inequality (see [6]).

Example 3.1 ([6]). *Let* $p = 193 = 7^2 + 12^2$, $\pi = 7 + 12i$, $x = 2 - 6i$, $y = 2 - 5i$, $z = 3 + 2i$. *Then* $(x - y)$ *mod* $\pi = -i$, $(x - z)$ *mod* $\pi = 6 + 4i$, $(z - y)$ *mod* $\pi = 1 + 7i$. *It follows that* $d'_M(x, y) = 1$, $d'_M(x, z) = 10$, $d'_M(y, z) = 8$, *which gives*

$$d'_M(x, z) > d'_M(x, y) + d'_M(y, z).$$

Therefore for Mannheim distance we will use the definition given in [6]. To simplify the notation, we denote both a Gaussian integer and its residue class in $\mathcal{G}/(\pi)$ by α.

Definition 3.1. Mannheim distance d_M between $\alpha, \beta \in \mathcal{G}$ (or $\mathcal{G}/(\pi)$) is

$$d_M(\alpha, \beta) = |a| + |b|,$$

where $a + bi$ is in the residue class of $\alpha - \beta$ modulo π with $|a| + |b|$ minimum.

Using Definition 3.1 for the elements in Example 3.1 we have $d_M(x, z) = 9$ since $6 + 4i \equiv -1 - 8i \pmod{\pi}$. The Mannheim metric is a Manhattan metric modulo a two-dimensional (2-D) grid.

Denote by $\mathcal{G}^*_\pi$ the set of the representatives of any residue class with minimum Mannheim weight according to Definition 3.1. The signal constellation for $\mathcal{G}^*_{5+2i}$ is presented in Figure 3.

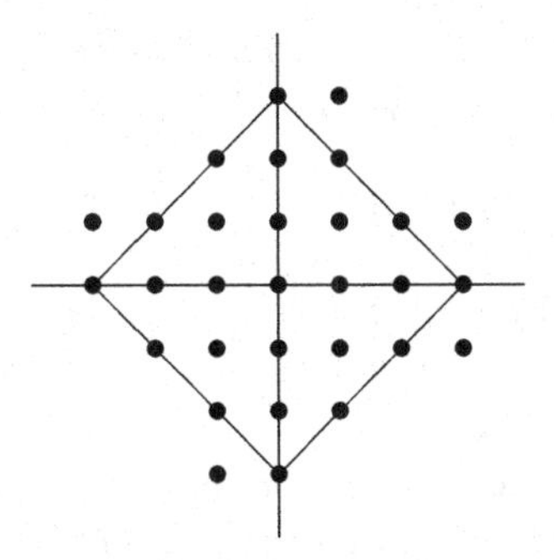

Figure 3. $\mathcal{G}^*_{5+2i}$.

Let $d_{\max}(\mathcal{G}^*_\pi)$ be the diameter of the constellation corresponding to $\mathcal{G}^*_\pi$ which is defined as the maximum Mannheim distance between two points. If $\pi = a+bi$ is a Gaussian prime with $N(\pi) = p$ where p is a rational prime satisfying $p \equiv 1 \pmod 4$, then $d_{\max}(\mathcal{G}^*_\pi) = \max\{|a|, |b|\} - 1$ (see [6]).

In [6], perfect codes of length one for the Mannheim metric have been characterized. But in this paper we consider codes of length $n \geq (p-1)/4$.

The Mannheim weight of the vector $x = (x_0, x_1, \ldots, x_{n-1})$ over $\mathcal{G}_\pi$ is given by

$$w_M(x) = \sum_{i=0}^{n-1} w_M(x_i),$$

and Mannheim distance between x and y is $d_M(x, y) = w_M(y - x)$. The Mannheim distance defines a metric in $\mathcal{G}^n_\pi$.

4. OMEC codes

A block code of length n over $\mathcal{G}_\pi$ is defined by

$$\mathcal{C} = \{c = (c_0, c_1, \ldots, c_{n-1}) : \; c_i \in \mathcal{G}_\pi, i = 0, 1, \ldots, n-1\}.$$

If $\mathcal{C}$ is a linear space over the field $\mathcal{G}_\pi$ it is called linear code. Minimum Mannheim distance of a code $\mathcal{C}$ is

$$d_M(\mathcal{C}) = \min\{w_M(c) | c \neq 0, c \in \mathcal{C}\}.$$

The OMEC (or One Mannheim Error Correcting) codes over the Gaussian integers are linear codes which are suited for QAM signal constellation (see [4]). They can correct errors of Mannheim weight 1.

Let $n = (p-1)/4$ and α be a primitive element of the field $\mathcal{G}_\pi$. An OMEC code of length n is a code over $\mathcal{G}_\pi$ with a parity-check matrix

$$H = (1 \ \ \alpha \ \ \alpha^2 \ \ \cdots \ \ \alpha^{n-1}). \tag{2}$$

This means that a vector $c \in \mathcal{G}_\pi^n$ is a codeword if and only if $Hc^T = 0$. The OMEC codes are linear codes with length n, dimension $n-1$ and minimum Mannheim distance $d_M = 3$.

Let c be a codeword and $r = c+e$ be the received vector. The syndrome of this vector is $Hr^T = He^T$. If $w_M(e) = 1$, then the vector-error e has only one nonzero entry (the Mannheim error) at position l, $0 \le l \le n-1$, which takes one of the four values $\pm 1, \pm i$. Hence $s = He^T = \pm\alpha^l$ or $\pm i\alpha^l$. Since $\alpha^{4n} = 1$, $\alpha^{2n} = -1$, $\{\alpha^n, \alpha^{3n}\} = \{\pm i\}$, then $s = \alpha^{l+tn}$, where $t = 0, 1, 2$, or 3. This gives us that $l = (\log_\alpha s) \bmod n$, so we can easily compute the location l of the error and its value which is $s\alpha^{-l} = \alpha^{tn}$.

Example 4.1. Let $p = 13$, $\pi = 3 + 2i$, $\alpha = 1 + i$, then $H = (1, 1+i, 2i)$. Assume that the received vector is $r = (1+i, i, -1+i)$. Then $s = Hr^T = -2 = \alpha^{11}$ and therefore $l = 11 \bmod 3 = 2$, $s\alpha^{-2} = \alpha^9 = i$. Hence $e = (0, 0, i)$ and the submitted codeword is $c = r - e = (1+i, i, -1)$.

This theory can be generalized to the length $n = (p^r - 1)/4$. For that we take a field of p^r elements generated by an irreducible polynomial $p(x)$ of degree r over the field $\mathcal{G}_\pi$. Denote this field by $\mathcal{G}_{p^r}$. Let α be a primitive element of $\mathcal{G}_{p^r}$ and

$$H = (1 \ \ \alpha \ \ \alpha^2 \ \ \cdots \ \ \alpha^{\frac{p^r-1}{4}-1}). \tag{3}$$

Definition 4.1. OMEC $[n, n-r, 3]$ codes are block codes over $\mathcal{G}_\pi$ defined by the matrix H given in (3) having length $n = (p^r - 1)/4$, dimension $k = n - r$ and minimum Mannheim distance $d_M = 3$ (see [4]).

Let us give an example.

Example 4.2. Let $p = 5$, $\pi = 2+i$ and $r = 2$. To construct $\mathcal{G}_{25}$ we use the primitive polynomial $p(x) = x^2 + x - i$. Hence $\mathcal{G}_{25} = \{ax + b,\ a, b \in \mathcal{G}_{2+i}\}$. We take $\alpha = x$ and for the matrix H instead of a polynomial $ax + b$ we take the vector-column $(a, b)^T$. It turns out that

$$H = \begin{pmatrix} 0 & 1 & -1 & 1+i & 1-i & -1 \\ 1 & 0 & i & -i & -1+i & 1+i \end{pmatrix}.$$

H defines a $[6, 4, 3]$ OMEC code. To decode the received vector $r = (1, 0, i, 1, 0, 0)$ we compute

$$s = Hr^T = \begin{pmatrix} 1 \\ -i \end{pmatrix} = \alpha^{14}.$$

Hence the position of the error is $l = 14 \bmod 6 = 2$ and its value is $s\alpha^{-2} = \alpha^{12} = -1$. Hence the submitted codeword is $c = (1, 0, 1 + i, 1, 0, 0)$.

The well known Hamming bound ([5]) states that each q-ary code $\mathcal{C}$ of length n and minimum distance d satisfies

$$|\mathcal{C}|V_q(t, n) \le q^n, \quad \text{where } t = \left\lfloor \frac{d-1}{2} \right\rfloor. \tag{4}$$

Here $V_q(t, n)$ denotes the number of vectors of weight $\le t$. The codes meeting this bound are called perfect.

Each vector over $\mathcal{G}_\pi$ of length n with Mannheim weight 1 has only one nonzero coordinate whose value is ± 1 or $\pm i$. Therefore in our case $V_p(1, n) = 4n$. If C is an OMEC code of length n and dimension $n - r$ over $\mathcal{G}_\pi$ then $q = p$, $d = 3$. We hence have $t = 1$, $|C| = p^{n-r}$, and get

$$|\mathcal{C}|V_q(t, n) = p^{n-r}(1 + 4n) \le p^n \quad \Rightarrow 1 + 4n \le p^r \quad \Rightarrow n \le \frac{p^r - 1}{4}.$$

It follows that the $[n, n - r, 3]$ OMEC codes are perfect.

5. Codes correcting errors of Mannheim weight ≥ 2

Let $\mathcal{C}$ be a code with parity-check matrix

$$H = \begin{pmatrix} 1 & \beta & \beta^2 & \dots & \beta^{n-1} \\ 1 & \beta^5 & \beta^{10} & \dots & \beta^{(n-1)5} \end{pmatrix},$$

where $\beta \in \mathcal{G}_{\pi^r}$, $\beta^n = i$. There is an one-to one correspondence between the set of vectors of length n over $\mathcal{G}_\pi$ and the set of all polynomials over the same fields with degrees $\le n - 1$ defined by

$$c = (c_0, c_1, \dots, c_{n-1}) \mapsto c(x) = c_0 + c_1 x + \dots + c_{n-1} x^{n-1}.$$

Then for the code $\mathcal{C}$ we have

$$c \in \mathcal{C} \iff c(\beta^{4k+1}) = 0 \quad \text{for} \quad k = 0, 1, \dots, t.$$

Thus there is a generator polynomial $g(x) \mid x^n - i$ such that

$$\mathcal{C} = \{c(x) = a(x)g(x) \bmod (x^n - i), \ a(x) \in \mathcal{G}_\pi[x]\}.$$

It is easy to see that if $c(x) \in \mathcal{C}$ then $xc(x) = ic_{n-1} + c_0 x + \cdots + c_{n-2}x^{n-1} \in \mathcal{C}$ and so $(ic_{n-1}, c_0, c_1, \ldots, c_{n-2}) \in \mathcal{C}$. We call such codes icyclic. The icyclic codes belong to the class of constacyclic codes. A code C is constacyclic if there is a constant a such that $(c_0, c_1, \ldots, c_{n-1}) \in C$ yields $(ac_{n-1}, c_0, c_1, \ldots, c_{n-2}) \in C$. For the icyclic codes the constant is $a = i$.

The code $\mathcal{C}$ determined by the parity check matrix H can correct errors of Mannheim weight 2. Let $c \in \mathcal{C}$ be the transmitted codeword, and $r = c + e$ be the received vector, where $w_M(e) \leq 2$. First we compute the syndrome $s = H \cdot r^T = \begin{pmatrix} s_1 \\ s_2 \end{pmatrix}$. Suppose there are 2 errors at positions l_1, l_2 with values $\beta^{L_1 - l_1}$, $\beta^{L_2 - l_2} \in \{\pm 1, \pm i\}$. It turns out that

$$s = \begin{pmatrix} s_1 \\ s_2 \end{pmatrix} = H \cdot e^T = \begin{pmatrix} \beta^{L_1} + \beta^{L_2} \\ \beta^{5L_1} + \beta^{5L_2} \end{pmatrix}.$$

Then we compute the error determinator polynomial

$$\sigma(z) = (z - \beta^{L_1})(z - \beta^{L_2}) = z^2 - s_1 z + \mu.$$

Using that $s_1 = \beta^{L_1} + \beta^{L_2}$ and $s_5 = \beta^{5L_1} + \beta^{5L_2}$, after some calculations we get $\mu^2 - s_1^2 \mu + \frac{s_1^5 - s_5}{5s_1} = 0$.

Example 5.1. $p = 17$, $r = 1$, $n = 4$, $\beta = \alpha = 1 + i$,

$$H = \begin{pmatrix} 1 & \beta & \beta^2 & \beta^3 \\ 1 & \beta^5 & \beta^{10} & \beta^{15} \end{pmatrix} = \begin{pmatrix} 1 & 1+i & 2i & -1-2i \\ 1 & -1+i & -2i & -2+i \end{pmatrix},$$

$g(x) = (x - \alpha)(x - \alpha^5) = x^2 - 2ix - 2$. Let $r = (2, -2i, -1 - i, 1)$ be the received vector. Then

$$s = H \cdot r^T = \begin{pmatrix} -1+i \\ 1+i \end{pmatrix} = \begin{pmatrix} \beta \\ \beta^5 \end{pmatrix} \quad \Longrightarrow \sigma(z) = z^2 - (-1+i)z + \mu.$$

Since $\pm(2 - i)$ are the roots of $\mu^2 + 2i\mu = 0$, we have $\sigma(z) = z^2 - (-1 + i)z + (\pm 2 - i)$. It follows that $z_1 = \beta^{L_1} = \beta$, $z_2 = \beta^{L_2} = \beta^6$, and therefore $L_1 = 1$, $L_2 = 6$. Hence

$$\beta^{1-l_1}, \beta^{6-l_2} \in \{1, \beta^4, \beta^8, \beta^{12}\}, \quad 0 \leq l_1, l_2 \leq 3.$$

It turns out that the first error is in position $l_1 = 1$ with value $\beta^0 = 1$, and the second is at $l_2 = 2$ with value $\beta^4 = i$. Thus we obtain

$$c = (2, -2i, -1 - i, 1) - (0, 1, i, 0) = (2, -1 - 2i, -1 - 2i, 1).$$

Unfortunately, this code can not correct single errors of Mannheim weight 2. This is not surprising as its minimum Mannheim distance is only 4. A necessary condition for a code $\mathcal{C}$ to correct all errors of Mannheim weight 2 is $d_M(\mathcal{C}) \geq 5$.

There are interesting open problems connected with the codes over Gaussian integers.

(1) To introduce and study inner products over $\mathcal{G}_\pi$ and corresponding orthogonality for codes. What about self-dual codes?
(2) To introduce and study other types of codes over $\mathcal{G}_\pi$.
(3) To prove bounds for the minimum Mannheim distance of a code.
(4) To study codes over other Euclidean rings (For example, codes over quadratic number fields are studied in [8]).

Bibliography

1. K. Conrad, *The Gaussian Integers*, preprint, `http://www.math.uconn.edu/~kconrad/blurbs/ugradnumthy/Zinotes.pdf`
2. G. Dresden and W. Dymàček, *Finding Factors of Factor Rings over the Gaussian Integers*, MAA Monthly **112**(2005), 602–611.
3. G. H. Hardy and E. M. Wright, *An Introduction to the Theory of Numbers*, 5nd edition, The Clarendon Press, Oxford Univ. Press, New York (1979).
4. K. Huber, *Codes over Gaussian integers*, IEEE Trans. Inf. Theory **40**(1994), 207–216.
5. J. MacWilliams and N.J.A. Sloane, *The Theory of Error-Correcting Codes*, North-Holland, Amsterdam 1977.
6. C. Martinez, R. Beivide, and E. Gabidulin, *Perfect codes for metrics induced by circulant graphs*, IEEE Trans. Inf. Theory **53**(2007), 3042–3052.
7. H. Morita, Ko Kamada, H. Kostadinov, A. J van Wijngaarden, *On single cross error correcting codes with minimum-energy signal constellations*, in Proc. of ISIT2007, Nice, France (2007), 26–30.
8. T. P. da N. Neto, J. C. Interlando, *Lattice Constellations and Codes From Quadratic Number Fields*, IEEE Trans. Inform. Theory **47**(2001), 1514–1527.

Received February 7, 2013

Proceedings of the 3rd International
Colloquium on Differential Geometry
and its Related Fields
Veliko Tarnovo, September 3–7, 2012

A SURVEY ON GENERALIZED LIOUVILLE MANIFOLDS

Midori GOTO

Faculty of Information Engineering, Fukuoka Institute of Technology,
3-30-1 Wajirohigashi, Higashi-ku,
Fukuoka, 811-0295, Japan
E-mail: m-gotou@fit.ac.jp

This is a survey on generalized Liouville manifolds which were studied in the joint work with K. Sugahara ([3]). Without assuming that the metric is positive definite, we define a generalized Liouville structure and generalized Liouville manifolds whose geodesic flows are completely integrable. We give many generalized Liouville manifolds and give a classification of them.

Keywords: Generalized Liouville structure; Liouville surface; Lorentzian metric; elliptic coordinate system; geodesic flow; symplectic structure.

1. Introduction

In the study of Riemannian manifolds, behaviors of geodesics have particularly significant meaning. However, there are less manifolds whose global behaviors of geodesics are known. So Liouville manifolds whose geodesic flows are completely integrable are important.

In Sections 2 and 3 we recall historical matters on elliptic coordinates and properties of Liouville line elements. In Section 4, generalized Liouville manifolds are defined. We find many Liouville surfaces, Lorentz-Liouville surfaces, Liouville manifolds and Lorentz-Liouville manifolds. Finally, we pose a conjecture which would suggest a classification of generalized Liouville manifolds.

By $C^\infty(M)$ we denote the set of all smooth functions on M.

2. Elliptic Coordinates

By the book, "JOSEPH LIOUVILLE 1809-1882" written by J. Lützen, we see that Liouville published three hundreds and eighty papers, see [9]. Among them three voluminous papers from 1846 and 1847 are concerned with a particular class of problems for which the equations of motion can be integrated. Let us quote several sentences from the book:

$\cdots$ they were inspired by a paper on geodesics on ellipsoids, namely by Jacobi's condensed (paper) $\cdots$ i.e., the use of elliptic coordinates allowed Jacobi to set forth the equation of the geodesics on an ellipsoid thereby bypassing its second-order differential equation in Cartesian coordinates, which has "such a complicated form that one is easily discouraged from any treatment of it". $\cdots$ Liouville started the investigation on 'mechanics'. He considered geodesics as paths of a particle moving freely on the surface. It is also Jacobi's idea, but Jacobi did not expressed in his published papers. Jacobi mentioned that the solution of Hamilton-Jacobi equation gives rise to the solution of the equation of motions. $\cdots$

The Jacobi's works have been introduced in the Darboux's lectures ([2]). Let us recall the elliptic coordinate system for later use (see [2] vol. II, n^0 459).

Let a, b, c be real numbers with $a > b > c$. Consider the surface $\Sigma(\sigma)$ defined by the equation

$$\frac{x^2}{a-\sigma} + \frac{y^2}{b-\sigma} + \frac{z^2}{c-\sigma} = 1.$$

Put

$$h(\sigma) = \frac{x^2}{a-\sigma} + \frac{y^2}{b-\sigma} + \frac{z^2}{c-\sigma} - 1.$$

We see that $h'(\sigma) > 0$, so the value of $h(\sigma)$ changes from $-\infty$ to ∞, and $\lim_{\sigma\to\pm\infty} h(\sigma) = 0$. Since $h(\sigma)$ is discontinuous at $\sigma = a, b$ and c, we obtain the following

Lemma 2.1. *For $x, y, z > 0$, there exists exactly one triplet $(\sigma_1, \sigma_2, \sigma_3)$ such that $h(\sigma_i) = 0, i = 1, 2, 3,$ and $a > \sigma_1 > b > \sigma_2 > c > \sigma_3$.*

Therefore we have a diffeomorphic correspondence between two sets

$$\{(x, y, z) \in \mathbb{R}^3 | \quad x > 0,\ y > 0,\ z > 0\}$$

and

$$\{(\sigma_1, \sigma_2, \sigma_3) \in \mathbb{R}^3 | \quad a > \sigma_1 > b > \sigma_2 > c > \sigma_3\}.$$

Let us denote by Φ the diffeomorphism of the second set to the first one. The triplet $(\sigma_1, \sigma_2, \sigma_3)$ is called the *elliptic coordinate system.* We see the following:

- $\Sigma(\sigma_1)$ is a hyperboloid of two sheets;
- $\Sigma(\sigma_2)$ is a hyperboloid of one sheet;

- $\Sigma(\sigma_3)$ is an ellipsoid.

Now, fix σ_3. Then a point $(x, y, z) \in \Sigma(\sigma_3)$ is determined by the two hyperboloids $\Sigma(\sigma_1)$ and $\Sigma(\sigma_2)$ passing through it. The curves $\sigma_1 = const$ and $\sigma_2 = const$ are the lines of curvature on the surface $\Sigma(\sigma_3)$.

Lemma 2.2. *Surfaces $\Sigma(\sigma_1)$, $\Sigma(\sigma_2)$ and $\Sigma(\sigma_3)$ are perpendicular to each other: $\Sigma(\sigma_1) \perp \Sigma(\sigma_3)$, $\Sigma(\sigma_2) \perp \Sigma(\sigma_3)$ and $\Sigma(\sigma_1) \perp \Sigma(\sigma_2)$.*

Proof. We note that the vector $\left(\dfrac{x}{a-\sigma}, \dfrac{y}{b-\sigma}, \dfrac{z}{c-\sigma}\right)$ is normal to $\Sigma(\sigma)$.

We shall prove the first relation. The other relations are obtained similarly. Using the following two equations

$$\frac{x^2}{a-\sigma_1} + \frac{y^2}{b-\sigma_1} + \frac{z^2}{c-\sigma_1} = 1 \quad \text{and} \quad \frac{x^2}{a-\sigma_3} + \frac{y^2}{b-\sigma_3} + \frac{z^2}{c-\sigma_3} = 1,$$

we have

$$(\sigma_1 - \sigma_3)\left(\frac{x^2}{(a-\sigma_1)(a-\sigma_3)} + \frac{y^2}{(b-\sigma_1)(b-\sigma_3)} + \frac{z^2}{(c-\sigma_1)(c-\sigma_3)}\right) = 0.$$

Hence it follows that

$$\left(\frac{x}{a-\sigma_1}, \frac{y}{b-\sigma_1}, \frac{z}{c-\sigma_1}\right) \cdot \left(\frac{x}{a-\sigma_3}, \frac{y}{b-\sigma_3}, \frac{z}{c-\sigma_3}\right) = 0.$$

Thus we obtain the desired relations. □

Let σ_3 be fixed. Then we put $\phi(\sigma_1, \sigma_2) = \Phi(\sigma_1, \sigma_2, \sigma_3) \in \Sigma(\sigma_3)$. The line element ds^2 on the ellipsoid $\Sigma(\sigma_3)$ is, due to Lemma 2.2, of the form

$$\begin{aligned} ds^2 &= \phi_*\left(\frac{\partial}{\partial\sigma_1}\right) \cdot \phi_*\left(\frac{\partial}{\partial\sigma_1}\right) d\sigma_1^2 + \phi_*\left(\frac{\partial}{\partial\sigma_2}\right) \cdot \phi_*\left(\frac{\partial}{\partial\sigma_2}\right) d\sigma_2^2 \\ &= \frac{\sigma_1-\sigma_2}{4}\left(\frac{\sigma_1-\sigma_3}{(a-\sigma_1)(b-\sigma_1)(c-\sigma_1)} d\sigma_1^2 - \frac{\sigma_2-\sigma_3}{(a-\sigma_2)(b-\sigma_2)(c-\sigma_2)} d\sigma_2^2\right) \end{aligned}$$

The line element of this type, namely

$$ds^2 = (U - V)(U_1^2 du^2 + V_1^2 dv^2), \tag{$*$}$$

where U, U_1 and V, V_1 are functions of a single valuable u and v respectively, is called the *Liouville line element.*

Nowadays the geodesic equations are expressed using the Christoffel symbols. (Elvin Christoffel was born at 1829 and died at 1900.) In the elliptic coordinates, the geodesics of the above line element $(*)$ are characterized elegantly as follows (see [2] vol. III, n^0 584; [7] p. 183–185; and [9]

p. 700–706): Choose a constant γ satisfying $U > \gamma > V$, where U lies in the interval (b, a) and V is in the interval (c, b). Then we replace $(*)$ by

$$\begin{aligned} ds^2 &= \left(\left(\sqrt{U-\gamma}\right)^2 + \left(\sqrt{\gamma - V}\right)^2\right)\left((U_1 du)^2 + (V_1 dv)^2\right) \\ &= \left(\sqrt{U-\gamma}\right)^2 (U_1 du)^2 + \left(\sqrt{\gamma-V}\right)^2 (V_1 dv)^2 \\ &\quad + 2U_1\sqrt{U-\gamma}\, du\, V_1\sqrt{\gamma-V}\, dv + \left(\sqrt{U-\gamma}\right)^2 (V_1 dv)^2 \\ &\quad + \left(\sqrt{\gamma-V}\right)^2 (U_1 du)^2 - 2U_1\sqrt{U-\gamma}\, du\, V_1\sqrt{\gamma-V}\, dv. \end{aligned}$$

It follows that

$$ds^2 = \left(U_1\sqrt{U-\gamma}\, du + V_1\sqrt{\gamma-V}\, dv\right)^2 + \left(V_1\sqrt{U-\gamma}\, dv - U_1\sqrt{\gamma-V}\, du\right)^2.$$

We introduce new coordinates u', v' by

$$\begin{aligned} du' &= U_1\sqrt{U-\gamma}\, du + V_1\sqrt{\gamma-V}\, dv, \\ dv' &= \frac{U_1}{\sqrt{U-\gamma}}\, du - \frac{V_1}{\sqrt{\gamma-V}}\, dv. \end{aligned}$$

In the new coordinates (u', v'), the line element is given by

$$ds^2 = du'^2 + (U-\gamma)(\gamma-V)dv'^2.$$

The u'-curves are geodesics and the v'-curves meet these geodesics orthogonally. The equation $dv'/dt = 0$ yields

Lemma 2.3. *In the elliptic coordinates (u, v), the geodesics are characterized by the equation*

$$\frac{U_1}{\sqrt{U-\gamma}}\, \dot{u} - \frac{V_1}{\sqrt{\gamma-V}}\, \dot{v} = 0$$

together with the condition $E = const$. Here γ is a constant with value in the interval (c, b) or in (b, a).

The elliptic coordinates in $\mathbb{R}^n, n > 3$, have been studied already, cf. [10].

The Liouville line elements have remarkable properties. Especially, the geodesic flow on a domain with a Liouville line element is completely integrable. To see that, we start with recalling related terminologies.

Let X be a vector field on a differentiable manifold M. A function F on M is called a *first integral* of X if and only if F is constant along the integral curves of X.

Let (M, g) be an n-dimensional Riemannian manifold. It is well-known that there is a canonical *symplectic structure* ω on the cotangent bundle T^*M of M, and the pair (T^*M, ω) is called the *symplectic manifold.*

Here the form ω is a closed and non-degenerate 2-form. Owing to the non-degeneracy of ω, the map assigning to a vector field X a function $i_X\omega = \omega(X, \cdot)$ is a bundle isomorphism of the tangent bundle TM to the cotangent bundle T^*M. Here, i_X denotes the interior derivation. Under the identification of TM and T^*M, for a function F on T^*M, there exists exactly one vector field X_F on M satisfying

$$i_{X_F}\omega = dF.$$

The vector field X_F is called the *Hamilton vector field* and F the *Hamilton function* or *Hamiltonian.* The Poisson bracket $\{F_1, F_2\}$ for functions F_1, F_2 on T^*M is defined by

$$\{F_1, F_2\} = \omega(X_{F_2}, X_{F_1})(= i_{X_{F_2}}\omega(X_{F_1})).$$

We denote by E the energy function on T^*M defined as a half of the square of the norm function. The geodesics on M are characterized as the projection onto M of integral curves of the Hamilton vector field X_E. Since $\{E, F\} = X_E(F)$, it follows that $\{E, F\} = 0$ if and only if F is a first integral of X_E (i.e., of the geodesic equation). The geodesic flow of M (i.e., the Hamiltonian system X_E) is given as a symplectic flow on T^*M generated by X_E. The geodesic flow on M is said to be *completely integrable* if there exist n first integrals $F_1 = E, F_2, \cdots, F_n$ of X_E on M, so that

(i) $\{F_i, F_j\} = 0$ for all $i, j = 1, \cdots, n$;
(ii) $dF_1 \wedge dF_2 \wedge \cdots \wedge dF_n \neq 0$ on an open dense subset of M.

It is well-known that the geodesic flows of ellipsoids possess n(= the dimension of the ellipsoid) first integrals which satisfy the above (i) and (ii). The first integrals are quadratic forms on each cotangent space and are simultaneously normalizable on each fibre.

3. Liouville Surfaces

In this section, we consider a Riemannian metric of the form

$$g = \big(f_1(x_1) + f_2(x_2)\big)(dx_1^2 + dx_2^2) \qquad (**)$$

where $x = (x_1, x_2)$ is a coordinate system, and f_i is a function of the single variable $x_i, (i = 1, 2)$. It is a Liouville line element. It is known that the surface which is equipped with the metric (**) has the following property. Let (x, ξ) be the canonical coordinate system on the cotangent bundle, and let

$$E = \frac{1}{2}\frac{1}{f_1(x_1) + f_2(x_2)}(\xi_1^2 + \xi_2^2)$$

be the energy function associated with the metric (**). Set

$$F = \frac{1}{f_1(x_1) + f_2(x_2)} \left(f_2(x_2)\xi_1^2 - f_1(x_1)\xi_2^2 \right).$$

Then the Poisson bracket $\{E, F\} = 0$ and the geodesic flow has the first integral F.

On the other hand, we have the following proposition.

Proposition 3.1 (Kiyohara [5]). *Let g be a Riemannian metric on a neighborhood U of a point $p \in \mathbb{R}^2$, and let $E \in C^\infty(T^*U)$ be the corresponding energy function. Assume that a function $F \in C^\infty(T^*U)$ satisfies the following three conditions:*

(1) $\{E, F\} = 0$;
(2) F_q *is a homogeneous polynomial of degree 2 for every* $q \in U$;
(3) $F_p \neq \mathbb{R}E_p$.

Then there is a coordinate system (x_1, x_2) on a (possibly smaller) neighborhood of p, and there are functions $f_i(x_i)(i = 1, 2)$ such that

$$g = \left(f_1(x_1) + f_2(x_2) \right) (dx_1^2 + dx_2^2)$$

and

$$F = \frac{1}{f_1(x_1) + f_2(x_2)} \left(f_2(x_2)\xi_1^2 - f_1(x_1)\xi_2^2 \right)$$

where $x = (x, \xi)$ is the associated canonical coordinates on the cotangent bundle. Furthermore, such coordinates (x_1, x_2) and functions f_1, f_2 are essentially unique.

In the next section we will see that the above Proposition, with some suitable modifications, is valid for Lorentzian metrics.

It is known that a surface with the Liouville line element, which was called a Liouville surface in the literature, is locally characterized as a 2-dimensional Riemannian manifold S whose geodesic flow has a first integral F which is a homogeneous polynomial of degree 2 on each fibre (see Darboux [2]).

In 1991, Kiyohara gave a definition of a (global) compact Liouville surface in terms of a first integral of the geodesic flow which is based on the above Proposition 3.1. He classified equivalence classes of compact Liouville surfaces completely. Soon after, non-compact Liouville surfaces were discussed in [4]. We shall recall the definition below. The above condition (3) is contained in the following condition (L3).

Definition 3.1 (cf. [4], [5], and [6]). A triplet (S, g, F) is called a *Liouville surface* if (S, g) is a 2-dimensional Riemannian manifold and F is a C^∞ function on the cotangent bundle T^*S satisfying the following conditions:

(L1) the Poisson bracket $\{E, F\}$ vanishes;
(L2) F_p is a homogeneous polynomial of degree 2 for any $p \in S$;
(L3) F is not of the form $rH + sE$, where $H \in C^\infty(T^*S)$ is fibrewise the square of a linear form, and $s, r \in \mathbb{R}$.

A point which does not satisfy (3) in the Proposition 3.1 is called a *singular point.*

Due to Kiyohara ([5]), the number of singular points gives rise to a classification of equivalent classes of Liouville surfaces as follows. Let (S, g, F) be a Liouville surface. Then the number of singular points is zero if S is diffeomorphic to a torus or a Klein bottle; it is 2 if S is diffeomorphic to a real projective plane $\mathbb{R}P^2$; it is 4 if S is difeomorphic to a sphere S^2; and there is no such a metric if the genus of S is greater than 1.

4. Definitions of Generalized Liouville Manifolds

Higher dimensional Liouville manifolds have been defined in [6] as follows:

Definition 4.1. Let (M, g) be an n-dimensional Riemannian manifold and let E be the coresponding energy function on the cotangent bundle T^*M. Let $\mathcal{F}$ be the n-dimensional vector space of C^∞ functions on T^*M which are homogeneous polynomials of degree 2 on each fibre. We denote by F_p the restriction of $F \in \mathcal{F}$ to the cotangent space T_p^*M at p. Then $\mathcal{F}$ is called a (*local*) *Liouville structure* if the following four conditions are satisfied:

(L1) $E \in \mathcal{F}$;
(L2) $\{F, H\} = 0$ holds for any $F, H \in \mathcal{F}$;
(L3) for every $p \in M$, F_p's are diagonalizable simultaneously;
(L4) at a point $p \in M$, $\dim\{F_p; F \in \mathcal{F}\} = n$.

When $\mathcal{F}$ is a (local) Liouville structure, a triplet $(M, g, \mathcal{F})$ is called a (*local*) *Liouville manifold.* If, furthermore, the metric g is geodesically complete, a triplet $(M, g, \mathcal{F})$ is called a *Liouville manifold.*

Remark 4.1. The above two conditions (L1) and (L2) imply that each function in $\mathcal{F}$ is a first integral of the geodesic flow of the Levi-Civita connection.

Including indefinite metrics, we generalize definitions of the (local) Liouville structure and the (local) Liouville manifold as follows.

Definition 4.2. Without assuming that the metric g is positive definite, when $\mathcal{F}$ satisfies the above four conditions (L1) – (L4) , we call $\mathcal{F}$ a (*local*) *generalized Liouville structure* and call a triplet $(M, g, \mathcal{F})$ a (*local*) *generalized Liouville manifold.* Furthermore, when g is geodesically complete, we call a triplet $(M, g, \mathcal{F})$ a *generalized Liouville manifold.*

In case g is a Riemannian metric, we simply call the triplet a *Riemann-Liouville manifold.* If g is a Lorentian metric, we call it a *Lorentz-Liouville manifold,* and so on. Moreover, in case $n = 2$, we call it a *generalized-Liouville surface*, a *Riemann-Liouville surface*, a *Lorentz-Liouville surface*, and so on, according to the situation of the metric g.

5. Lorentz-Liouville Surfaces

The following theorem is the Lorentzian version of Proposition 3.1.

Theorem 5.1 ([3]). *Let g be a Lorentzian metric defined in a neighborhood U of a point $p \in \mathbb{R}^2$, and let E be the energy function associated with g. Suppose that a C^∞ function F on the cotangent bundle T^*U satisfies the following conditions:*

(1) $\{F, H\} = 0$;
(2) *F_q is a homogeneous polynomial of degree 2 for each $q \in U$;*
(3) $F_q \neq \mathbb{R}E_p$;
(4) *for each point q, E_q and F_q are simultaneously diagonalizable.*

Then, in a neighborhood of p, there are coordinate system (x_1, x_2) and functions $f_i(x_i)(i = 1, 2)$ such that

$$g = \big(f_1(x_1) + f_2(x_2)\big)(dx_1^2 - dx_2^2)$$

and

$$F = \frac{1}{f_1(x_1) + f_2(x_2)}\big(f_2(x_2)\xi_1^2 + f_1(x_1)\xi_2^2\big).$$

Here, (x, ξ) is the canonical coordinate system on the cotangent bundle. Furthermore, these coordinate system (x_1, x_2) and functions $f_i(x_i)(i = 1, 2)$ are essentially unique.

Since we can prove that all the points on any geodesic starting at a singular point in the null direction turn out to be singular points, we obtain the following theorem. It characterizes Lorentz-Liouville surfaces.

Theorem 5.2 ([3]). *There are no singularities on any Lorentz-Liouville surface.*

Hence, we have the following classification of Lorentz-Liouville surfaces.

Theorem 5.3. *The universal covering space of a Lorentz-Liouville surface is diffeomorphic to a rectangular domain of* $\mathbb{R}^2$. *The metric* g *is of the form*

$$g = \big(f_1(x_1) + f_2(x_2)\big)(dx_1^2 - dx_2^2)$$

globally with respect to the canonical coordinates of $\mathbb{R}^2$.

6. Quadratic Surfaces in $\mathbb{R}^3$

In this section, we prove that quadratic surfaces in $\mathbb{R}^3$ admit generalized Liouville structures with respect to the natural indefinite metric.

Let (x_0, x_1, x_2) be the natural coordinates of $\mathbb{R}^3$. Let $\epsilon = (\epsilon_0, \epsilon_1, \epsilon_2)$ be one of $(1,1,1), (1,1,-1)$ and $(1,-1,-1)$. For each ϵ we consider the inner product of $\mathbb{R}^3$ defined by

$$g_\epsilon = \epsilon_0 dx_0^2 + \epsilon_1 dx_1^2 + \epsilon_2 dx_2^2.$$

For real numbers $a_0 < a_1 < a_2$ and $\lambda \neq \{a_0, a_1, a_2\}$, put

$$Q_\lambda := \frac{\epsilon_0 x_0^2}{a_0 - \lambda} + \frac{\epsilon_1 x_1^2}{a_1 - \lambda} + \frac{\epsilon_2 x_2^2}{a_2 - \lambda}.$$

According to the signature of ϵ_i's, the valuables $\lambda_0 < \lambda_1 < \lambda_2$ have domains as follows:

(0) for $(\epsilon_0, \epsilon_1, \epsilon_2) = (1,1,1)$,

$$\lambda_0 \in (-\infty, a_0],\ \lambda_1 \in [a_0, a_1],\ \lambda_2 \in [a_1, a_2];$$

(1) for $(\epsilon_0, \epsilon_1, \epsilon_2) = (1,1,-1)$,

$$\lambda_0 \in (-\infty, a_0],\ \lambda_1 \in [a_0, a_1],\ \lambda_2 \in [a_2, \infty);$$

(2) for $(\epsilon_0, \epsilon_1, \epsilon_2) = (1,-1,-1)$,

$$\lambda_0 \in (-\infty, a_0],\ \lambda_1 \in [a_1, a_2],\ \lambda_2 \in [a_2, \infty).$$

Then, using the equations

$$Q_{\lambda_0} = Q_{\lambda_1} = Q_{\lambda_2} = 1,$$

we obtain

$$x_0^2 = -\frac{(\lambda_0 - a_0)(\lambda_1 - a_0)(\lambda_2 - a_0)}{\epsilon_0(a_1 - a_0)(a_2 - a_0)};$$

$$x_1^2 = -\frac{(\lambda_0 - a_1)(\lambda_1 - a_1)(\lambda_2 - a_1)}{\epsilon_1(a_0 - a_1)(a_2 - a_1)};$$

$$x_2^2 = -\frac{(\lambda_0 - a_2)(\lambda_1 - a_2)(\lambda_2 - a_2)}{\epsilon_2(a_0 - a_2)(a_1 - a_2)}.$$

Thus we have a coordinate system $(\lambda_0, \lambda_1, \lambda_2)$ for each quadrant of $\mathbb{R}^3$, as in Section 3. However, since ellipsoids may not be realized for all the value of ϵ, the name "elliptic coordinates" is not suitable. We call it the *quadratic coordinates.*

In the quadratic coordinates, the metric g_ϵ is of the form

$$\begin{aligned} g_\epsilon = &- \frac{(\lambda_0 - \lambda_1)(\lambda_0 - \lambda_2)}{4(\lambda_0 - a_0)(\lambda_0 - a_1)(\lambda_0 - a_2)} d\lambda_0^2 \\ &- \frac{(\lambda_1 - \lambda_0)(\lambda_1 - \lambda_2)}{4(\lambda_1 - a_0)(\lambda_1 - a_1)(\lambda_1 - a_2)} d\lambda_1^2 \\ &- \frac{(\lambda_2 - \lambda_0)(\lambda_2 - \lambda_1)}{4(\lambda_2 - a_0)(\lambda_2 - a_1)(\lambda_2 - a_2)} d\lambda_2^2. \end{aligned}$$

Here we note that ϵ does not appear in the expression, because the domains of λ_i's $(i = 0, 1, 2)$ are determined according to the signatures of ϵ_i's.

Theorem 6.1 ([3]). *Let us take $\lambda_0 \in (-\infty, a_0)$ and fix it. Then the surface $Q_{\lambda_0} = 1$ is a Liouville surface with respect to the first fundamental form g_{λ_0} which is induced from the inner product g_ϵ of $\mathbb{R}^3$. In particular, in case $\epsilon = (\epsilon_0, \epsilon_1, \epsilon_2) = (1, 1, -1)$, it is a Lorentz-Liouville surface.*

Theorem 6.2 ([3]). *According to the signature of ϵ_1, we choose λ_1 from the interval (a_0, a_1) or from (a_1, a_2) and fix it. Then the surface $Q_{\lambda_1} = 1$ is a Liouville surface with respect to the first fundamental form which is induced from the inner product g_ϵ of $\mathbb{R}^3$.*

Theorem 6.3 ([3]). *According to the signature of ϵ_2, we choose λ_2 from the interval (a_1, a_2) or from (a_2, ∞) and fix it. Then the surface $Q_{\lambda_2} = 1$ is a Liouville surface with respect to the first fundamental form which is induced from the inner product g_ϵ of $\mathbb{R}^3$.*

Thus the classification of the Lorentz-Liouville surfaces has been done.

7. Further Results

Like the same way as we see in the previous section, we can prove that quadratic hypersurfaces in $\mathbb{R}^{n+1}$ have the generalized-Liouville structures with respect to the canonical indefinite metric.

An $(n+1)$-dimensional hyperbolic surfaces is realized in the upper half space of the $(n+1)$-dimensional Cartesian space. If the upper half space is equipped with the usual metric or an indefinite hyperbolic metric, we can prove that quadratic hypersurfaces which are restricted to the domain admit the generalized-Liouville structure.

Theorem 7.1. *For a quadratic hypersurface in an $(n+1)$-dimensional Cartesian space, we can express its first integrals explicitly and show that it admits a generalized-Liouville structure with respect to the canonical indefinite metric, including the Minkowski metric, as well as the usual Euclidean metric.*

It is known that ellipsoids in an $(n+1)$-dimensional Euclidean space are Liouville manifolds (cf. [10]). In the 2-dimensional case, using the elliptic coordinates, geodesic equations and their first integrals have been investigated explicitly in [8].

Theorem 7.2. *Quadratic hypersurfaces in a Cartesian $(n+1)$-space, restricted to the half space equipped with the usual hyperbolic metric and an indefinite hyperbolic metric, admit generalized-Liouville structures with respect to the first fundamental form.*

So far, we could find many generalized-Liouville manifolds. Now, let us try to classify them.

Definition 7.1. Let $(M, \mathcal{F})$ be a generalized Liouville manifold of dimension n. Let $\mathcal{F}^*$ be the set of all F in $\mathcal{F} - \{0\}$ such that $\{p \in M : F_p = 0\}$ is a closed submanifold of codimension 2. For $F, F' \in \mathcal{F}$, we write $F \sim F'$ if there exists some $c \in \mathbb{R}$ such that $F = cF'$. The number $r = n - \#(\mathcal{F}/\sim)$ is called the *rank* of $(M, \mathcal{F})$.

The rank of Riemannian-Liouville manifolds of dimension n is equal to or less than $(n-1)$. The proper Riemannian-Liouville manifolds of rank one have been classified in [7].

We pose the following conjecture:

Conjecture 7.1. *The rank of a generalized–Liouville manifold of dimension n is equal to or less than $(n-2)$ if its metric is indefinite.*

In order to classify generalized–Liouville manifolds, we have to classify Liouville manifolds of rank two.

Bibliography

1. A. Besse, *Manifolds all of whose Geodesics are Closed*, Springer-Verlag, 1978.
2. G. Darboux, *Leçons sur la théorie générale des surfaces* vol. I -IV, Gauthier-Villars, 1972.
3. M. Goto & K. Sugahara, *Generalized Liouville Manifolds*, forthcoming.
4. M. Igarashi, K. Kiyohara & K. Sugahara, *Noncompact Liouville surfaces*, J. Math. Soc. Japan **45** (1993), 459-479.
5. K. Kiyohara, *Compact Liouville surfaces*, J. Math. Soc. Japan **43** (1991), 555-591.
6. K. Kiyohara, *Two Classes of Riemannian Manifolds Whose Geodesic Flows Are Integrable.* Memoirs of AMS, vol. **130**, 1997.
7. W. Klingenberg, *Lectures on closed geodesics*, Springer, Berlin- Heidelberg, 1978.
8. W. Klingenberg, *Riemannian Manifolds*, 2nd ed., Walter de Gruyter, Berlin-New York, 1982.
9. J. Lützen, *JOSEPH LIOUVILLE 1809-1882, Master of Pure and Applied Mathematics*, Studies in the History of Mathematics and Physical Sciences **15**, Springer-Verlag, Berlin-Heidelberg, 1990.
10. A. Thimm, *Integrabilität beim Geodätischen Fluß*, Bonner Mathematische Schriften Nr. **103**, Bonn 1978.

Received December 17, 2012
Revised February 9, 2013

Proceedings of the 3rd International
Colloquium on Differential Geometry
and its Related Fields
Veliko Tarnovo, September 3–7, 2012

THE RELATIVISTIC NON-LINEAR QUANTUM DYNAMICS FROM THE $\mathbb{C}P^{N-1}$ GEOMETRY

Dedicated to my dear wife Ludmila
for her constant patience and support.

Peter LEIFER

Medical Physics and Informatics, Crimea State University,
Simferopol, 95006 Crimea, Ukraine
E-mail: peter.leifer@gmail.com
www.csmu.edu.ua

Einstein's program of the unified field theory transformed nowadays to the "theory of everything" (TOE) requiring new primordial elements and relations between them. Definitely, they must be elements of the quantum nature. One of most fundamental quantum elements are pure quantum states. Their basic relations are defined by the geometry of the complex projective Hilbert space. In the framework of such geometry all physical concepts should be formulated and derived in the natural way. Analysis following this logic shows that inertia and inertial forces are originated not in spacetime but in the space of quantum states since they are generated by the deformation of quantum states as a reaction on an external interaction or self-interaction. It turns out that the $\mathbb{C}P^{N-1}$ geometry intrinsically connected with such deformations. Therefore, such geometry should clarify the old problem of inertial mass (dynamical mass generation), and, on the other hand, leads to the law of environment independence (invariance) of the quantum numbers of quantum particles. The conservation law of proper energy-momentum following from this invariance has been applied to self-interacting quantum Dirac's electron.

Keywords: Complex projective space; Cartan decomposition; Coset space; Flexible quantum reference frame; Local dynamical variables; Affine gauge fields; Conservation law; Quantum formulation of the inertia principle; Quasi-linear PDE's; Jacobi vector fields.

1. Introduction

The second quantization method is the basis of the QFT [9]; it presents the top of the linear approach to the essentially non-linear problem of quantum interaction and self-interaction. This formal apparatus realizes the physical concept of the corpuscles-wave duality. It is well known, however, that this universality should be broken for interacting quantum fields [5]. Divergences and necessity of the renormalization procedure during solution of the typical problems of QED is the most acute consequence of this method [9]. These problems stem partly from the formal, "stiff" method of quantization being applied to the dynamical variables and to the Fourier components of wave function. However, ideally, the quantization might be realized by the soliton-like lumps currying discrete portions of the mass, charge, spin, etc. Such a way requires new non-linear wave equations derived from unknown "first principles" connected with unified theory of quantum interactions. Modern attempts to unify initially electromagnetic, weak and strong interactions are based upon obvious observation that gravity is much weaker than other fundamental forces. But such approach is similar to splashing water together with a kid since these attempts close the way to understand the source of inertia and gravity - a mass.

I would like discuss here intrinsic unification of general relativity and non-linear quantum dynamics based on the eigen-dynamics of pure quantum state in $\mathbb{C}P^{N-1}$ leading to the geometric formulation of non-Abelian gauge fields $\vec{P} = P^i(x,\pi)\frac{\partial}{\partial \pi^i} + c.c. = P^{\mu}(x)\Phi^i_{\mu}(\pi)\frac{\partial}{\partial \pi^i} + c.c.$ carrying the self-interaction [15, 17, 18]. These gauge fields are considerably differ from the Yang-Mills fields $\hat{A}_{\mu}(x) = A^{\alpha}_{\mu}(x)\hat{\lambda}_{\alpha}$ since from the technical point of view the fixed matrices $\hat{\lambda}_{\alpha} \in \mathfrak{su}(N)$ have been replaced by the smooth state-dependent vector fields on $\mathbb{C}P^{N-1}$. Principally, in the base of such approach lies idea of the priority of internal $SU(N)$ symmetry and its breakdown with subsequent classification of quantum motions in $\mathbb{C}P^{N-1}$ over spacetime symmetries [19–21]. This idea leads to the quantum formulation of the inertia principle in the space state of the internal degrees of freedom, not in spacetime [15]. Namely, such "matter field" like electron is represented by a dynamical process of the geodesic motion in $\mathbb{C}P^{N-1}$ and the geodesic variations shall be related to the gauge fields. Then the first order quasi-linear PDE's field equations for electron itself taking the place of the boundary conditions for the variation problem and Jacobi fields play the role of the generalized affine non-Abelian state-dependent gauge field associated with the gauge group $H = U(1) \times U(N)$. How one may identify this field with electroweak field will be discussed elsewhere. I will be concentrated here on the problem of the rest field mass of self-interacting electron and the origin of the inertial forces. In order to do this I will discuss initially motivation leading to new formulation of the inertia principle.

Clarification

0) The Einstein's convention of the summation in identical co- and contra-variant indexes has been used.
1) Flexible quantum setup (FQS) is anholonomic reference frame $A^\mu(x)\Phi^i_\mu(\pi)\frac{\partial}{\partial\pi^i}$ in $\mathbb{C}P^{N-1}$ whose spacetime coefficient functions $A^\mu(x)$ realizing a quantum setup "tuning" by variation of these components.
2) Local dynamical variables (LDV's) are vector fields $\Phi^i_\mu(\pi)\frac{\partial}{\partial\pi^i} + c.c.$ on $\mathbb{C}P^{N-1}$ corresponding to the $SU(N)$ generators [4].
3) Superrelativity means physical equivalence of any conceivable quantum setup, i.e. the quantum numbers of "elementary" particles like mass, charge, spin, etc., are the same anywhere. This invariance grantees the self-identity of quantum particles in any ambient.
4) Self-identity means the conservation of the fundamental LDV's corresponding mentioned quantum numbers.
5) The conservation of LDV's may be expressed by the affine parallel transport of LDV's in $\mathbb{C}P^{N-1}$.
6) Coefficient functions $\Phi^i_\mu(\pi)$ of the $SU(N)$ generators acting on quantum states in $\mathbb{C}P^{N-1}$ replace classical force vector fields acting on material point with charge, mass, etc. Being multiplied and contracted with potentials $P^\mu(x)$ they comprise vector field of proper energy-momentum giving the rate of the quantum state variation $\frac{d\pi^i}{d\tau} = \frac{c}{\hbar}P^\mu\Phi^i_\mu(\pi)$. Their divergency $L_\lambda = \frac{\partial\Phi^n_\lambda(\pi)}{\partial\pi^n} + \Gamma^m_{mn}\Phi^n_\lambda(\pi)$ may be treated as non-Abelian charges. From the formal point of view it comes out that the projective Hilbert space $\mathbb{C}P^{N-1}$ serves as base of the principle fiber bundle with $SU(N)$ structure group instead of the spacetime with Poincare group and its representations as in the traditional QFT.
7) No connections of this theory with the commonly used $\mathbb{C}P^{N-1}$ sigma models in low-dimensional space-time. $\mathbb{C}P^{N-1}$ compact manifold serves as base manifold; 4D space-time arises in the frame fibre bundle.

2. Intrinsic unification of relativity and quantum principles

The localization problem in spacetime mentioned above and deep difficulties of divergences in quantum field theory (QFT) insist to find a new primordial quantum element instead of the classical material point and its probabilistic quantum counterpart. I will use unlocated quantum state of a system - a specific quantum motion [10] as such primordial element. Quantum states of single quantum particles may be represented by vectors $|\Psi\rangle$, $|\Phi\rangle, \ldots$ of linear functional Hilbert space $\mathcal{H}$ with countable or even finite dimensions since these states related to the internal degrees of freedom like spin, charge, etc. It is important to note that the correspondence between a quantum state and its vectors representation in $\mathcal{H}$ is not isomorphic. It is rather homomorphic, when a full equivalence class of proportional vectors, so-called rays $\{\Psi\} = z|\Psi\rangle$, where $z \in \mathcal{C} \setminus \{0\}$ corresponds to the one quantum state $|\Psi\rangle$. The rays of quantum states may be represented by points of complex projective Hilbert space $\mathbb{C}P^{\infty}$ or its finite dimension subspace $\mathbb{C}P^{N-1}$. Points of $\mathbb{C}P^{N-1}$ represent generalized coherent states (GCS) that will be used thereafter as fundamental physical concept instead of material point. This space will be treated as the *space of "unlocated quantum states"* as the analog of the *"space of unlocated shapes"* [27]. We are dealing with the lift of the quantum dynamics from $\mathbb{C}P^{N-1}$ into the *space of located quantum states.* That is, the variance between Shapere & Wilczek construction and our scheme is that the dynamics of unlocated quantum states should be represented by the motions of the localizable 4D "field-shell" in dynamical spacetime, whereas the configuration space is a Hilbert space, i.e. functional space containing the "field-shell" solutions of quasi-linear field equations (see below).

Two simple observations serve as the basis of the intrinsic unification of relativity and quantum principles. The first observation concerns interference of quantum states in a fixed quantum setup.

A. The linear interference of pure quantum states (amplitudes) shows the symmetries relative spacetime transformations of whole setup. This interference has been studied in "standard" quantum theory. Such symmetries reflects, say, the *first order of relativity*: the physics is same if any *complete setup* subject (kinematical, not dynamical!) shifts, rotations, boosts as whole in a single Minkowski space-time. According to our notes given some times ago [15, 16] one should add to this list a freely falling quantum setup (super-relativity).

The second observation concerns a dynamical "deformation" of some quantum setup.

B. If one dynamically changes the setup configuration or its "environment", then the pure state (amplitude of an event) will be generally changed. Nevertheless there is a different type of tacitly assumed deep symmetry that may be formulated on the intuitive level as the invariance of physical properties of "quantum particles", i.e. the invariance of their quantum numbers like mass, spin, charge, etc., relative variation of the quantum state due to the ambient variance. This means that the physical properties expressed by intrinsic dynamical variables of, say, electrons in two different setups S_1 and S_2 are the same. It is close to the nice Fock's idea of "relativity to observation devices" but it will be realized in infinitesimal form as following:

One postulates that the invariant content of this physical properties may be kept if one makes the infinitesimal variation of some "flexible quantum setup" reached by a small variation of some fields by adjustment of tuning devices.

A new concept of local dynamical variable (LDV) [21] should be introduced for the realization of the "flexible quantum setup". This construction is naturally connected with methods developed in studying geometric phase.[3,23] I seek, however, conservation laws for LDV's in the quantum state space following from the new formulation of the inertia principle [15, 18]. Details of plausible reasonings leading to such formulation may be found in the e-print [14]. The one of the most serious conclusion is as follows: *coset transformations on the quantum state space serves as a new geometric counterpart of quantum interaction or self-interaction* [15–19].

3. Affine non-Abelian gauge potential

A new gauge structure in the framework of the both classical mechanics of deformable bodies and the "geometric phase" in quantum mechanics has been discovered last time. These gauge fields lead mainly to Coriolis terms (see nice review [23] and corresponding references therein). There is a question: is it possible to use some variational principle leading automatically to the dynamical nature of the inertial mass? I will apply the variational principle (in the form of affine parallel transport of the vector field of $SU(N)$ generators defining infinitesimal transformation in the state space) since the origin of the inertial mass will be associated with proper quantum motion in the quantum state space, not in the spacetime. This affine parallel transport closely connected with the non-Abelian gauge invariance.

The local Abelian gauge invariance was ordinary connected with the invariance of the Maxwell equations. Yang-Mills fields serve as local non-Abelian gauge fields in the Standard Model (SM). New type of the non-Abelian gauge fields arises under the conservation of the LDV's of quantum particles. Say, stability of the solution of characteristic equations (see below) under the Jacobi geodesic variation in $\mathbb{C}P^3$ ensures the self-conservation of the electron due to parallel transport of the internal quantum energy-momentum.[15,17] Thereby, one has the coupling electrons by the Jacobi gauge vector fields.

It is interesting that already about 20 years ago so-called "magnetic

Jacobi fields" have been used as the variation of trajectories of classical particles on Kähler manifolds [1]. Deformed geodesics treated as trajectory of classical particles in the Jacobi magnetic fields obey generalized Jacobi equations with additional term depends on strength of an uniform magnetic field and the sectional curvature of a complex projective space. Quantum dynamics requires essentially different approach similar to the geometric phase ideology.

The geometric phase is an intrinsic property of the family of eigenstates. There are in fact a set of local dynamical variables (LDV) that like the geometric phase intrinsically depends on eigenstates. For us these are interesting vector fields $\xi^k(\pi^1_{(j)},...,\pi^{N-1}_{(j)}) : \mathbb{C}P^{N-1} \to \mathcal{C}$ associated with the reaction of quantum state $\pi^i_{(j)}$ on the action of internal "unitary field" $\exp(i\epsilon\lambda_\sigma)$ given by Φ^i_σ. Notation is defined in the equation (12) later. In view of future discussion of infinitesimal unitary transformations, it is useful to compare *velocity* of variation of the Berry's phase

$$\dot{\gamma}_n(t) = -\mathbf{A}_n(\mathbf{R})\dot{\mathbf{R}}, \tag{1}$$

where $\mathbf{A}_n(\mathbf{R}) = \Im\langle n(\mathbf{R})|\nabla_{\mathbf{R}} n(\mathbf{R})\rangle$ with the affine parallel transport of the vector field $\xi^k(\pi^1,...,\pi^{N-1})$ given by the equations

$$\frac{d\xi^i}{d\tau} = -\Gamma^i_{kl}\xi^k\frac{d\pi^l}{d\tau}. \tag{2}$$

The parallel transport of Berry is similar but it is not identical to the affine parallel transport. The last one is the fundamental because this agrees with Fubini-Study "quantum metric tensor" G_{ik^*} in the base manifold $\mathbb{C}P^{N-1}$. The affine gauge field given by the connection

$$\Gamma^i_{mn} = \frac{1}{2}G^{ip^*}\left(\frac{\partial G_{mp^*}}{\partial \pi^n} + \frac{\partial G_{p^*n}}{\partial \pi^m}\right) = -\frac{\delta^i_m\pi^{n^*} + \delta^i_n\pi^{m^*}}{1+\sum|\pi^s|^2} \tag{3}$$

is of course more close to the Wilczek-Zee non-Abelian gauge fields [28] where the Higgs potential has been replaced by the affine gauge potential (3) whose shape in the case $\mathbb{C}P^1$ is depicted in Fig. 1. It is involved in the affine parallel transport of LDV's [20–22] which agrees with the Fubini-Study metric (9).

The transformation law of the connection forms $\Gamma^i_k = \Gamma^i_{kl}d\pi^l$ in $\mathbb{C}P^{N-1}$ under the differentiable transformations of local coordinates $\Lambda^i_m = \frac{\partial \pi^i}{\partial \pi'^m}$ is as follows:

$$\Gamma'^i_k = \Lambda^i_m\Gamma^m_j\Lambda^{-1j}_k + d\Lambda^i_s\Lambda^{-1s}_k. \tag{4}$$

It is similar to the well known transformations of non-Abelian fields. However the physical sense of these transformations is quite different. Namely:

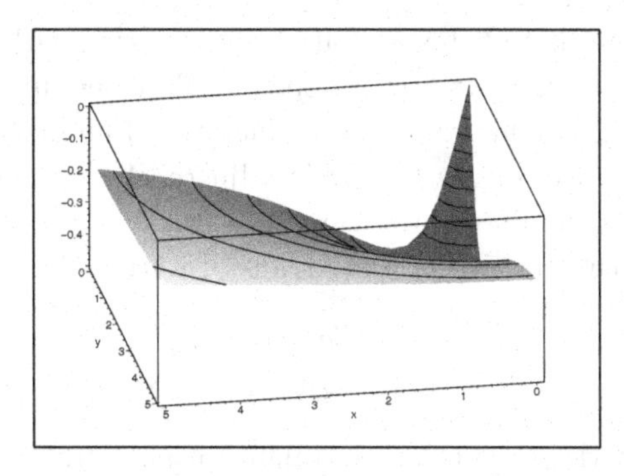

Figure 1. The shape of the gauge potential associated with the affine connection in CP(1): $\Gamma = -2\frac{|\pi|}{1+|\pi|^2}, \pi = u + iv$.

the Cartan's moving reference frame takes here the place of "flexible quantum setup", whose motion refers to itself with infinitesimally close coordinates. Thus we will be rid of necessity in "second particle" [2] as an external reference frame.

4. Dynamical spacetime

The distance between two quantum states of an electron in $\mathbb{C}P^3$ is given by the Fubini-Study invariant interval $dS_{F.-S.} = G_{ik*}d\pi^i d\pi^{k*}$. The speed of the interval variation is given by the equation

$$\left(\frac{dS_{F.-S.}}{d\tau}\right)^2 = G_{ik*}\frac{d\pi^i}{d\tau}\frac{d\pi^{k*}}{d\tau} = \frac{c^2}{\hbar^2}G_{ik*}(\Phi^i_\mu P^\mu)(\Phi^{k*}_\nu P^{\nu*}) \tag{5}$$

relative "quantum proper time" τ where energy-momentum vector field $P^\mu(x)$ obeys field equations that will be derived later. This internal dynamics of *"unlocated quantum states"* in $\mathbb{C}P^{N-1}$ should be expressed by the *quantum states cum location* in "dynamical spacetime" coordinates x^μ assuming that variation of coordinates δx^μ arise due to the transformations of Lorentz reference frame that involved in the covariant derivative $\frac{\delta P^\nu}{\delta \tau} = (\frac{\partial P^\nu}{\partial x^\mu} + \Gamma^\nu_{\mu\lambda}P^\lambda)\frac{\delta x^\mu}{\delta \tau}$ in dynamical spacetime (DST). Such procedure may be called "inverse representation" [15, 17, 18, 21] since this intended to represent quantum motions in $\mathbb{C}P^{N-1}$ by "quantum Lorentz transformation" in DST as it will be described below.

Since there is no a possibility to use classical physical reference frame comprising usual clock and solid scales on the deep quantum level, I will use the "field frame" from the four components of the vector field of the proper energy-momentum $P^\mu = (\frac{\hbar\omega}{c}, \hbar\vec{k})$ instead. This means that the period T

and the wave length λ of the oscillations associating with an electron's field are identified with flexible (state-dependent) scales in the DST. Thereby, the local Lorentz "field frame" is in fact the 4-momentum tetrad whose components may be locally (in $\mathbb{C}P^3$) adjusted by state-dependent "quantum boosts" and "quantum rotations".

It is convenient to take Lorentz transformations in the following form

$$\begin{aligned} ct' &= ct + (\vec{x}\vec{a}_Q)\delta\tau \\ \vec{x'} &= \vec{x} + ct\vec{a}_Q\delta\tau + (\vec{\omega}_Q \times \vec{x})\delta\tau \end{aligned} \tag{6}$$

where I put for the parameters of quantum acceleration and rotation the definitions $\vec{a}_Q = (a_1/c, a_2/c, a_3/c)$, $\vec{\omega}_Q = (\omega_1, \omega_2, \omega_3)$ [24] in order to have for the "proper quantum time" τ the physical dimension of time. The expression for the "4-velocity" V^μ is as follows

$$V^\mu_Q = \frac{\delta x^\mu}{\delta\tau} = (\vec{x}\vec{a}_Q, ct\vec{a}_Q + \vec{\omega}_Q \times \vec{x}). \tag{7}$$

The coordinates x^μ of an imaging point in the dynamical spacetime serve here merely for the parametrization of the energy-momentum distribution in the "field shell" described by quasi-linear field equations [15, 18] that will be derived below.

5. Self-interacting quantum electron

Since the spicetime priority is replaced by the priority of the state space, operators corresponding quantum dynamical variables should be expressed not in the terms spacetime coordinates like spatial, temporal differentials or angles but in terms of coordinates of state vectors. Furthermore, deleting redundant common multiplier, one may use local projective coordinates of rays. These local state space coordinates will be arguments of the local dynamical variables represented by the vector fields on $\mathbb{C}P^{N-1}$ [21].

Further, one needs the invariant classification of quantum motions [19]. This invariant classification is the quantum analog of classical conditions of inertial and accelerated motions. They are rooted into the global geometry of the dynamical group manifold. Namely, the geometry of $G = SU(N)$, the isotropy group $H = U(1)\times U(N-1)$ of the pure quantum state and the coset $G/H = SU(N)/S[U(1) \times U(N-1)] = \mathbb{C}P^{N-1}$ as geometric counterpart of the self-interaction, play an essential role in the classification of quantum motions [19].

In order to formulate the quantum (internal) energy-momentum conservation law in the state space, let us discuss the local eigen-dynamics of quantum system with finite quantum degrees of freedom N. It will be realized below in the model of self-interacting quantum electron where spin/charge degrees of freedom in $\mathbb{C}^4$ have been taken into account [17]. The LDV's like the energy-momentum and should be expressed in terms of the projective local coordinates π^k, $1 \leq i,k,j \leq N-1$ of quantum state $|\Psi\rangle = \psi^a|a\rangle$, $1 \leq a \leq N$, where ψ^a is a homogeneous coordinate on $\mathbb{C}P^{N-1}$

$$\pi^i_{(j)} = \begin{cases} \dfrac{\psi^i}{\psi^j}, & \text{if } 1 \leq i < j, \\ \dfrac{\psi^{i+1}}{\psi^j}, & \text{if } j \leq i < N, \end{cases} \tag{8}$$

since $SU(N)$ acts effectively only on the space of rays, i.e. on equivalent classes relative the relation of equivalence of quantum states distanced by a non-zero complex multiplier. LDV's will be represented by linear combinations of $SU(N)$ generators in local coordinates of $\mathbb{C}P^{N-1}$ equipped with the Fubini-Study metric [13]

$$G_{ik^*} = [(1 + \sum |\pi^s|^2)\delta_{ik} - \pi^{i^*}\pi^k](1 + \sum |\pi^s|^2)^{-2}. \tag{9}$$

Hence the internal dynamical variables and their norms should be state-dependent, i.e. local in the state space [4]. These local dynamical variables realize a non-linear representation of the unitary global $SU(N)$ group in the Hilbert state space $\mathbb{C}^N$. Namely, N^2-1 generators of $G = SU(N)$ may be divided in accordance with the Cartan decomposition: $[B,B] \in H, [B,H] \in B, [B,B] \in H$. The $(N-1)^2$ generators

$$\Phi^i_h \frac{\partial}{\partial \pi^i} + c.c. \in H, \quad 1 \leq h \leq (N-1)^2 \tag{10}$$

of the isotropy group $H = U(1) \times U(N-1)$ of the ray (Cartan sub-algebra) and $2(N-1)$ generators

$$\Phi^i_b \frac{\partial}{\partial \pi^i} + c.c. \in B, \quad 1 \leq b \leq 2(N-1) \tag{11}$$

are the coset $G/H = SU(N)/S[U(1) \times U(N-1)]$ generators realizing the breakdown of the $G = SU(N)$ symmetry of the generalized coherent states (GCS's). Here Φ^i_σ, $\;1 \leq \sigma \leq N^2-1$ are the coefficient functions of the generators of the non-linear $SU(N)$ realization as follows:

$$\Phi^i_\sigma = \lim_{\epsilon \to 0} \epsilon^{-1} \left\{ \frac{[\exp(i\epsilon\hat{\lambda}_\sigma)]^i_m \psi^m}{[\exp(i\epsilon\hat{\lambda}_\sigma)]^j_m \psi^m} - \frac{\psi^i}{\psi^j} \right\} = \lim_{\epsilon \to 0} \epsilon^{-1}\{\pi^i(\epsilon\hat{\lambda}_\sigma) - \pi^i\}. \tag{12}$$

Thereby each of the N^2-1 generators $\hat{\lambda}_\sigma$ may be represented by vector fields $\vec{G}_\sigma$ comprising the coefficient functions Φ^i_σ contracted with the corresponding partial derivatives $\frac{\partial}{\partial \pi^i} = \frac{1}{2}(\frac{\partial}{\partial \Re \pi^i} - i\frac{\partial}{\partial \Im \pi^i})$ and $\frac{\partial}{\partial \pi^{*i}} = \frac{1}{2}(\frac{\partial}{\partial \Re \pi^i} + i\frac{\partial}{\partial \Im \pi^i})$ as follows:

$$\vec{G}_\sigma = \Phi^i_\sigma \frac{\partial}{\partial \pi^i} + \Phi^{*i}_\sigma \frac{\partial}{\partial \pi^{*i}}. \tag{13}$$

There are a lot of attempts to build speculative model of electron as extended compact object in existing spacetime, see for example [12]. The model of the extended electron proposed here is quite different. Self-interacting quantum electron is *a periodic motion of quantum degrees of freedom along closed geodesics* γ obeying equation

$$\nabla_{\dot{\gamma}}\dot{\gamma} = 0 \tag{14}$$

in the projective Hilbert state space $\mathbb{C}P^3$. Namely, it is assumed that the motion of spin/charge degrees of freedom comprises of stable attractor in the state space, whereas its "field-shell" in dynamical space-time arises as a consequence of the local conservation law of the proper energy-momentum vector field. *This conservation law leads to PDE's whose solution give the distribution of energy-momentum in DST that keeps motion of spin/charge degrees of freedom along geodesic in* $\mathbb{C}P^3$. The periodic motion of quantum spin/charge degrees of freedom generated by the coset transformations from $G/H = SU(4)/S[U(1) \times U(3)] = \mathbb{C}P^3$ will be associated with inertial "mechanical mass" and the gauge transformations from $H = U(1) \times U(3)$ rotates closed geodesics in $\mathbb{C}P^3$ as whole. These transformations will be associated with Jacobi fields corresponding mostly to the electromagnetic energy.

In order to built the LDV corresponding to the internal energy-momentum of relativistic quantum electron we shall note that the matrices

$$\hat{\gamma}_0 = \begin{pmatrix} 1 & 0 & 0 & 0 \\ 0 & 1 & 0 & 0 \\ 0 & 0 & -1 & 0 \\ 0 & 0 & 0 & -1 \end{pmatrix}, \quad \hat{\gamma}_1 = \begin{pmatrix} 0 & 0 & 0 & -i \\ 0 & 0 & -i & 0 \\ 0 & i & 0 & 0 \\ i & 0 & 0 & 0 \end{pmatrix},$$
$$\hat{\gamma}_2 = \begin{pmatrix} 0 & 0 & 0 & -1 \\ 0 & 0 & 1 & 0 \\ 0 & 1 & 0 & 0 \\ -1 & 0 & 0 & 0 \end{pmatrix}, \quad \hat{\gamma}_3 = \begin{pmatrix} 0 & 0 & -i & 0 \\ 0 & 0 & 0 & i \\ i & 0 & 0 & 0 \\ 0 & -i & 0 & 0 \end{pmatrix}, \tag{15}$$

originally introduced by Dirac [11] may be represented as linear combinations of the "standard" $SU(4)$ λ-generators [7]

$$\hat{\gamma}_0 = \hat{\lambda}_3 + \frac{1}{3}\left[\sqrt{3}\hat{\lambda}_8 - \sqrt{6}\hat{\lambda}_{15}\right], \quad \hat{\gamma}_1 = \hat{\lambda}_2 + \hat{\lambda}_{14},$$
$$\hat{\gamma}_2 = \hat{\lambda}_1 - \hat{\lambda}_{13}, \qquad \hat{\gamma}_3 = -\hat{\lambda}_5 + \hat{\lambda}_{12}. \tag{16}$$

Since any state $|S\rangle$ has the isotropy group $H = U(1) \times U(N)$, only the coset transformations $G/H = SU(N)/S[U(1) \times U(N-1)] = \mathbb{C}P^{N-1})$ effectively act in $\mathbb{C}^N$. One should remember, however, that the concrete representation of hermitian matrices belonging to subsets h or b (as defined above) depends on a priori chosen vector (all "standard" classification of the traceless matrices of Pauli, Gell-Mann, etc., is based on the vector $(1,0,0,...,0)^T$). *The Cartan's decomposition of the algebra* $\mathfrak{su}(N)$ *is unitary invariant and I will use it instead of the Foldy-Wouthuysen decomposition in "even" and "odd" components.*

Infinitesimal variations of the proper energy-momentum evoked by interaction charge-spin degrees of freedom (implicit in $\hat{\gamma}^\mu$) that may be expressed in terms of local coordinates π^i since there is a diffeomorphism between the space of the rays $\mathbb{C}P^3$ and the $SU(4)$ group sub-manifold of the coset transformations $G/H = SU(4)/S[U(1) \times U(3)] = \mathbb{C}P^3$ and the isotropy group $H = U(1) \times U(3)$ of some state vector. It will be expressed by the combinations of the $SU(4)$ generators $\hat{\gamma}_\mu$ of unitary transformations that will be defined by an equation arising under infinitesimal variation of the energy-momentum

$$\Phi^i_\mu(\pi) = \lim_{\epsilon \to 0} \epsilon^{-1} \left\{ \frac{[\exp(i\epsilon\hat{\gamma}_\mu)]^i_m \psi^m}{[\exp(i\epsilon\hat{\gamma}_\mu)]^j_m \psi^m} - \frac{\psi^i}{\psi^j} \right\} = \lim_{\epsilon \to 0} \epsilon^{-1}\{\pi^i(\epsilon\hat{\gamma}_\mu) - \pi^i\}, \tag{17}$$

arose in a nonlinear local realization of $SU(4)$ [17]. Here $\psi^m, 1 \le m \le 4$ are the ordinary bi-spinor amplitudes. The twelve coefficient functions $\Phi^i_\mu(\pi)$ in the map $U_1 : \{\psi_1 \neq 0\}$ are as follows:

$$\begin{aligned}
&\Phi^1_0(\pi) = 0, && \Phi^2_0(\pi) = -2i\pi^2, && \Phi^3_0(\pi) = -2i\pi^3;\\
&\Phi^1_1(\pi) = \pi^2 - \pi^1\pi^3, && \Phi^2_1(\pi) = -\pi^1 - \pi^2\pi^3, && \Phi^3_1(\pi) = -1 - (\pi^3)^2;\\
&\Phi^1_2(\pi) = i(\pi^2 + \pi^1\pi^3), && \Phi^2_2(\pi) = i(\pi^1 + \pi^2\pi^3), && \Phi^3_2(\pi) = i(-1 + (\pi^3)^2);\\
&\Phi^1_3(\pi) = -\pi^3 - \pi^1\pi^2, && \Phi^2_3(\pi) = -1 - (\pi^2)^2, && \Phi^3_3(\pi) = \pi^1 - \pi^2\pi^3.
\end{aligned} \tag{18}$$

Now I will define the Γ-vector field

$$\vec{\Gamma}_\mu = \Phi^i_\mu(\pi^1, \pi^2, \pi^3)\frac{\partial}{\partial \pi^i} \tag{19}$$

and then the internal energy-momentum operator will be defined as the *functional vector field*

$$P^\mu \vec{\Gamma}_\mu \Psi(\pi^1, \pi^2, \pi^3) = P^\mu(x)\Phi^i_\mu(\pi^1, \pi^2, \pi^3)\frac{\partial}{\partial \pi^i}\Psi(\pi^1, \pi^2, \pi^3) + c.c. \tag{20}$$

acting on the "total wave function", where the ordinary 4-momentum $P^\mu = (\frac{E}{c} - \frac{e}{c}\phi, \vec{P} - \frac{e}{c}\vec{A}) = (\frac{\hbar\omega}{c} - \frac{e}{c}\phi, \hbar\vec{k} - \frac{e}{c}\vec{A})$ (not operator-valued) should be identified with the solution of quasi-linear "field-shell" PDE's for the contravariant components of the energy-momentum tangent vector field in $\mathbb{C}P^3$

$$P^i(x,\pi) = P^\mu(x)\Phi^i_\mu(\pi^1,\pi^2,\pi^3), \tag{21}$$

where $P^\mu(x)$ is energy-momentum distribution that comprise of the "field-shell" of the self-interacting electron.

One sees that infinitesimal variation of the internal energy-momentum is represented by the operator of partial differentiation in complex local coordinates π^i with corresponding coefficient functions $\Phi^i_\mu(\pi^1,\pi^2,\pi^3)$. Then the single-component "total wave function" $\Psi(\pi^1,\pi^2,\pi^3)$ should be studied in the framework of new quasi-linear PDE's [17, 18]. There are of course four such functions $\Psi(\pi^1_{(1)},\pi^2_{(1)},\pi^3_{(1)})$, $\Psi(\pi^1_{(2)},\pi^2_{(2)},\pi^3_{(2)})$, $\Psi(\pi^1_{(3)},\pi^2_{(3)},\pi^3_{(3)})$, $\Psi(\pi^1_{(4)},\pi^2_{(4)},\pi^3_{(4)})$ – one function in each local map.

Since the least action principle is correct only in average that is clear from Feynman's summation of quantum amplitudes a more deep principle should be used for the derivation of fundamental quantum equations of motion. The quantum formulation of the inertia law has been used [15]. The "field-shell" equations are derived as the consequence of the conservation law of the proper energy-momentum [17, 18].

What the inertia law means for quantum system and its states? Formally the classical inertia principle is tacitly accepted in the package with relativistic invariance. But we already saw that the problem of identification and localization of quantum particles in classical spacetime is problematic and therefore they require clarification and mathematically correct formulation. I assumed that quantum version of the inertia law may be formulated as follows:

The inertial quantum motion of the quantum system is expressed as a self-conservation of its local dynamical variables like proper energy-momentum, spin, charge, etc. in the quantum state space, not in spacetime.

The conservation law of the energy-momentum vector field in $\mathbb{C}P^3$ during inertial evolution will be expressed by the equation of the affine parallel transport

$$\frac{\delta[P^\mu(x)\Phi^i_\mu(\pi)]}{\delta\tau} = 0, \tag{22}$$

which is equivalent to the following system of four coupled quasi-linear

PDE's for the dynamical spacetime distribution of the energy-momentum "field-shell" of the quantum state

$$V^{\mu}_{Q}(\frac{\partial P^{\nu}}{\partial x^{\mu}} + \Gamma^{\nu}_{\mu\lambda}P^{\lambda}) = -\frac{c}{\hbar}(\Gamma^{m}_{mn}\Phi^{n}_{\mu}(\pi) + \frac{\partial \Phi^{n}_{\mu}(\pi)}{\partial \pi^{n}})P^{\nu}P^{\mu}, \tag{23}$$

and ordinary differential equations for relative amplitudes giving in fact the definition of the proper energy-momentum P^{μ}

$$\frac{d\pi^{k}}{d\tau} = \frac{c}{\hbar}\Phi^{k}_{\mu}P^{\mu}. \tag{24}$$

These equations serve as the *equations of characteristics* for the linear "super-Dirac" equation

$$i\{P^{\mu}\Phi^{i}_{\mu}(\pi)\frac{\partial \Psi}{\partial \pi^{i}} + c.c.\} = mc\Psi \tag{25}$$

that agrees with ODE

$$i\hbar\frac{d\Psi}{d\tau} = mc^{2}\Psi \tag{26}$$

for single total state function Ψ of free but self-interacting quantum electron "cum location" moving in DST like free material point with the rest mass m.

Simple relation given by the Fubini-Study metric for the square of the frequency associated with the velocity traversing geodesic line during spin/charge variations in $\mathbb{C}P^{3}$ sheds the light on the mass problem of self-interacting electron. Taking into account the "off-shell" dispersion law

$$\begin{aligned}\frac{\hbar^{2}}{c^{2}}\Omega^{2} &= \frac{\hbar^{2}}{c^{2}}G_{ik*}\frac{d\pi^{i}}{d\tau}\frac{d\pi^{k*}}{d\tau} = G_{ik*}(\Phi^{i}_{\mu}P^{\mu})(\Phi^{k*}_{\nu}P^{\nu*}) \\ &= (G_{ik*}\Phi^{i}_{\mu}\Phi^{k*}_{\nu})P^{\mu}P^{\nu*} = G_{\mu\nu}P^{\mu}P^{\nu*} = m^{2}c^{2},\end{aligned} \tag{27}$$

one has

$$i\frac{d\Psi}{d\tau} = \frac{mc^{2}}{\hbar}\Psi = \Psi\sqrt{G_{ik*}\frac{d\pi^{i}}{d\tau}\frac{d\pi^{k*}}{d\tau}} = \Psi\frac{1}{d\tau}\sqrt{dS^{2}_{F.-S.}} = \pm\Omega\Psi \tag{28}$$

and, therefore has,

$$\Psi(T) = \Psi(0)e^{\pm iS_{F.-S.}} = \Psi(0)e^{\pm i\int_{0}^{T}\Omega d\tau}. \tag{29}$$

Note that $\Omega(\tau) = \frac{m(\tau)c^{2}}{\hbar}$ depends on the proper time τ in our theory since the energy-momentum is dynamically distributed in the diffuse mass "off-shell" (27). The metric tensor of the local DST in the vicinity of electron $G_{\mu\nu} = G_{ik*}\Phi^{i}_{\mu}\Phi^{k*}_{\nu}$ is state-dependent, therefore the gravity in the vicinity of the electron generated by the coset transformations. All this and the general coordinate invariance do not considered in this paper since the spacetime curvature is the effect of the second order [24] in comparison with Coriolis contribution to the pseudo-metric [23] in boosted and rotated state dependent Lorentz reference frame.

The system of quasi-liner PDE's (23) following from the conservation law has been shortly discussed under strong simplification assumptions [17, 18]. The theory of such quasi-liner PDE's equations is well known [8]. One has the quasi-linear PDE's system with the identical principle part V_Q^μ for which we will build ODE's system of characteristics

$$\begin{aligned}
\frac{dx^\nu}{d\tau} &= V_Q^\nu, \\
\frac{dP^\nu}{d\tau} &= -V_Q^\mu \Gamma_{\mu\lambda}^\nu P^\lambda - \frac{c}{\hbar}(\Gamma_{mn}^m \Phi_\mu^n(\gamma) + \frac{\partial \Phi_\mu^n(\gamma)}{\partial \pi^n}) P^\nu P^\mu, \\
\frac{d\pi^k}{d\tau} &= \frac{c}{\hbar}\Phi_\mu^k P^\mu, \\
\frac{d\Psi}{d\tau} &= -i\frac{mc^2}{\hbar}\Psi.
\end{aligned} \tag{30}$$

In order to provide integration one should to find self-consistent solutions for state-dependent "quantum boosts" $\vec{a}_Q$ and "quantum rotations" $\vec{\omega}_Q$ involved in the "four velocity" V_Q^μ. See details in the e-print [14].

6. Solutions of the "field-shell" quasi-linear PDE's

Let me shortly discuss the system of characteristic equations

$$\frac{dP^\nu}{d\tau} = -V_Q^\mu \Gamma_{\mu\lambda}^\nu P^\lambda - \frac{c}{\hbar}(\Gamma_{mn}^m \Phi_\mu^n(\gamma) + \frac{\partial \Phi_\mu^n(\gamma)}{\partial \pi^n}) P^\nu P^\mu \tag{31}$$

of the PDE's system with the identical principle part V_Q^μ

$$V_Q^\mu \frac{\partial P^\nu}{\partial x^\mu} = -V_Q^\mu \Gamma_{\mu\lambda}^\nu P^\lambda - \frac{c}{\hbar}(\Gamma_{mn}^m \Phi_\mu^n(\pi) + \frac{\partial \Phi_\mu^n(\pi)}{\partial \pi^n}) P^\nu P^\mu. \tag{32}$$

Writing the characteristic equations (32) in symmetrical form with $S^\nu = -V_Q^\mu \Gamma_{\mu\lambda}^\nu P^\lambda - \frac{c}{\hbar}(\Gamma_{mn}^m \Phi_\mu^n(\pi) + \frac{\partial \Phi_\mu^n(\pi)}{\partial \pi^n}) P^\nu P^\mu$,

$$\begin{aligned}
\frac{dx^0}{V_Q^0} &= \frac{dx^1}{V_Q^1} = \frac{dx^2}{V_Q^2} = \frac{dx^3}{V_Q^3} \\
&= \frac{dP^0(x)}{S^0} = \frac{dP^1(x)}{S^1} = \frac{dP^2(x)}{S^2} = \frac{dP^3(x)}{S^3} \\
&= \frac{d\pi^1}{\frac{c}{\hbar}P^\mu(x)\Phi_\mu^1} = \frac{d\pi^2}{\frac{c}{\hbar}P^\mu(x)\Phi_\mu^2} = \frac{d\pi^3}{\frac{c}{\hbar}P^\mu(x)\Phi_\mu^3} \\
&= \frac{i\hbar d\Psi(x,\pi,P)}{mc^2\Psi(x,\pi,P)} = d\tau
\end{aligned} \tag{33}$$

one may integrate each combination of the equations given above. For example the first pair of equations

$$\frac{dx^0}{V_Q^0} = \frac{dP^0(x)}{S^0} \tag{34}$$

may be integrated taking into account $V_Q^0 = (\vec{a}\vec{x})$ as follows:

$$\int dx^0 = ct + cT = \int \frac{V_Q^0}{S^0} dP^0(x). \tag{35}$$

The right integral has a form

$$\int \frac{V_Q^0}{S^0} dP^0(x) = \hbar L_\alpha x^\alpha \int \frac{\sqrt{1+As}ds}{B(1+\sqrt{1+As})+Cs^2}, \tag{36}$$

where $s = P^0$, $A = \frac{4L_\alpha x^\alpha}{\hbar}$, $B = -\hbar L_\alpha P^\alpha$, $C = 2L_0 L_\alpha x^\alpha$. The substitution $q = \sqrt{1+As}$ leads to the integral from rational integrand as follows:

$$\begin{aligned} ct + cT &= 2A\hbar L_\alpha x^\alpha \int \frac{q^2 dq}{Cq^4 - 2Cq^2 + A^2Bq + A^2B + C} \\ &= \hbar L_\alpha x^\alpha \frac{1}{AB}[2\ln(q+1) \\ &\quad + \sum_{s=1}^{3} \frac{(-R_s^2 + R_s(2+(A^2B/C)) - 1 - (A^2B/C))\ln(q-R_s)}{3R_s^2 - 2R_s - 1}], \end{aligned} \tag{37}$$

where R_1, R_2, R_3 are the roots of the polynomial $Z^3 - Z^2 - Z + 1 + A^2B/C = 0$ and $A^2B/C = -\frac{8L_\alpha x^\alpha L_\alpha P^\alpha}{\hbar L_0}$. Therefore

$$\begin{aligned} \frac{ABc(t+T)}{\hbar L_\alpha x^\alpha} &= \ln(q+1)^2 \\ &\quad + \frac{N_1}{D_1}\ln(q-R_1) + \frac{N_2}{D_2}\ln(q-R_2) + \frac{N_3}{D_3}\ln(q-R_3), \end{aligned} \tag{38}$$

i.e.

$$\exp\left(\frac{-4c(t+T)L_\alpha P^\alpha}{\hbar}\right) = (q+1)^2(q-R_1)^{\frac{N_1}{D_1}}(q-R_2)^{\frac{N_2}{D_2}}(q-R_3)^{\frac{N_3}{D_3}}, \tag{39}$$

where $N_s = -R_s^2 + R_s(2+(A^2B/C)) - 1 - (A^2B/C)$, $D_s = 3R_s^2 - 2R_s - 1$. One will see that after substitution of these expressions all integrations are elementary containing rational fractions, logarithms and arctan/arctanh only. Their shapes are very complicated and the problem of functionally independent invariants did not solved up to now. This solution represents the lump of self-interacting electron in the co-moving Lorentz reference frame.

Up to now we dealt with the field configuration shaping, say, extended self-interaction electron itself. "Field-shell" equations gave the distribution of the proper energy-momentum vector field $P^{\mu}(x)\Phi^i_{\mu}(\pi)$ in the tetrad whose four vectors $P^{\mu}(x)$ are functions over the DST. This means that geodesic motion of spin/charge degrees of freedom have been lifted into the frame fibre bundle over $\mathbb{C}P^3$. No words were told, however, about the interaction between electrons. Next paragraph contains a draft dedicated to possible solution of this problem.

7. State-dependent Jacobi gauge fields

The quantum dynamics of spin/charge degrees of freedom of the single self-interacting quantum electron goes along geodesic in $\mathbb{C}P^3$. The lift of this geodesic into the frame fiber bundle leads to the first order PDE's system. It is reasonable assume that interaction of two electrons may deform the geodesic and therefore the lift of the deformed geodesic will be deformed too together with field equations. Adachi and coauthors discussed already so-called Jacobi magnetic fields (closed Kähler 2-form) on Kähler manifolds and particularly on $\mathbb{C}P^N$ [1]. One needs, however, the quantum theory of interaction of the extended (non-local) electrons.

I will be concentrated here on the basic variations of the geodesic, namely those that generated by the isotropy group $H = U(1) \times U(3)$. This group is considered as gauge group transforming one electron motion along geodesic γ_1 to the motion of second electron along geodesic γ_2. Thereby these variations may be connected with pure Jacobi vector fields on $\mathbb{C}P^3$.

I will treat the correction for the stationary solution of (31) $J^i(x,\pi) = p^{\mu}(x)\Phi^i_{\mu}(\pi)$ as Jacobi vector field, i.e. solution of the Jacobi equation

$$\nabla_{\dot{\gamma}}\nabla_{\dot{\gamma}}J + R(\dot{\gamma}, J)\dot{\gamma} = 0. \tag{40}$$

This requirement puts additional restriction on the components of anholonomic frame $\Phi^i_{\mu}(\pi)$ in $CP(3)$ that obey to the Duffing type equation with cubic non-linearity

$$\begin{aligned}\ddot{\Phi}^i_{\mu} + 2\omega\dot{\Phi}^i_{\mu} + \omega^2\Phi^i_{\mu} &= -R^i_{klm*}\Phi^l_{\mu}\dot{\pi}^k\dot{\pi}^{m*}\\ &= -\frac{c^2}{\hbar^2}R^i_{klm*}\Phi^l_{\mu}\Phi^k_{\nu}P^{\nu}\Phi^{m*}_{\lambda}P^{\lambda*},\end{aligned} \tag{41}$$

where $R^i_{klm*} = \delta^i_k G_{lm*} + \delta^i_l G_{km*}$ is the curvature tensor of the $\mathbb{C}P^3$.

Complicated equations for Φ^i_{μ} requires detailed investigation but solution of the Jacobi equation for $J^i(x,\pi) = p^{\mu}(x)\Phi^i_{\mu}(\pi)$ is well known and this is very easy [4]. One may distinguish two kinds of Jacobi fields: tangent Jacobi vector field $J_{tang}(\pi) = (a_i\tau + b_i)U^i(\pi)$ giving initial frequencies traversing the geodesic and the initial phases, and the normal Jacobi vector field $J_{norm}(\pi) = [c_i \sin(\sqrt{\kappa}\tau) + d_i \cos(\sqrt{\kappa}\tau)]U^i(\pi)$ showing deviation from one geodesic to another [4]. There are of course the continuum forms of the geodesic variations but for us only *internal gauge fields* are interesting. We will use only narrow class of such variations: geodesic to geodesic. It is well known that the isotropy group $H = U(1) \times U(N-1)$ rotates geodesic [13] whose generators with corresponding coefficient functions Φ^i_h may be identified with the normal Jacobi vector fields. Thereby, two invariantly separated motions of quantum state have been taken into account:

1) *along a tangent Jacobi vector field*, i.e. "free motion" of spin/charge degrees of freedom along the geodesic line in $\mathbb{C}P^3$ generated by the coset transformations $G/H = SU(4)/S[U(1) \times U(3)] = \mathbb{C}P^3$ (oscillation of a massive mode in the vicinity of a minimum of the affine gauge potential across its valley) and,
2) deviation of geodesic motion *in the direction of the normal Jacobi vector field transversal to the reference geodesic* generated by the isotropy group $H = U(1) \times U(3)$ (oscillation of the massless mode along the valley of the affine gauge potential).

These oscillators cannot be of course identified with the Fourier components oscillators of a pure electromagnetic field. But this deformation of geodesic in the base manifold $CP(3)$ induces the nine-parameter deformation of the "field-shell" quasi-linear PDE's, described by the Jacobi fields in $\mathbb{C}P^3$ playing the role of non-Abelian electromagnetic-like field carrier of interaction between electrons. Thus it should contain the fine structure constant $\alpha = \frac{e^2}{\hbar c}$. Then holomorphic sectional curvature of $\mathbb{C}P^3$ assumed in this paper equal "1" must be somehow related to the α. However, the logical way leading to this connection is unclear yet.

8. Conclusion and future outlook

This theory has the program character and should be treated as preliminary framework for the fundamental problems of the grand unification. It is clearly that gigantic volume of work is left for future. I would like formulate here very basic and general principles applied to the single quantum relativistic self-interacting electron in the elementary discussion. The generalization of this theory on a different kind of fermions and their interactions is only under investigation and will not be mentioned presently.

Analysis of the localization problem in QFT and all consequences of its formal apparatus like divergences, unnecessary particles, etc., shows that we should have the realistic physical theory. Such a theory requires *intrinsic unification* of quantum principles based on the fundamental concept of quantum states and the principle of relativity ensures the physical equivalence of any conceivable quantum setup. Realization of such program evokes the necessity of the state-dependent affine gauge field in the state space that acquires reliable physical basis under the quantum formulation of the inertia law (self-conservation of local dynamical variables of quantum particle during inertial motion). Representation of such affine gauge field in dynamical spacetime has been applied to the relativistic extended self-interacting Dirac's electron [15, 17, 18].

The Fubini-Study metric in $\mathbb{C}P^{N-1}$ is the positive definite metric in the base manifold. Whereas the Lorentz metric is the indefinite pseudometric $h_{\mu\nu}$ in the "vertical" sub-space [23] generated by the gauge isotropy sub-group $H = S[U(1) \times U(N-1)]$ in the frame fibre bundle. The metric tensor $G_{\mu\nu}$, i.e. the gravity in the vicinity of the electron generated by the coset transformations, and the general coordinate invariance does not considered in this paper since the spacetime curvature is the effect of the second order [24] in comparison with Coriolis contribution to the pseudometric in boosting and rotating state dependent Lorentz reference frame. The application to the general relativity is a future problem.

The second quantization of the gauge fields has not been discussed in this paper. The Hilbert space has an indefinite metric in the gauge theories. We need the gauge fixing to quantize the gauge fields. This may carry out in consistent with the fundamental symmetry $\mathbb{C}P^{N-1}$. It will be realized by the choice of the section of the frame fiber bundle giving by the boundary conditions of the "field-shell" PDE's for the proper energy-momentum P^μ (23) and the initial conditions for the components of the anholonomic frame Φ^i_μ for (41). Details of this gauge fixing will be reported elsewhere.

Dirac's relativistic equations of electron save mass on-shell condition under the introduction of two internal quantum degrees of freedom by matrices belonging to the $\mathfrak{su}(4)$. This theory perfectly fits to the electron in the external electromagnetic field. The Dirac equation has positive and negative energy solutions, which is interpreted as a particle and an antiparticle. In order to make solutions the positive energy, we have to consider the Dirac sea or the second quantization. But in the Section 5, we described the Dirac's single self-interacting quantum electron where tangent state dependent vector fields to $\mathbb{C}P^3$ replace the Dirac's matrices. The fields in the vicinity of the electron and their equations of motion have been derived from the fundamental representation of quantum states motions in the $\mathbb{C}P^3$

geometry. I follow Dirac's ideas [12] trying to find the mass spectrum of the electron's generation. This problem is formulated but it does not solved yet.

In this paper, only $SU(N)$ group generically connected with quantum state space geometry is discussed. But, of course, all Lie groups are possible as gauge groups as it was dictated by physical experiments.

Acknowledgements

I am very grateful to Professor T. Adachi for his invitation to ICDG-2012 and interesting discussions. I would like express especial gratitude to the referee for his big efforts for the clarification of some dark places of my presentation and correction of my English.

Bibliography

1. T. Adachi, *A comparison theorem on magnetic Jacobi fields*, Proc. Edinburgh Math. Soc., **40**, 293-308, (1997).
2. J. Anandan & Y. Aharonov, *Geometric quantum phase and angles*, Phys. Rev. D, **38**, (6) 1863-1870 (1988).
3. M.V. Berry, *The Quantum Phase, Five Years After*, in *Geometric Phases in Physics*, World Scientific. 1989.
4. A.L. Besse, *Manifolds all of whose Geodesics are Closed*, Springer-Verlag, Berlin, Heidelberg, New Yourk, (1978).
5. D.I. Blochintzev, *Whether always the "duality" of waves and particles does exist?*, Uspechy Phys. Nauk, **XLIV**, No.1, 104-109 (1951).
6. N. Bohr, *On the constitution of Atoms and Molecules*, Phil. Mag. **26**, 1-25 (1913).
7. F.E. Close, *An introduction to quarks and partones*, 438, Academic Press, London, NY, San Francisco, (1979).
8. R. Courant & D. Hilbert, *Methods of Mathematical Physics*, **2**, Partial Differential Equations, 830, Wiley, (1989).
9. P.A.M. Dirac, *Lectures on quantum field theory*, Yeshiva University, NY, (1967).
10. P.A.M. Dirac, *The principls of quantum mechanics*, Fourth Edition, Oxford, At the Clarebdon Press, 1958.
11. P.A.M. Dirac, *The quantum theory of the electron*, Proc. Royal. Soc. A **117**, 610-624 (1928).
12. P.A.M. Dirac, *An extensible model of the electron*, Proc. Royal. Soc. A **268**, 57-67 (1962).
13. S. Kobayashi & K. Nomizu, *Foundations of Differential Geometry, V. II*, Interscience Publishers, New York-London-Sydney, (1969).
14. P. Leifer, *The role of the $CP(N-1)$ geometry in the intrinsic unification of the general relativity and QFT*, arXiv:1209.6389v2.
15. P. Leifer, *The quantum content of the inertia law and field dynamics*, arXiv:1009.5232v1.

16. P. Leifer, *Inertia as the "threshold of elasticity" of quantum states*, Found. Phys. Lett., **11**, (3) 233 (1998).
17. P. Leifer, *Self-interacting quantum electron*, arXiv:0904.3695v4.
18. P. Leifer, T. Massalha, *Field-shell of the self-interacting quantum electron*, Annales de la Fondation Louis de Broglie, **36**, 29-51 (2011).
19. P. Leifer, *Superrelativity as an Element of a Final Theory*, Found. Phys. **27**, (2) 261 (1997).
20. P. Leifer, *Objective quantum theory based on the $CP(N-1)$ affine gauge field*, Annales de la Fondation Louis de Broglie, **32**, (1) 25-50 (2007).
21. P. Leifer, *State-dependent dynamical variables in quantum theory*, JETP Letters, **80**, (5) 367-370 (2004).
22. P. Leifer, *An affine gauge theory of elementary particles*, Found. Phys. Lett., **18**, (2) 195-204 (2005).
23. R.G. Littlejohn & M. Reinsch, *Gauge fields in the separation of rotations and internal motions in the n-body problem*, Rev. Mod. Physics, **69**, No.1, 213-275 (1997).
24. C.W. Misner, K.S. Thorne & J.A. Wheeler, *Gravitation*, W.H.Freeman and Company, San Francisco, 1279 (1973).
25. T.D. Newton & E.P. Wigner, *Localized States for Elementary Systems*, Rev. Mod. Phys., **21**, No.3, 400-406 (1949).
26. E. Schrödinger, *Quantisierung als Eigenwertprobleme*, Ann. Phys. **79**, 361-376 (1926).
27. A. Shapere & F. Wilczek, *Gauge Kinamatics of Deformable Bodies*, in *Geometric Phases in Physics*, World Science, (1988).
28. F. Wilczek & A. Zee, *Appearance of gauge structure in simple dynamical systems*, Phys. Rev. Lett., **52**, (24) 2111-2114 (1984).

Received December 12, 2012
Revised February 24, 2013

Proceedings of the 3rd International
Colloquium on Differential Geometry
and its Related Fields
Veliko Tarnovo, September 3–7, 2012

WAVE EQUATIONS, INTEGRAL TRANSFORMS, AND TWISTOR THEORY ON INDEFINITE GEOMETRY

Fuminori NAKATA

Faculty of Human Development and Culture, Fukushima University,
Kanayagawa, Fukushima, 960-1296, Japan
E-mail: fnakata@educ.fukushima-u.ac.jp

This article is a survey of recent results on LeBrun-Mason type twistor theory. We summarize two types of results given by the author, one is the twistor theory for Tod-Kamada metric and the other is its analogy on $\mathbb{R}^4$. We introduce several integral transforms to write down these correspondences explicitly. We obtain consequences concerning solutions to wave equations by making use of these integral transforms. A general method of twistor construction on the indefinite geometry is also summarized.

Keywords: Twistor theory; Pseudo-Riemannian geometry; Wave equation; Integral transform.

1. Introduction

Twistor theory, originated by R. Penrose [12], brought a rich harvest on mathematical physics and developed into the gauge theory and so on. In the method of Penrose, we establish correspondences between a twistor space (complex manifold satisfying certain conditions) and a space-time (certain differential geometric structure, e.g. self-dual conformal 4-manifold, Einstein-Weyl 3-manifold). In this correspondence, a space-time is recovered as a parameter space of a family of rational curves (i.e. $\mathbb{CP}^1$) on a twistor space (see Atiyah et al [1], Hitchin [3]). On the other hand, relatively recently, another type of twistor theory is obtained by C. LeBrun and L. J. Mason. In this theory, a space-time is recovered from a family of holomorphic disks. The following is the typical result on the LeBrun-Mason type twistor correspondences.

Theorem 1.1 (LeBrun-Mason [7]). *There is a natural one-to-one correspondence between*

- *self-dual Zollfrei conformal structure* $[g]$ *on* $S^2 \times S^2$ *of signature* $(--$

++), and

- *pair $(\mathbb{CP}^3, P)$, where P is an embedded $\mathbb{RP}^3$*

on the neighborhood of the standard objects.

Here an indefinite conformal structure $[g]$ is called Zollfrei if and only if all the maximal null geodesics are closed. Notice that the Zollofrei condition is a global condition defined only on the indefinite geometry. In this theorem, the self-dual manifold $(S^2 \times S^2, [g])$ is recovered from the twistor space $(\mathbb{CP}^3, P)$ as the space of holomorphic disks on $\mathbb{CP}^3$ with the boundaries lying on P. In this article, we give a sketch of this construction (Section 4). We remark that the Zollfrei condition is so strong that it gives a strong restriction to the topology on self-dual manifolds. In fact, LeBrun and Mason showed that a self-dual Zollfrei manifold is homeomorphic to either $S^2 \times S^2$ or $(S^2 \times S^2)/\mathbb{Z}_2$.

Following Theorem 1.1, the author proved that Tod-Kamada metrics are Zollfrei, and determined their twistor spaces explicitly (Nakata [10]). Here Tod-Kamada metrics are S^1-invariant indefinite self-dual metrics on $S^2 \times S^2$ which are firstly obtained by K. P. Tod [13] and also investigated by H. Kamada[4] independently. Moreover, as a consequence of the twistor theory for Tod-Kamada metric, the author obtained results concerning a wave equation and integral transforms. Though we can deal with these consequences without twistor theory, it seems difficult to find these results without a background of twistor correspondence.

The author also established another twistor correspondence, that is, a correspondence for indefinite self-dual conformal structure on $\mathbb{R}^4$ (Nakata [11]). While we need the Zollfrei condition in $S^2 \times S^2$ case, we can relax this condition in $\mathbb{R}^4$ case, and instead we need to assume an extra condition concerning rapidly decreasingness. This $\mathbb{R}^4$ case is the simplest non-compact case, and to establish twistor correspondences on other non-compact manifolds is a future problem.

The organization of this article is as follows. In Section 2, we explain a consequence of twistor theory for Tod-Kamada metrics, without going to the detail of the twistor theory. Here we study the geometry of small circles on S^2. We introduce integral transforms and explain that the solutions to the wave equation on the de Sitter 3-space are simply obtained by using these integral transforms.

Another parallel story is introduced in Section 3. We study the geometry of circles on a cylinder. The results explained in this section is a consequence of the twistor correspondence for an indefinite self-dual metric on $\mathbb{R}^4$.

In Section 4, we summarize a general theory of LeBrun-Mason twistor correspondence. We explain the way to construct indefinite space-times (self-dual 4-manifolds, Einstein-Weyl 3-manifolds) from twistor spaces by using the family of holomorphic disks, and we do not deal with the converse correspondence in this article.

In Section 5, twistor correspondence for Tod-Kamada metric is explained, and in Section 6 the outline of the twistor correspondence on $\mathbb{R}^4$ is shown.

2. Small circles on the 2-sphere

Let M be the set of oriented small circles on the unit sphere $S^2 \subset \mathbb{R}^3$. If we define a domain $\Omega_{(t,y)} \subset S^2$ for each $(t, y) \in \mathbb{R} \times S^2$ by

$$\Omega_{(t,y)} := \{u \in S^2; \langle u, y\rangle > \tanh t\},$$

then each oriented small circle is uniquely written as $\partial\Omega_{(t,y)}$. Hence M is identified with $\mathbb{R} \times S^2$. Further, we equip $M \cong \mathbb{R} \times S^2$ with an indefinite metric

$$g = -dt^2 + \cosh^2 t \cdot g_{S^2},$$

where g_{S^2} is the standard metric on S^2. This metric g is called the de Sitter metric, and the semi-Riemannian manifold (M, g) is called the de Sitter 3-space. We write $(M, g) = (S^3_1, g_{S^3_1})$ from now on following the standard notation.

On the above construction of the de Sitter 3-space, there is a nice geometric correspondence as follows. Notice the Apollonian circles on S^2; one of them is the family of small circles containing two common points on S^2, and this family corresponds to a space-like geodesic on $(S^3_1, g_{S^3_1})$. Another is the circles containing one common point and being tangent to each other, and this family corresponds to a null geodesic. The third type corresponds to a time-like geodesic.

Now we define two integral transforms $R, Q : C^\infty(S^2) \to C^\infty(S^3_1)$ so that for $h \in C^\infty(S^2)$

$$Rh(t, y) = \frac{1}{2\pi}\int_{\partial\Omega_{(t,y)}} h\, dS^1 \tag{1}$$

$$Qh(t, y) = \frac{1}{2\pi}\int_{\Omega_{(t,y)}} h\, dS^2, \tag{2}$$

where dS^1 is the standard measure on the small circle $\partial\Omega_{(t,y)}$ of the total length 2π and dS^2 is the standard measure on the unit sphere S^2. Notice that $Rh(t, y)$ is nothing but the mean value of h along the circle $\partial\Omega(t, y)$.

The transform Q is related with the wave equation as follows. Let $*$ be the Hodge operator on S^3_1 and Δ_{S^2} be the Laplace operator on S^2. The wave equation on the de Sitter space S^3_1 is given by

$$\square V := *d*dV = 0 \tag{3}$$

or precisely

$$\left(-\frac{\partial^2}{\partial t^2} - 2\tanh t\,\frac{\partial}{\partial t} + (\cosh t)^{-2}\Delta_{S^2}\right)V = 0. \tag{4}$$

If we introduce a function space

$$C^\infty_*(S^2) = \left\{h \in C^\infty(S^2)\ ;\ \int_{S^2} hdS^2 = 0\right\}$$

then we obtain the following.

Theorem 2.1 (Nakata [10], Proposition 2.6 & Theorem 3.1). *For each $h \in C^\infty_*(S^2)$, the function $V = Qh \in C^\infty(S^3_1)$ satisfies the conditions:*

- *V solves the wave equation $\square V = 0$, and*
- *$V(t, y) \to 0$ and $\partial_t V(t, y) \to 0$ as $t \to \pm\infty$ uniformly for $y \in S^2$.*

Conversely, if $V \in C^\infty(S^3_1)$ satisfies these conditions, then there exists a unique function $h \in C^\infty_*(S^2)$ satisfying $V = Qh$.

Sketch of the proof. For given $h \in C^\infty_*(S^2)$, there exists a unique function $\tilde{h} \in C^\infty_*(S^2)$ satisfying $h = -\Delta_{S^2}\tilde{h}$. We can show following formulas directly by using the divergence formula:

$$\frac{\partial}{\partial t}R\tilde{h}(t, y) = -Q\Delta_{S^2}\tilde{h}(t, y), \tag{5}$$

$$\frac{\partial}{\partial t}Q\tilde{h}(t, y) = -(\cosh t)^{-2}R\tilde{h}(t, y). \tag{6}$$

Now we claim that the operators R and Δ_{S^2} are commutative. Actually both operators are $SO(3)$-equivariant, and this insists that both operators are diagonalized with respect to the spherical harmonics. Hence they commute.

Then we obtain

$$\frac{\partial^2}{\partial t^2}R\tilde{h} = -\frac{\partial}{\partial t}Q\Delta_{S^2}\tilde{h} = (\cosh t)^{-2}R\Delta_{S^2}\tilde{h} = (\cosh t)^{-2}\Delta_{S^2}R\tilde{h},$$

that is

$$\left(-\frac{\partial^2}{\partial t^2} + (\cosh t)^{-2}\Delta_{S^2}\right)R\tilde{h} = 0. \tag{7}$$

Applying ∂_t, we obtain $\Box Qh = 0$.

The converse is proved by using the uniqueness theorem for the solution of hyperbolic equations. See Nakata [10] for the detail. $\Box$

3. Circles on the cylinder

We can establish a similar theory as Section 2 on the two-dimensional cylinder

$$\mathcal{C} = \{(\omega, v) \in \mathbb{R}^2 \times \mathbb{R} \mid |\omega| = 1\} \simeq S^1 \times \mathbb{R}$$

instead of S^2. Let M be the set of circles on $\mathcal{C}$ which are cut out by planes on $\mathbb{R}^3 = \mathbb{R}^2 \times \mathbb{R}$. Then M is globally coordinated by $(t, x) \in \mathbb{R} \times \mathbb{R}^2$ so that the corresponding circle $C_{(t,x)}$ is cut out by the plane $\{(\omega, v) \in \mathbb{R}^2 \times \mathbb{R} \mid v = t + \langle \omega, x \rangle\}$, that is,

$$C_{(t,x)} = \{(\omega, v) \in \mathcal{C} \mid v = t + \langle \omega, x \rangle\}. \tag{8}$$

Now we equip $M \simeq \mathbb{R} \times \mathbb{R}^2$ with a flat Lorentz metric

$$g = -dt^2 + dx_1^2 + dx_2^2 \qquad (x = (x_1, x_2)). \tag{9}$$

With respect to this Lorentz metric g, the geometry on M is nicely related to the geometry on $\mathcal{C}$ in the following way.

First, for each point $p = (\omega, v) \in \mathcal{C}$ let $\Pi_p \subset M$ be the set of circles passing through p, i.e.

$$\Pi_p = \{(t, x) \in M \mid p \in C_{(t,x)}\} = \{(t, x) \in M \mid v = t + \langle \omega, x \rangle\}.$$

Then Π_p is a null plane on (M, g), i.e. the metric g degenerates on Π_p. Since every null plane on M is written in this way by a unique $p \in \mathcal{C}$, the cylinder $\mathcal{C}$ is identified with the set of null planes on M.

Next we consider a family of planes on $\mathbb{R}^3$ with common axis. If this axis intersects with $\mathcal{C}$ at two distinct points p and q, then this family cut out circles passing through common points p and q. In this case, these circles correspond to a straight line on M written as $\Pi_p \cap \Pi_q$. This is a space-like geodesic on (M, g). If the axis is tangent to $\mathcal{C}$ at p, then we obtain a family of planar circles which are mutually tangent at p. In this case, the corresponding line on M is a null geodesic contained in the null plane Π_p. If the axis is apart from $\mathcal{C}$, then we obtain a family of circles which folliate the cylinder $\mathcal{C}$. This family corresponds to a time-like geodesic.

Now we introduce an integral transform $R' : C^\infty(\mathcal{C}) \to C^\infty(M)$ so that for a given function $h(\omega, v) \in C^\infty(\mathcal{C})$

$$R'h(t,x) = \frac{1}{2\pi}\int_{C_{(t,x)}} h\, d\theta = \frac{1}{2\pi}\int_{|\omega|=1} h\left(\omega, t + \langle \omega, x\rangle\right) d\theta \qquad (\omega = e^{i\theta})$$
$$= \frac{1}{2\pi}\int_0^{2\pi} h(e^{i\theta}, t + x_1\cos\theta + x_2 \sin\theta) d\theta. \tag{10}$$

A significant property of the transform R' is that its image $u = R'h$ satisfies the wave equation

$$-\frac{d^2u}{dt^2} + \frac{d^2u}{dx_1^2} + \frac{d^2u}{dx_2^2} = 0. \tag{11}$$

Here notice that R' has a non-trivial kernel. Actually, if the function $h(\omega, v)$ is independent of v, then $R'h(t,x)$ is constant and is equal to the constant term of the Fourier expansion of $h = h(\omega)$ which can vanish for non-trivial h. One natural way to avoid such obvious kernel is to replace the function space $C^\infty(\mathcal{C})$ with the rapidly decreasing functions $\mathcal{S}(\mathcal{C})$, where $\mathcal{S}(\mathcal{C})$ is the space of smooth functions h on $\mathcal{C}$ which, for any integers $k, l \geq 0$ and any differential operator D on S^1, satisfy

$$\sup_{(\omega,v)\in\mathcal{C}} \left|(1+|v|^k)\frac{\partial^l}{\partial v^l}(Dh)(\omega,v)\right| < \infty.$$

Under some assumptions for rapidly decreasingness, we can establish the inverse transform of R' by making use of the Radon transform, which we summarize here (see Helgason [2] for the detail). Let us notice the initial plane $M_0 = \{(t,x) \in M \mid t = 0\} \simeq \mathbb{R}^2$. Each null plane $\Pi_{(\omega,v)} \subset M$ corresponding to $(\omega, v) \in \mathcal{C}$ intersects with M_0 by a straight line

$$l_{(\omega,v)} = \Pi_{(\omega,v)} \cap M_0 = \{x \in M_0 \mid \langle \omega, x\rangle = v\}.$$

Since $l_{(\omega,v)} = l_{(-\omega,-v)}$, the cylinder $\mathcal{C}$ is also identified with the double cover of the set of straight lines on M_0.

Let $\mathcal{S}(M_0)$ be the space of rapidly decreasing functions on M_0. Here a smooth function $f(x)$ on M_0 is called rapidly decreasing if and only if f satisfies, for any integers $k, l, m \geq 0$,

$$\left||x|^k \partial_{x_1}^l \partial_{x_2}^m f(x)\right| < \infty.$$

For given function $f(x) \in \mathcal{S}(M_0)$, its Radon transform $\hat{f}(\omega, v) \in \mathcal{S}(\mathcal{C})$ is defined by

$$\hat{f}(\omega, v) = \int_{l_{(\omega,v)}} f\, dm, \tag{12}$$

where dm is the Euclid measure on the line $l_{(\omega,v)}$. On the other hand, for a function $h(\omega,v) \in C^\infty(\mathcal{C})$, its dual Radon transform $\check{h}(x) \in C^\infty(M_0)$ is defined by

$$\check{h}(x) = \frac{1}{2\pi}\int_{|\omega|=1} h(\omega,\langle\omega,x\rangle)d\theta = R'h(0,x).$$

The inversion formula of the Radon transform for $f(x) \in \mathcal{S}(M_0)$ is

$$f = \frac{1}{2i}\left(\mathcal{H}_v\partial_v\hat{f}\right)^\vee, \quad \text{where} \quad \mathcal{H}_v h(\omega,v) = \frac{i}{\pi}\,\mathrm{pv}\int_{-\infty}^{\infty}\frac{h(\omega,\nu)}{\nu - v}d\nu. \tag{13}$$

The operator $\mathcal{H}_v$ is called the Hilbert transform.

Now the inverse of R' is given in the following way.

Theorem 3.1 (Nakata [11], Theorem 2.1). *Suppose $u(t,x) \in C^\infty(M)$ satisfies the following conditions:*

- *u solves the wave equation* (11),
- *$f_0(x) = u(0,x)$ and $f_1(x) = u_t(0,x)$ are rapidly decreasing functions for $x \in \mathbb{R}^2$.*

Then we can write $u = R'h$, where $h \in \mathcal{S}(\mathcal{C})$ is defined by

$$h = \frac{1}{4\pi i}\mathcal{H}_v(\partial_v\hat{f}_0 + \hat{f}_1). \tag{14}$$

This theorem is easily proved by the inversion formula (13) and the uniqueness theorem for the solutions to a hyperbolic PDE (see Helgason [2]).

4. General method on the twistor correspondence

The results in the previous sections are naturally obtained from LeBrun-Mason type twistor theory. In this section, we summarize the general method of the LeBrun-Mason theory.

4.1. *Partial indices of a holomorphic disk*

Let us denote $\mathbb{D} = \{z \in \mathbb{C} \mid |z| \le 1\}$. Let (Z,P) be the pair of a complex n-manifold Z and a totally real submanifold P. A holomorphic disk on (Z,P) is the image $(D,\partial D)$ of a continuous map $(\mathbb{D},\partial\mathbb{D}) \to (Z,P)$ which is holomorphic on the interior of $\mathbb{D}$.

Now, we recall the notion of the partial indices of a holomorphic disk $(D,\partial D)$ on (Z,P) (see LeBrun [6]). We write $\mathcal{N}$ the complex normal bundle of D in Z and $\mathcal{N}_\mathbb{R}$ the real normal bundle of ∂D in P. We have $\mathcal{N}_\mathbb{R} \subset \mathcal{N}|_{\partial D}$

by definition. We can construct a virtual holomorphic vector bundle $\widehat{\mathcal{N}}$ on the double $\mathbb{CP}^1 = D \cup_{\partial D} \overline{D}$ by patching $\mathcal{N} \to D$ and $\overline{\mathcal{N}} \to \overline{D}$ so that $\mathcal{N}_{\mathbb{R}}$ coincides. Then we can write $\widehat{\mathcal{N}} = \mathcal{O}(k_1) \oplus \cdots \oplus \mathcal{O}(k_{n-1})$, and we call the $(n-1)$-tuple of integers $(k_1, \cdots, k_{n-1})$ the partial indices of the holomorphic disk D. The partial indices are uniquely determined by D up to permutation.

4.2. *Construction of* $(--++)$ *self-dual space*

A Riemannian conformal 4-manifold is called self-dual if and only if $W_- \equiv 0$, where W_- is the anti-self-dual part of the Weyl curvature tensor. Quite similarly, we can define self-duality for an indefinite conformal manifold $(M, [g])$ of signature $(--++)$. Such indefinite self-dual manifolds are, at least locally, obtained by the twistor method in the following way.

Let $\mathcal{T}$ be a complex 3-manifold and let $\mathcal{T}_{\mathbb{R}}$ be a totally real submanifold on $\mathcal{T}$. Suppose that there exists a family of holomorphic disks on $(\mathcal{T}, \mathcal{T}_{\mathbb{R}})$ smoothly parametrized by a real 4-manifold M. Such a family is described by the double fibration:

$$M \xleftarrow{\mathfrak{g}} (\mathcal{Z}, \mathcal{Z}_{\mathbb{R}}) \xrightarrow{(\mathfrak{f}, \mathfrak{f}_{\mathbb{R}})} (\mathcal{T}, \mathcal{T}_{\mathbb{R}}), \tag{15}$$

where $\mathfrak{g}$ is a smooth $(\mathbb{D}, \partial\mathbb{D})$-bundle on M, $\mathfrak{f}$ is a smooth map which is holomorphic along each fiber of $\mathfrak{g}$, and $\mathfrak{f}_{\mathbb{R}}$ is the restriction of $\mathfrak{f}$. For each $x \in M$, let us write $D_x = \mathfrak{f}(\mathfrak{g}^{-1}(x))$ which is the holomorphic disk corresponding to x. Now we assume the following conditions:

A1) the differential $\mathfrak{f}_*$ is of full-rank on $\mathcal{Z} \setminus \mathcal{Z}_{\mathbb{R}}$ (i.e. $\mathfrak{f}$ is locally diffeomorphic on $\mathcal{Z} \setminus \mathcal{Z}_{\mathbb{R}}$),
A2) the differential $(\mathfrak{f}_{\mathbb{R}})_*$ is of full-rank, and
A3) for each $x \in M$, the partial indices of D_x is $(1, 1)$.

In this setting, the following holds (see Section 10 of LeBrun-Mason[7]).

Proposition 4.1. *There exists a unique smooth self-dual conformal structure $[g]$ of signature $(--++)$ on M such that for each $p \in \mathcal{T}_{\mathbb{R}}$ the set $\mathfrak{S}_p = \{x \in M \mid p \in \partial D_x\}$ gives a β-surface for $[g]$.*

Here a β-surface is a two-dimensional submanifold on which the conformal structure vanishes and of which the tangent bivector is anti-self-dual everywhere.

4.3. *Construction of* $(-++)$ *Einstein-Weyl structure*

There is a similar construction for three-dimensional Einstein-Weyl structure (see Hitchin [3], LeBrun-Mason [8], Nakata [9]), which is sometimes called the minitwistor correspondence. An Einstein-Weyl structure on a manifold N is the pair $([g], \nabla)$ of a conformal structure $[g]$ and an affine connection ∇ on N satisfying the compatibility condition (Weyl condition) $\nabla g \propto g$ and the Einstein-Weyl condition $R_{(ij)} \propto g_{ij}$ where $R_{(ij)}$ is the symmetrized Ricci tensor of ∇. In particular, if g is an Einstein metric and ∇ is the Levi-Civita connection of g, then $([g], \nabla)$ is Einstein-Weyl.

Let $\mathcal{S}$ be a complex surface and $\mathcal{S}_{\mathbb{R}}$ be a totally real submanifold on $\mathcal{S}$. Suppose that there exists a family of holomorphic disks on $(\mathcal{S}, \mathcal{S}_{\mathbb{R}})$ smoothly parametrized by a real 3-manifold N, which is described by the following diagram:

$$N \overset{\mathfrak{g}}{\longleftarrow} (\mathcal{W}, \mathcal{W}_{\mathbb{R}}) \overset{(\mathfrak{g}, \mathfrak{g}_{\mathbb{R}})}{\longrightarrow} (\mathcal{S}, \mathcal{S}_{\mathbb{R}}). \tag{16}$$

Let us assume the conditions:

B1) the differential $\mathfrak{g}_*$ is of full-rank on $\mathcal{W} \setminus \mathcal{W}_{\mathbb{R}}$,
B2) the differential $(\mathfrak{g}_{\mathbb{R}})_*$ is of full-rank, and
B3) for each $x \in N$, the partial index of the disk $D_x = \mathfrak{g}(\mathfrak{g}^{-1}(x))$ is 2.

Then the following holds (see Nakata [9]).

Proposition 4.2. *There exists a unique smooth Einstein-Weyl structure* $([g], \nabla)$ *of signature* $(-++)$ *on* N *so that for each* $p \in \mathcal{S}_{\mathbb{R}}$ *the set* $\mathfrak{S}_p = \{x \in N \mid p \in \partial D_x\}$ *is a totally geodesic null surface on* N.

Here a surface on an indefinite conformal manifold $(N, [g])$ is called null if and only if $[g]$ degenerate on it.

5. Twistor theory for Tod-Kamada metric

In this section, we summarize the twistor correspondence for Tod-Kamada metric. Tod-Kamada metrics are indeinite self-dual metrics on $S^2 \times S^2$ of signature $(--++)$ which are firstly given by K. P. Tod [13] and are also studied by H. Kamada [4]. The author determined the twistor space corresponding to Tod-Kamada metric in the sense of LeBrun-Mason theory (see Nakata [10]).

5.1. *Model case*

Let us start from the twistor space $(\mathbb{CP}^3, \mathbb{RP}^3)$, where

$$\mathbb{RP}^3 = \{[z_0 : z_1 : z_2 : z_3] \in \mathbb{CP}^3 \mid z_3 = \bar{z}_0,\ z_2 = \bar{z}_1\}. \tag{17}$$

Notice that $\mathbb{RP}^3$ coincides with the fixed point set of the antiholomorphic involution $\sigma : \mathbb{CP}^3 \to \mathbb{CP}^3$ defined by

$$\sigma([z_0 : z_1 : z_2 : z_3]) = [\bar{z}_3 : \bar{z}_2 : \bar{z}_1 : \bar{z}_0].$$

Each σ-invariant complex line on $\mathbb{CP}^3$ intersects with $\mathbb{RP}^3$ along a circle, so this line is divided into two holomorphic disks on $(\mathbb{CP}^3, \mathbb{RP}^3)$. The family of such holomorphic disks is parametrized by $S^2 \times S^2$, and we can check that this family satisfies the conditions A1) to A3) in Section 4.2. Hence by Proposition 4.1 we obtain an indefinite self-dual conformal structure on $S^2 \times S^2$, which turns out to be the class of the standard product metric $(-g_{S^2}) \times g_{S^2}$. This correspondence is described by the following double fibration:

$$S^2 \times S^2 \quad \overset{\mathfrak{g}}{\longleftarrow} \quad (\mathcal{Z}, \mathcal{Z}_{\mathbb{R}}) \quad \overset{(\mathfrak{f}, \mathfrak{f}_{\mathbb{R}})}{\longrightarrow} \quad (\mathbb{CP}^3, \mathbb{RP}^3). \tag{18}$$

Next let us introduce a holomorphic $\mathbb{C}^*$-action on $\mathbb{CP}^3$ by

$$\mu \cdot [z_0 : z_1 : z_2 : z_3] = [\mu z_0 : \mu z_1 : z_2 : z_3] \qquad (\mu \in \mathbb{C}^*).$$

If $\mu \in U(1)$, then μ-action commutes with the involution σ, so it preserves $\mathbb{RP}^3$. Hence this action can be understood as a $(\mathbb{C}^*, U(1))$-action on $(\mathbb{CP}^3, \mathbb{RP}^3)$. If we take the quotient of the free part of this action, we obtain the minitwistor space $(\mathbb{CP}^1 \times \mathbb{CP}^1, S^2)$. On the other hand, the above $U(1)$-action on $(\mathbb{CP}^3, \mathbb{RP}^3)$ induces an action on the set of holomorphic disks, hence we obtain a $U(1)$-action on $S^2 \times S^2$. The quotient of the free part of this action is $S^3_1 \simeq \mathbb{R} \times S^2$. In this way we obtain the double fibration

$$S^3_1 \quad \longleftarrow \quad (\mathcal{W}, \mathcal{W}_{\mathbb{R}}) \quad \longrightarrow \quad (\mathbb{CP}^1 \times \mathbb{CP}^1, S^2) \tag{19}$$

as a free quotient of (18). On this double fibration, the conditions B1) to B3) are satisfied. Hence by Proposition 4.2, we obtain an Einstein-Weyl structure on S^3_1 which turns out to be the one induced from the de Sitter metric. The geometry on the small circles on S^2 explained in Section 2 is obtained from the double fibration (19) by noticing the boundaries of holomorphic disks on $(\mathbb{CP}^1 \times \mathbb{CP}^1, S^2)$.

5.2. *Tod-Kamada metric*

Now we introduce Tod-Kamada metric. Let (V, A) be the pair of a function $V \in C^\infty(S^3_1)$ and a 1-form $A \in \Omega^1(S^3_1)$. For each pair (V, A) satisfying

$V > 0$ everywhere, we define an S^1-invariant indefinite metric $g_{(V,A)}$ on $S^1 \times S^3_1$ by

$$g_{(V,A)} = -V^{-1}(ds + A)^2 + V g_{S^3_1}, \tag{20}$$

where s is the coordinate on S^1-direction.

It is known that the indefinite metric $g_{(V,A)}$ on $S^1 \times S^3_1$ is self-dual if and only if (V, A) satisfies the monopole equation

$$*dV = dA \tag{21}$$

(see Kamada [4], LeBrun [5]). We remark that if (21) holds, then V satisfies the wave equation $\Box V = *d*dV = 0$. Noticing that $S^2 \times S^2$ is a natural compactification of $S^1 \times S^3_1$, we obtain the following.

Proposition 5.1 (Kamada [4]). *For each solution (V, A) to the equation (21) satisfying $V > 0$, the indefinite metric $\bar{g}_{(V,A)} = (\cosh s)^{-2} g_{(V,A)}$ extends smoothly to the self-dual metric on $S^2 \times S^2$ if and only if there exist smooth functions F_+ and F_- on $\mathbb{R} \times S^2$ in variables r^2, q^2 and y such that*

$$\begin{aligned} V(t,y) &= 1 + r^2 F_-(r^2, y) \qquad r = e^t \qquad \text{and} \\ V(t,y) &= 1 + q^2 F_+(q^2, y) \qquad q = e^{-t}, \end{aligned} \tag{22}$$

as $r \to +0$ and as $q \to +0$ respectively.

By construction, any Tod-Kamada metric $\bar{g}_{(V,A)}$ has an S^1-symmetry. When $(V, A) = (1, 0)$, the metric $\bar{g}_{(1,0)}$ coincides with the standard product metric. Hence Tod-Kamada metrics are S^1-invariant deformations of product metric on $S^2 \times S^2$ parametrized by monopoles.

The author proved that every smooth monopole (V, A) satisfying the condition (22) are essentially generated from a single smooth function h on S^2 as follows (Theorem 5.1). Let $\check{*}$ and $\check{d}$ are the Hodge operator and the exterior derivative of S^2-direction of $S^3_1 \cong \mathbb{R} \times S^2$. Two monopoles (V, A) and (V', A') are called gauge equivalent if and only if there exists a function $\phi \in C^\infty(S^3_1)$ such that $(V', A') = (V, A + d\phi)$. If (V, A) and (V', A') are gauge equivalent, then the metric $g_{(V,A)}$ and $g_{(V',A')}$ are isometric. Recall that $R : C^\infty(S^2) \to C^\infty(S^3_1)$ is the integral transform defined in (1), and that $C^\infty_*(S^2) = \{h \in C^\infty(S^2) : \int_{S^2} h dS^2 = 0\}$.

Theorem 5.1 (Nakata [10], Corollary 4.8). *For each function $h \in C^\infty_*(S^2)$ satisfying $|\partial_t Rh(t,y)| < 1$, the pair*

$$V = 1 + \partial_t Rh, \qquad A = -\check{*}\,\check{d}\,Rh \tag{23}$$

satisfies the conditions in Proposition 5.1, hence we obtain a self-dual metric $\bar{g}_{(V,A)}$ *on* $S^2 \times S^2$*. Conversely, if* (V', A') *is a solution of the monopole equation* (21) *satisfying* $V' > 0$ *and*

$$V'(t,y) \to 1, \quad V'_t(t,y) \to 0 \quad \text{as } t \to \pm\infty \quad \text{uniformly for } y \in S^2, \tag{24}$$

then there is a unique (V, A) *in the gauge equivalent class of* (V', A') *and a unique* $h \in C^\infty_*(S^2)$ *such that* (23) *holds.*

Notice that Kamada's condition (22) is replaced by the condition (24).

5.3. *Twistor space of Tod-Kamada metric*

As already mentioned, Tod-Kamada metric is considered as an S^1-invariant deformation of the standard product metric on $S^2 \times S^2$. So it would be natural to expect that the twistor space corresponding to the Tod-Kamada metric is obetained as an $U(1)$-invariant deformation of $(\mathbb{CP}^3, \mathbb{RP}^3)$. Following this observation, we define a deformation of $\mathbb{RP}^3$ in $\mathbb{CP}^3$ by

$$P_h := \left\{ [z_0 : z_1 : z_2 : z_3] \in Z \;\middle|\; z_3 = \bar{z}_0 e^{-h(z_1/z_0)},\ z_2 = \bar{z}_1 e^{-h(z_1/z_0)} \right\}, \tag{25}$$

where h is a smooth function on $\mathbb{CP}^1 \cong S^2$. Notice that $P_h = \mathbb{RP}^3$ if $h \equiv 0$. Since P_h depends on h essentially up to constant, we assume $h \in C^\infty_*(S^2)$. Then the following holds.

Theorem 5.2 (Nakata [10], Theorem 7.1). *For each* $h \in C^\infty_*(S^2)$ *satisfying* $|\partial_t Rh(t,y)| < 1$*, the corresponding Tod-Kamada metric* $\bar{g}_{(V,A)}$ *defined in Theorem 5.1 is Zollfrei. Moreover the following objects correspond to each other*

- *Tod-Kamada metric* $\bar{g}_{(V,A)}$*, and*
- *the pair* $(\mathbb{CP}^3, P_h)$

in the sense of Theorem 1.1.

The proof is obtained by writing down holomorphic disks explicitly. See Nakata [10] for the detail.

6. Twsitor theory for indefinite self-dual metric on $\mathbb{R}^4$

As already mentioned, the results for small circles on S^2 (Section 2) is connected with the twistor theory for Tod-Kamada metric. On the other hand, the results for circles on the cylinder (Section 3) is connected with another twistor theory. In this Section, we show just the outline of this theory (See Nakata [11] for the detail).

6.1. *Model case*

Let H be the degree 1 holomorphic line bundle over $\mathbb{CP}^1$ and we put $Z = H \oplus H$. The total space Z can be embedded in $\mathbb{CP}^3$ as

$$Z = \{[y_0 : y_1 : y_2 : y_3] \in \mathbb{CP}^3 \mid (y_0, y_1) \neq (0,0)\}. \tag{26}$$

We define a totally real submanifold $Z_\mathbb{R}$ in Z by

$$Z_\mathbb{R} = \left\{ [e^{-\frac{i\theta}{2}} : e^{\frac{i\theta}{2}} : \eta : \bar{\eta}] \in \mathbb{CP}^3 \;\middle|\; (e^{i\theta}, \eta) \in S^1 \times \mathbb{C} \right\}. \tag{27}$$

The space $(Z, Z_\mathbb{R})$ fits to the story of Section 4.2, so we obtain an indefinite self-dual 4-manifold which turns out to be the flat Minkowski space $\mathbb{R}^{2,2}$. This correspondence is described by the double fibration

$$\mathbb{R}^{2,2} \quad \overset{\mathfrak{g}}{\longleftarrow} \quad (\mathcal{Z}, \mathcal{Z}_\mathbb{R}) \quad \overset{(\mathfrak{f}, \mathfrak{f}_\mathbb{R})}{\longrightarrow} \quad (Z, Z_\mathbb{R}). \tag{28}$$

Next we introduce a $(\mathbb{C}, \mathbb{R})$-action on $(Z, Z_\mathbb{R})$ by

$$\mu \cdot [y_0 : y_1 : y_2 : y_3] = [y_0 : y_1 : y_2 - i\nu y_1 : y_3 + i\nu y_0] \qquad (\nu \in \mathbb{C} \text{ or } \mathbb{R}). \tag{29}$$

This action is free, and let the quotient space $(\mathcal{S}, \mathcal{C})$. Then $\mathcal{S}$ is diffeomorphic to the degree 2 line bundle H^2, and the totally real submanifold $\mathcal{C}$ is diffeomorphic to the cylinder $S^1 \times \mathbb{R}$.

The $\mathbb{R}$-action (29) induces an $\mathbb{R}$-action on $\mathbb{R}^{2,2}$, and its quotient is the flat Lorentz space $\mathbb{R}^{1,2}$. This Lorentz space $\mathbb{R}^{1,2}$ is also considered as the set of holomorphic disks on $(\mathcal{S}, \mathcal{C})$. This correspondence is described by the double fibration

$$\mathbb{R}^{1,2} \quad \longleftarrow \quad (\mathcal{W}, \mathcal{W}_\mathbb{R}) \quad \longrightarrow \quad (\mathcal{S}, \mathcal{C}). \tag{30}$$

6.2. *Deformation of the twistor correspondence*

Now we deform the twistor space $(Z, Z_\mathbb{R})$ respecting the $\mathbb{R}$-action. We notice that, the $(\mathbb{C}, \mathbb{R})$-quotient $(Z|_\mathcal{C}, Z_\mathbb{R}) \to \mathcal{C}$ is considered as a trivial $(\mathbb{C}, \mathbb{R})$-bundle, that is we can trivialize

$$Z|_\mathcal{C} = \{(p, \nu) \in \mathcal{C} \times \mathbb{C}\},$$
$$Z_\mathbb{R} = \{(p, \nu) \in \mathcal{C} \times \mathbb{R}\} = \{(p, \nu) \in Z|_\mathcal{C} \mid \mathrm{Im}(\nu) = 0\}.$$

Now, for each smooth function $h(p) \in C^\infty(\mathcal{C})$, we define a deformation $\mathcal{P}_h$ of $Z_\mathbb{R}$ by

$$\mathcal{P}_h = \{(p, \nu) \in Z|_\mathcal{C} \mid \mathrm{Im}(\nu) = h(p)\}. \tag{31}$$

In this way, we obtain deformed twistor spaces $(Z, \mathcal{P}_h)$ parametrized by functions $h(p) \in C^\infty(\mathcal{C})$. The twistor correspondence for this twistor space is written down in the following way.

We use a global coordinate (t, x_1, x_2) on $\mathbb{R}^{1,2}$ so that the flat Lorentz metric is written as

$$\underline{g} = -dt^2 + dx_1^2 + dx_2^2.$$

Let $\check{d}$ and $\check{*}$ be the exterior derivative and the Hodge operator along the (x_1, x_2)-direction of $\mathbb{R}^{1,2}$. We defined the integral transform $R' : C^\infty(\mathcal{C}) \to C^\infty(\mathbb{R}^{1,2})$ as (10).

Theorem 6.1 (Nakata [11], Theorem 5.1). *For each $h \in C^\infty(\mathcal{C})$, there is a smooth $\mathbb{R}^4$-family of holomorphic disks on $(Z, \mathcal{P}_h)$. Moreover if $\partial_t R'h < 1$, a natural self-dual conformal structure on the parameter space $\mathbb{R}^4 = \mathbb{R} \times \mathbb{R}^{1,2}$ is induced and is represented by the metric*

$$g_{(V,A)} = -V^{-1}(ds + A)^2 + V\underline{g}, \tag{32}$$

where (V, A) is the pair of the function $V \in C^\infty(\mathbb{R}^{1,2})$ and the 1-form $A \in \Omega^1(\mathbb{R}^{1,2})$ defined by

$$V = 1 - \partial_t R'h \qquad \text{and} \qquad A = \check{*}\check{d}R'h. \tag{33}$$

Similar to the de Sitter case (Section 5.2), the pair (V, A) obtained as (33) saisfies the monopole equation $*dV = dA$ which is an equivalent condition to the self-duality of the indefinite metric (32). The converse correspondence of Theorem 6.1 is also established under the rapidly decreasing conditions as in Theorem 3.1. We omit the detail of this direction.

Bibliography

1. M. F. Atiyah, N. Hitchin, I. M. Singer, *Self-duality in Four-dimensional Riemannian Geometry*, Proc. R. Soc. Lond. **A.362**(1978), 425–461.
2. S. Helgason: *The Radon Transform (Second Edition)*, Progress in Mathematics vol.5, Birkhäusar (1999).
3. N. J. Hitchin, *Complex manifolds and Einstein's equations*, Twistor Geometry and Non-Linear Systems, Lecture Notes in Mathematics vol. 970 (1982).
4. H. Kamada, *Compact Scalar-flat Indefinite Kähler Surfaces with Hamiltonian S^1-Symmetry*, Comm. Math. Phys. **254**(2005), 23–44.
5. C. LeBrun, *Explicit self-dual metrics on* $\mathbb{CP}_2\sharp\cdots\sharp\mathbb{CP}_2$, J. Diff. Geom. **34**(1991), 223–253.
6. C. LeBrun, *Twistors, Holomorphic Disks, and Riemann Surfaces with Boundary*, Perspectives in Riemannian geometry, CRM Proc. Lecture Notes, 40, Amer. Math. Soc. Providence, RI, (2006), 209–221.
7. C. LeBrun, L. J. Mason, *Nonlinear Gravitons, Null Geodesics, and Holomorphic Disks*, Duke Math. J. **136**(2007), 205–273.
8. C. LeBrun, L. J. Mason, *The Einstein-Weyl Equations, Scattering Maps, and Holomorphic Disks*, Math. Res. Lett. **16**(2009) 291-301.

9. F. Nakata, *A construction of Einstein-Weyl spaces via LeBrun-Mason type twistor correspondence*, Comm. Math. Phys. **289**(2009) 663-699.
10. F. Nakata, *Wave equations and LeBrun-Mason correspondence* Trans. Amer. Math. Soc. **364**(2012), 4763–4800.
11. F. Nakata, *An integral transform on a cylinder and the twistor correspondence* Diff. Geom. and its Appl. **30**(2012), 428–437.
12. R. Penrose, *Nonlinear gravitons and curved twistor theory*, Gen. Rel. Grav. **7**(1976), 31–52.
13. K. P. Tod, *Indefinite conformally-ASD metric on* $S^2 \times S^2$: Further advances in twistor theory. Vol.III. Chapman & Hall/CRC (2001) 61–63, reprinted from Twistor Newsletter **36**(1993).

Received December 4, 2012
Revised March 28, 2013

Proceedings of the 3rd International
Colloquium on Differential Geometry
and its Related Fields
Veliko Tarnovo, September 3–7, 2012

A DYNAMICAL SYSTEMATIC ASPECT OF HOROCYCLIC CIRCLES IN A COMPLEX HYPERBOLIC SPACE

Toshiaki ADACHI *

Department of Mathematics, Nagoya Institute of Technology,
Nagoya 466-8555, Japan
E-mail: adachi@nitech.ac.jp

In this note we attempt to reconsider circles of positive geodesic curvature on a complex hyperbolic space by use of magnetic fields on real hypersurfaces. In particular, we show that every horocyclic circle is a trajectory for some Sasakian magnetic field on a horosphere. We show also that the same property holds for horocyclic helices of proper order 3 each of which is an orbit of one parameter family of isometries and lies on a totally geodesic complex hyperbolic plane.

Keywords: Sasakian magnetic fields; Trajectories; Circles; Horospheres; Essential Killing helices; Extrinsic shapes.

1. Introduction

A smooth curve γ parameterized by its arclength is said to be a *helix* of proper order d if it satisfies the following system of equations of Frenet-Serre type

$$\nabla_{\dot{\gamma}} Y_j = -k_{j-1} Y_{j-1} + k_j Y_{j+1} \qquad (j = 1, \ldots, d)$$

with positive constants $k_1, \ldots, k_{d-1}$ and an orthonormal frame field $\{Y_1 = \dot{\gamma}, Y_2, \ldots, Y_d\}$ along γ. Here, we set Y_0, Y_{d+1} to be null vector fields along γ and $k_0 = k_d = 0$. These positive constants and the frame field are called the *geodesic curvatures* and the *Frenet frame* of γ, respectively. When $d = 2$, it is also called a *circle* of positive geodesic curvature. In their paper [13], Nomizu-Yano studied circles from the viewpoint of submanifold theory. It is needless to say that geodesics play quite an important role in the study of Riemannian manifolds. Since circles are simple curves next to geodesics

*The author is partially supported by Grant-in-Aid for Scientific Research (C) (No. 24540075) Japan Society of Promotion Science.

in the sense of Frenet formula, they should play an important role, mainly in the study of submanifolds (see [14, 15], for example). But as we need information on their accelerations it is not so easy to treat them.

If we restrict ourselves to helices on a Kähler manifold with complex structure J, we can consider invariants other than their geodesic curvatures. For a helix γ with frame field $\{Y_1, \dots, Y_d\}$, we set $\tau_{ij} = \langle Y_i, JY_j \rangle$ $(1 \le i < j \le d)$ and call them its *complex torsions*. Circles of complex torsion $\tau_{12} = \pm 1$ can be interpreted as trajectories for Kähler magnetic fields (see §3). We are hence interested in interpreting other circles from such a dynamical systematic point of view. In this note we give a light on horocyclic circles and helices of proper order 3 on a complex hyperbolic space. We show that every horocyclic circle is the extrinsic shape of a trajectory for some Sasakian magnetic field on a horosphere. This result corresponds to the fact that every horocyclic circle on a real hyperbolic space is a geodesic on a horosphere in this space. Also, we correct an error in [3] concerning properties of trajectories on a horosphere in a complex hyperbolic space.

2. Circles on a complex space form

We shall start by recalling properties of circles on a complex hyperbolic space $\mathbb{C}H^n$. A smooth curve γ parameterized by its arclength is said to be *closed* if there is a positive t_p satisfying $\gamma(t + t_p) = \gamma(t)$ for all t. The minimum positive t_p with this property is called the *length* of γ. When γ is not closed, we say it is open. In this case we cinsider its length infinity. Since $\mathbb{C}H^n$ is not compact we have two kinds of open curves, unbounded open curves and bounded open curves. We call a curve γ parameterized by its arclength unbounded in both directions if both $\gamma([0,\infty))$ and $\gamma((-\infty, 0])$ are unbounded sets. As $\mathbb{C}H^n$ is a typical example of Hadamard manifolds, we can compactify it by taking its ideal boundary $\partial\mathbb{C}H^n$ formed by asymptotic classes of geodesics. For a curve γ which is unbounded in both directions, we set $\gamma(\infty) = \lim_{t\to\infty} \gamma(t)$, $\gamma(-\infty) = \lim_{t\to-\infty} \gamma(t) \in \partial\mathbb{C}H^n$ if they exist, and call them points at infinity.

We define a function $\nu : (0, \infty) \to [0, 1]$ by

$$\nu(k) = \begin{cases} 0, & \text{if } 0 < k < 1, \\ 2(k^2-1)^{3/2}/(3\sqrt{3}k), & \text{if } 1 \le k \le 2, \\ 1, & \text{if } k > 2. \end{cases}$$

On a complex hyperbolic space $\mathbb{C}H^n(-4)$ of constant holomorphic sectional curvature -4, circles of positive geodesic curvature have the following properties (see [7]).

1) A circle of geodesic curvature k is bounded if and only if either $k > 2$ or the absolute value of its complex torsion satisfies $|\tau_{12}| < \nu(k)$. Otherwise, it is unbounded in both directions.
2) An unbounded circle γ of geodesic curvature k has a single point at infinity ($\gamma(\infty) = \gamma(-\infty)$) and lies on a horosphere if $1 \leq k \leq 2$ and $|\tau_{12}| = \nu(k)$. Otherwise, it has two distinct points at infinity ($\gamma(\infty) \neq \gamma(-\infty)$).
3) A circle of complex torsion $\tau_{12} = \pm 1$ and of geodesic curvature $k > 2$ is closed. Its length is $2\pi/\sqrt{k^2 - 4}$.
4) A circle of null complex torsion and of geodesic curvature $k > 1$ is closed. Its length is $2\pi/\sqrt{k^2 - 1}$.
5) A bounded circle of complex torsion $0 < |\tau_{12}| < 1$ is closed if and only if there exist relatively prime positive integers p, q satisfying $p > q$ and $3\sqrt{3}k\tau = \pm 2(k^2 - 1)^{3/2}\tau_{p,q}$, where $\tau_{pq} = q(9p^2 - q^2)(3p^2 + q^2)^{-3/2}$. In this case its length is given as $\delta(p, q)\pi\sqrt{(3p^2 + q^2)/\{3(k^2 - 1)\}}$, where $\delta(p, q) = 2$ when the product pq is even and $\delta(p, q) = 1$ when pq is odd.

We call a circle on $\mathbb{C}H^n(-4)$ *horocyclic* if its geodesic curvature and its complex torsion satisfy $1 \leq k \leq 2$ and $|\tau_{12}| = \nu(k)$.

We say two smooth curves γ_1, γ_2 parameterized by arclength are *congruent* to each other if there exist an isometry φ of the base manifold and a constant t_c satisfying $\gamma_2(t) = \varphi \circ \gamma_1(t + t_c)$ for all t. In case we can take $t_c = 0$, we say that they are congruent to each other in strong sense. On a complex hyperbolic space $\mathbb{C}H^n$, two circles of positive geodesic curvature are congruent to each other in strong sense if and only if they have the same geodesic curvature and the same absolute value of complex torsion. Thus the moduli space, which is the set of all congruence classes, of circles of positive geodesic curvature coincides set-theoretically with the band $(0, \infty) \times [0, 1]$.

It is clear that two closed circles have the same length if they are congruent to each other. Thus we can define a length function of the moduli space of circles. Taking into account of properties of circles on $\mathbb{C}H^n(-4)$, we see that we have a lamination $\mathcal{F} = \{\mathcal{F}_\mu\}_{0 \leq \mu \leq 1}$ on the moduli space of bounded circles whose leaves are set-theoretically maximal with respect to the property that the length function is continuous (see [2]). Leaves of this lamination is given as

$$\mathcal{F}_\mu = \begin{cases} \{[\gamma_{k,0}] \mid k > 1\}, & \text{if } \mu = 0, \\ \{[\gamma_{k,\tau}] \mid 3\sqrt{3}k\tau(k^2 - 1)^{-3/2} = 2\mu,\ 0 < \tau < 1\}, & \text{if } 0 < \mu < 1, \\ \{[\gamma_{k,1}] \mid k > 2\}, & \text{if } \mu = 1, \end{cases}$$

where $[\gamma_{k,\tau}]$ denotes the congruence class containing a circle of geodesic curvature k and of complex torsion τ.

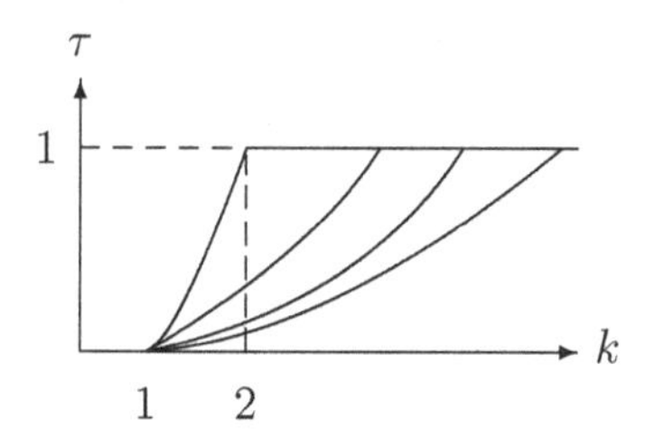

Figure 1. Lamination on the moduli space of bounded circles on $\mathbb{C}H^n(-4)$

As we can see in Figure 1, the lamination has singularities only on the leaf of circles of complex torsion ± 1. The author hence doubts whether all circles on a Kähler manifold are of the same kind. From the viewpoint of submanifold theory, Chen-Maeda [10] studied their "type-number". But they did not treat all curves having the same type-number.

3. Sasakian magnetic fields

As a generalization of static magnetic fields on a Euclidean 3-space, a closed 2-form $\mathbb{B}$ on a Riemannian manifold M is said to be a *magnetic field.* We define a skew symmetric operator $\Omega_{\mathbb{B}} : TM \to TM$ by $\mathbb{B}(v, w) = \langle v, \Omega_{\mathbb{B}}(w)\rangle$ for arbitrary v, $w \in T_pM$ at an arbitrary point $p \in M$. We say a smooth curve γ parameterized by its arclength to be a *trajectory* for $\mathbb{B}$ if it satisfies the differential equation $\nabla_{\dot\gamma}\dot\gamma = \Omega_{\mathbb{B}}(\dot\gamma)$. Given a unit tangent vector $u \in UM$ we denote by γ_u the trajectory for $\mathbb{B}$ with initial vector $\dot\gamma_u(0) = u$. We can then define a magnetic flow $\mathbb{B}\varphi_t : UM \to UM$ by $\mathbb{B}\varphi_t(u) = \dot\gamma_u(t)$. When $\mathbb{B}$ is the trivial 2-form, its magnetic flow is the geodesic flow. When a magnetic field is given, then its constant multiples are also magnetic fields. Considering their magnetic flows, we may say that these magnetic flows form a perturbation of the geodesic flow.

On a Kähler manifold with complex structure J, we have a natural closed 2-form $\mathbb{B}_J$ which is called the Kähler form. Its constant multiple $\mathbb{B}_\kappa = \kappa\mathbb{B}_J$ is said to be a *Kähler magnetic field.* Since we have $\Omega_{\mathbb{B}_\kappa} = \kappa J$, a smooth curve γ parameterized by its arclength is a trajectory for $\mathbb{B}_\kappa$ if it satisfies $\nabla_{\dot\gamma}\dot\gamma = \kappa J\dot\gamma$. As J is parallel, we see it is a circle of geodesic curvature $|\kappa|$ and of complex torsion $-\mathrm{sgn}(\kappa)$, where $\mathrm{sgn}(\alpha)$ of a constant α denotes its signature. Thus we can treat circles of complex torsion ± 1 on a Kähler manifold quite similarly as we treat geodesics (see [1, 4] and its references).

We are hence interested in circles of complex torsion τ_{12} with $|\tau_{12}| < 1$. We here take a real hypersurface M of a Kähler manifold $\widetilde{M}$. On this real hypersurface M we have a natural contact metric structure $(\phi, \xi, \eta, \langle\ ,\ \rangle)$ which is induced by the complex structure J and Riemannian metric $\langle\ ,\ \rangle$ of the ambient manifold $\widetilde{M}$. Taking a unit normal (local) vector field $\mathcal{N}$ of M in $\widetilde{M}$ we have $\xi = -J\mathcal{N}$, $\eta(v) = \langle v, \xi\rangle$ and $\phi(v) = Jv - \eta(v)\mathcal{N}$ for each $v \in TM$. By use of this structure we can define a closed 2-form $\mathbb{F}_\phi$ by $\mathbb{F}_\phi(v, w) = \langle v, \phi w\rangle$ for v, $w \in T_pM$ at an arbitrary point $p \in M$ (see [9]). We say a constant multiple $\mathbb{F}_\kappa = \kappa\mathbb{F}_\phi$ $(\kappa \in \mathbb{R})$ of this canonical 2-form to be a *Sasakian magnetic field*. We use this terminology even if the base manifold is not Sasakian.

Since we have $\Omega_{\mathbb{F}_\kappa} = \kappa\phi$, a smooth curve γ parameterized by its arclength is a trajectory for $\mathbb{F}_\kappa$ if it satisfies $\nabla_{\dot\gamma}\dot\gamma = \kappa\phi\dot\gamma$. Being different from trajectories for Kähler magnetic fields, as ϕ is not parallel, its geodesic curvature $|\kappa|\sqrt{1 - \langle\dot\gamma, \xi_\gamma\rangle^2}$ depends on velocity vectors. Hence it is not even a helix in general. To calculate its geodesic curvatures we set $\rho_\gamma = \langle\dot\gamma, \xi_\gamma\rangle$ and call it the *structure torsion* of γ. Denoting the connection of $\widetilde{M}$ by $\widetilde{\nabla}$ and the shape operator of M in $\widetilde{M}$ by A, we have the following formulas of Gauss and Weingarten

$$\widetilde{\nabla}_X Y = \nabla_X Y + \langle AX, Y\rangle\mathcal{N}, \qquad \widetilde{\nabla}_X\mathcal{N} = -AX \tag{3.1}$$

for vector fields X, Y tangent to M. Since these show that $\nabla_X\xi = \phi AX$, by taking account that A is symmetric and ϕ is skew-symmetric, we find the differential of ρ_γ is given as

$$\rho_\gamma' = \langle\dot\gamma, \xi\rangle' = \langle\kappa\phi\dot\gamma, \xi\rangle + \langle\dot\gamma, \phi A\dot\gamma\rangle = \frac{1}{2}\langle\dot\gamma, (\phi A - A\phi)\dot\gamma\rangle.$$

In $\mathbb{C}H^n$ we have a typical real hypersurfaces, which are called hypersurfaces of type (A). They are horospheres, geodesic spheres and tubes around $\mathbb{C}H^\ell$ $(1 \le \ell < n)$ each of whose shape operator A and characteristic tensor ϕ are simultaneously diagonalizable $(A\phi = \phi A)$. Thus, on these real hypersurfaces, every trajectory for a Sasakian magnetic field has constant structure torsion.

4. Extrinsic circular trajectories

We here study trajectories for Sasakian magnetic fields on a horosphere *HS* in a complex hyperbolic space $\mathbb{C}H^n(-4)$. When we take a smooth curve γ

on *HS*, we can regard it as a curve in $\mathbb{C}H^n$. That is, we consider a curve $\tilde{\gamma} = \iota \circ \gamma$ with an isometric embedding $\iota : HS \to \mathbb{C}H^n$. We call this curve the *extrinsic shape* of γ. When the extrinsic shape of a curve is a circle of positive geodesic curvature, we shall call this curve *extrinsic circular.*

As the shape operator A of a horosphere *HS* in $\mathbb{C}H^n(-4)$ satisfies $Av = v$ and $A\xi = 2\xi$ for every tangent vector v of *HS* with $v \perp \xi$, by use of Gauss and Weingarten formulas (3.1) we have

$$\widetilde{\nabla}_{\dot{\gamma}}\dot{\gamma} = \nabla_{\dot{\gamma}}\dot{\gamma} + \langle A\dot{\gamma}, \dot{\gamma}\rangle \mathcal{N} = \kappa\phi\dot{\gamma} + (1+\rho_\gamma^2)\mathcal{N}, \tag{4.1}$$

and

$$\begin{aligned}\widetilde{\nabla}_{\dot{\gamma}}\{\kappa\phi\dot{\gamma} + (1+\rho_\gamma^2)\mathcal{N}\} &= \widetilde{\nabla}_{\dot{\gamma}}\{\kappa J\dot{\gamma} + (1-\kappa\rho_\gamma+\rho_\gamma^2)\mathcal{N}\} \\ &= \kappa J\widetilde{\nabla}_{\dot{\gamma}}\dot{\gamma} - (1-\kappa\rho_\gamma+\rho_\gamma^2)A\dot{\gamma} \\ &= -\{\kappa^2(1-\rho_\gamma^2) + (1+\rho_\gamma^2)^2\}\dot{\gamma} + (\kappa+\rho_\gamma)(1+\rho_\gamma^2-\kappa\rho_\gamma)(\rho_\gamma\dot{\gamma}-\xi).\end{aligned} \tag{4.2}$$

Thus, γ is extrinsic circular if and only if either $\rho_\gamma = \pm 1$ or $\rho_\gamma \neq \pm 1$ and $(\kappa+\rho_\gamma)(1+\rho_\gamma^2-\kappa\rho_\gamma) = 0$. In both cases, the geodesic curvature and the complex torsion are given as

$$k_1 = \sqrt{\kappa^2(1-\rho_\gamma^2) + (1+\rho_\gamma^2)^2}, \quad \tau_{12} = -\frac{1}{k_1}\{\kappa(1-\rho_\gamma^2) + \rho_\gamma(1+\rho_\gamma^2)\}$$

and the Frenet frame is given as $\{\dot{\gamma}, (\kappa\phi\dot{\gamma} + (1+\rho_\gamma^2)\mathcal{N})/k_1\}$.

For curves on a horosphere *HS* we shall employ some terminologies as for their extrinsic shapes. For example, for a curve on *HS* its points at infinity mean the points at infinity of its extrinsic shape.

Proposition 4.1. *Let γ be a trajectory for a Sasakian magnetic field $\mathbb{F}_\kappa$ with $\rho_\gamma = \pm 1$ on a horosphere HS in $\mathbb{C}H^n(-4)$. Its extrinsic shape is a circle of geodesic curvature 2 and of complex torsion ∓ 1, where double signs take the opposite signatures. Hence it is unbounded in both directions and has a single point at infinity.*

Proposition 4.2 ([3]). *Let γ be a trajectory for a Sasakian magnetic field $\mathbb{F}_\kappa$ with $\rho_\gamma \neq \pm 1$ on a horosphere HS in $\mathbb{C}H^n(-4)$.*

(1) *It is extrinsic circular if and only if $\kappa = \rho_\gamma + \rho_\gamma^{-1}$ or $\kappa = -\rho_\gamma$.*
(2) *When $\kappa = \rho_\gamma + \rho_\gamma^{-1}$, its extrinsic shape is a trajectory for the Kähler magnetic field $\mathbb{B}_\kappa$. That is, it is a circle of geodesic curvature $|\kappa|$ and of complex torsion $-\mathrm{sgn}(\kappa)$.*
(3) *When $\kappa = -\rho_\gamma$, its extrinsic shape is a circle of geodesic curvature $k_1 = \sqrt{1+3\rho_\gamma^2}$ and of complex torsion $\tau_{12} = -2\rho_\gamma^3\big/\sqrt{1+3\rho_\gamma^2}$.*

Corollary 4.1. *A Sasakian magnetic field* $\mathbb{F}_\kappa$ *on HS does not have extrinsic circular trajectories of structure torsion* $\rho \neq \pm 1$ *if and only if* $1 \leq |\kappa| < 2$.

We now study properties of trajectories on *HS*. When $\kappa = \rho_\gamma + \rho_\gamma^{-1}$and $|\rho_\gamma| < 1$, as we have $|\kappa| > 2$ and the extrinsic shape is a trajectory for a Kähler magnetic field, we obtain the following.

Corollary 4.2. *An extrinsic circular trajectory* γ *for* $\mathbb{F}_\kappa$ *on HS in* $\mathbb{C}H^n(-4)$ *with* $\kappa = \rho_\gamma + \rho_\gamma^{-1}$ *and* $\rho_\gamma \neq \pm 1$ *is closed. In this case its length is* $2\pi/\sqrt{\kappa^2-4}\ \big(= 2\pi|\rho_\gamma|/(1-\rho_\gamma^2)\big)$.

Next we study the case that $\kappa = -\rho_\gamma$ and $|\rho_\gamma| < 1$. In this case, we see $|\tau_{12}| < 1$ and $\nu(k_1) = 2|\rho_\gamma|^3/\sqrt{1+3\rho_\gamma^2} = |\tau_{12}|$. Thus, being different from the case that $\kappa = \rho_\gamma + \rho_\gamma^{-1}$, we find that γ is unbounded in both directions.

Proposition 4.3. *An extrinsic circular trajectory* γ *for* $\mathbb{F}_\kappa$ *on HS in* $\mathbb{C}H^n(-4)$ *with* $\kappa = -\rho_\gamma$ *is unbounded in both directions, hence has a single point at infinity.*

Remark 4.1. We may consider that the case $\rho_\gamma = \pm 1$ is included in the case (3) of Proposition 4.2. Because trajectories of structure torsion ± 1 are geodesic on *HS* and do not depend on strengths of Sasakian magnetic fields, we consider $\kappa = \mp 1$ in this case.

We now study the converse of the third assertion in Proposition 4.2. We shall show that all circles on $\mathbb{C}H^n$ with single point at infinity are extrinsic shapes of trajectories for Sasakian magnetic fields. To show this we here make mention of horospheres briefly for the sake of readers' convenience. A horosphere *HS* is a level set of a Busemann function b_σ on $\mathbb{C}H^n$ which is defined by $b_\sigma(q) = \lim_{t\to\infty}\big\{t - d(q,\sigma(t))\big\}$ for a geodesic σ of unit speed on $\mathbb{C}H^n$. If $HS = b_\sigma^{-1}(t_0)$, then we may say that it is the "limit" of geodesic spheres $\big\{S_r(\sigma(t_0-r)) \bigm| r > 0\big\}$ of radius r centered at $\sigma(t_0-r)$ as r goes to infinity. Thus, for given a unit tangent vector v at an arbitrary point $p \in \mathbb{C}H^n$ we have a unique horosphere which passes through p and is the level set of the Busemann function defined by the geodesic of initial vector v. We shall call v an "inward" unit normal vector of this horosphere.

Theorem 4.1. *Let* γ *be a circle of positive geodesic curvature* k *and of complex torsion* τ *on a complex hyperbolic space* $\mathbb{C}H^n(-4)$.

(1) *If γ is unbounded and has a single point at infinity, then it is an extrinsic shape of a trajectory for some Sasakian magnetic field on some horosphere. When $|\tau| < 1$, the choices of the horosphere and the Sasakian magnetic field are unique.*

(2) *If γ is a bounded circle of complex torsion ± 1, then it is also the extrinsic shape of a trajectory for some Sasakian magnetic field on some horosphere.*

Proof. (1) Since γ is unbounded and has a single point at infinity, we see $1 \leq k \leq 2$ and $|\tau| = \nu(k)$. We set $\rho = -\mathrm{sgn}(\tau)\sqrt{(k^2-1)/3}$ and $v = \big(kY(0) + \rho J\dot{\gamma}(0)\big)/(1+2\rho^2)$, where we set $kY = \widetilde{\nabla}_{\dot{\gamma}}\dot{\gamma}$. We consider a horosphere *HS* which makes v to be the inward unit normal vector at $\gamma(0)$. Then its inward unit normal $\mathcal{N}$ satisfies

$$\langle \dot{\gamma}(0), -J\mathcal{N}_{\gamma(0)}\rangle = \frac{-k\tau+\rho}{1+2\rho^2} = \rho.$$

If we take a trajectory for a Sasakian magnetic field $\mathbb{F}_{-\rho}$ on *HS* with initial vector $\dot{\gamma}(0)$, its structure torsion is hence ρ and its extrinsic shape $\hat{\gamma}$ is a circle of geodesic curvature $\sqrt{1+3\rho^2} = k$ satisfying

$$\widetilde{\nabla}_{\dot{\hat{\gamma}}}\dot{\hat{\gamma}}(0) = -\rho\phi\dot{\gamma}(0) + (1+\rho^2)\mathcal{N}_{\gamma(0)} = -\rho J\dot{\gamma}(0) + (1+2\rho^2)\mathcal{N}_{\gamma(0)} = kY(0).$$

Thus we find $\hat{\gamma}$ coincides with γ.

We next show the uniqueness. Suppose that the extrinsic shape of a trajectory ς for a Sasakian magnetic field $\mathbb{F}_\kappa$ on a horosphere M coincides with γ. Following Remark 4.1 we have $\kappa = -\rho_\varsigma$ by Proposition 4.2 and have

$$\begin{cases} \dot{\varsigma} = \dot{\varsigma} - \rho_\varsigma\xi_M - \rho_\varsigma J\mathcal{N}_M, \\ kY = \kappa\phi\dot{\varsigma} + (1+\rho_\varsigma^2)\mathcal{N}_M = \kappa J\big(\dot{\varsigma} - \rho_\varsigma\xi_M\big) + (1+\rho_\varsigma^2)\mathcal{N}_M, \end{cases}$$

where $\mathcal{N}_M$ denotes the inward unit normal of M and $\xi_M = -J\mathcal{N}_M$. We hence obtain $\mathcal{N}_M = (kY - \kappa J\dot{\varsigma})/(1+\rho_\varsigma^2 - \kappa\rho_\varsigma)$, which shows that M coincides with the above horosphere *HS*. This guarantees that ς coincides with the above trajectory because $\dot{\varsigma}(0) = \dot{\gamma}(0)$.

(2) In this case we have $k > 2$. We take a unit tangent vector $w \in U_{\gamma(0)}\mathbb{C}H^n$ which is orthogonal to both $\dot{\gamma}(0)$ and $J\dot{\gamma}(0)$. We set $\rho = \mp\big(k - \sqrt{k^2-4}\big)/2$, where the double sign takes the opposite signature of that of the complex torsion ± 1 of the circle. We consider a horosphere *HS* which makes the vector $\sqrt{1-\rho^2}\,w + \rho J\dot{\gamma}(0)$ to be the inward unit normal vector at $\gamma(0)$. Then its inward unit normal $\mathcal{N}$ satisfies $\langle \dot{\gamma}(0), -J\mathcal{N}_{\gamma(0)}\rangle = \rho$. If we take a trajectory for a Sasakian magnetic field $\mathbb{F}_{\mp k}$ on *HS* with initial vector $\dot{\gamma}(0)$, its structure torsion is hence ρ. As $\rho + \rho^{-1} = \mp k$, its extrinsic

shape $\hat{\gamma}$ is a circle of geodesic curvature k and of complex torsion ± 1 by Proposition 4.2. We hence get the conclusion. □

Remark 4.2. In Theorem 4.1, the choices of horospheres and Sasakian magnetic fields for circles of complex torsion $|\tau| = 1$ are as follows.

(1) For each unbounded circle of complex torsion $|\tau| = 1$, the choice of the horosphere is unique by the proof of (1) in Theorem 4.1. But the extrinsic circular trajectory is a geodesic on this horosphere and does not depend on the choice of Sasakian magnetic fields.
(2) For each bounded circle, we have infinitely many choice of horospheres, because $\rho \neq \pm 1$ and unit tangent vectors $w \in U_{\gamma(0)}\mathbb{C}H^n$ which are orthogonal to both $\dot{\gamma}(0)$ and $J\dot{\gamma}(0)$ form a unit sphere S^{2n-2} $\left(\subset U_{\gamma(0)}\mathbb{C}H^n\right)$ in the proof of (2) in Theorem 4.1. But once we choose a horosphere, the choice of Sasakian magnetic field is unique.

By Theorem 4.1, we can regard the boundary of the moduli space of bounded circles in the moduli space of circles on $\mathbb{C}H^n$ can be obtained as a set of congruence classes of extrinsic circular trajectories on a horosphere. Here, on congruence of trajectories for Sasakian magnetic fields as curves on a horosphere, we have the following.

Lemma 4.1 ([3]). *Let γ_1, and γ_2 are trajectories for Sasakian magnetic fields $\mathbb{F}_{\kappa_1}$ and $\mathbb{F}_{\kappa_2}$ on a horosphere HS in $\mathbb{C}H^n$, respectively. They are congruent to each other if and only if one of the following conditions holds:*

i) $|\rho_{\gamma_1}| = |\rho_{\gamma_2}| = 1$,
ii) $\rho_{\gamma_1} = \rho_{\gamma_2} = 0$ *and* $|\kappa_1| = |\kappa_2|$,
iii) $0 < |\rho_{\gamma_1}| = |\rho_{\gamma_2}| < 1$ *and* $\kappa_1\rho_{\gamma_1} = \kappa_2\rho_{\gamma_2}$.

By a similar argument we can show that every bounded circle on $\mathbb{C}H^n$ is the extrinsic shape of some trajectory for a Sasakian magnetic field on some geodesic sphere (c.f. [6]). We shall discuss this in somewhere in future.

5. Other trajectories for Sasakian magnetic fields

In order to complete our study on extrinsic shapes of trajectories for Sasakian magnetic fields, we show other cases in this section. To explain extrinsic shapes of trajectories we prepare some terminologies of helices. We call a smooth curve *Killing* if it is generated by some Killing vector field. That is, it is an orbit of a one-parameter family of isometries. On $\mathbb{C}H^n$, a helix is Killing if and only if all its complex torsions are constant (see [12]).

Circles on $\mathbb{C}H^n$ are always Killing, but not necessarily for helices of proper order greater than 2. We say a helix of proper order 2ℓ or $2\ell-1$ on $\mathbb{C}H^n$ to be *essential* if it lies on a totally geodesic $\mathbb{C}H^\ell$ in $\mathbb{C}H^n$.

A helix of proper order 3 on $\mathbb{C}H^n$ is essential and Killing if and only if their geodesic curvatures and complex torsions satisfy

$$\tau_{12} = \pm k_1\Big/\sqrt{k_1^2+k_2^2}, \quad \tau_{13}=0, \quad \tau_{23} = \pm k_2\Big/\sqrt{k_1^2+k_2^2},$$

where double signs take the same signatures (see [4, 11]). A helix of proper order 4 on $\mathbb{C}H^n$ is essential and Killing if and only if their geodesic curvatures and complex torsions satisfy $\tau_{13}=\tau_{24}=0$ and satisfy one of the following;

$$\text{i)}\ \tau_{12}=\tau_{34}=\pm\frac{k_1+k_3}{\sqrt{k_2^2+(k_1+k_3)^2}}, \qquad \tau_{23}=\tau_{14}=\pm\frac{k_2}{\sqrt{k_2^2+(k_1+k_3)^2}},$$

$$\text{ii)}\ \tau_{12}=-\tau_{34}=\pm\frac{k_1-k_3}{\sqrt{k_2^2+(k_1-k_3)^2}}, \qquad \tau_{23}=-\tau_{14}=\pm\frac{k_2}{\sqrt{k_2^2+(k_1-k_3)^2}}.$$

In each of the above, double sings take the same signatures (see [4, 11]).

We continue the calculation in §4 for the case $\rho_\gamma \neq \pm 1$, $\kappa \neq \rho_\gamma + \rho_\gamma^{-1}$ and $\kappa \neq -\rho_\gamma$. For a trajectory γ for a Sasakian magnetic field $\mathbb{F}_\kappa$ on a horosphere *HS* we have (4.1) and (4.2). We put

$$k_2 = \frac{1}{k_1}\big|(\kappa+\rho_\gamma)(1+\rho_\gamma^2-\kappa\rho_\gamma)\big|\sqrt{1-\rho_\gamma^2}, \quad Y_2 = \frac{1}{k_1}\{\kappa\phi\dot\gamma + (1+\rho_\gamma^2)\mathcal{N}\}.$$

By using (3.1) again, we have

$$\begin{aligned}\widetilde\nabla_{\dot\gamma}(\rho_\gamma\dot\gamma-\xi) &= (\kappa\rho_\gamma-1)\phi\dot\gamma + \rho_\gamma(\rho_\gamma^2-1)\mathcal{N} \\ &= -\sqrt{1-\rho_\gamma^2}\,\mathrm{sgn}\big((\kappa+\rho_\gamma)(1+\rho_\gamma^2-\kappa\rho_\gamma)\big)k_2Y_2 \\ &\qquad + \frac{2\kappa\rho_\gamma-1-\rho_\gamma^2}{k_1^2}\{(1+\rho_\gamma^2)\phi\dot\gamma - \kappa(1-\rho_\gamma^2)\mathcal{N}\}.\end{aligned}$$

Thus the extrinsic shape of γ is a helix of proper order 3 when $2\kappa\rho_\gamma = 1+\rho_\gamma^2$ holds. As we have

$$\widetilde\nabla_{\dot\gamma}\{(1+\rho_\gamma^2)\phi\dot\gamma - \kappa(1-\rho_\gamma^2)\mathcal{N}\} = -(2\kappa\rho_\gamma-1-\rho_\gamma^2)(\rho_\gamma\dot\gamma-\xi),$$

we can conclude that the extrinsic shape is a helix of proper order 4 in general case by setting

$$Y_3 = \operatorname{sgn}\big((\kappa+\rho_\gamma)(1+\rho_\gamma^2-\kappa\rho_\gamma)\big)\frac{\rho_\gamma\dot\gamma-\xi}{\sqrt{1-\rho_\gamma^2}},$$

$$Y_4 = \operatorname{sgn}\big((2\kappa\rho_\gamma-1-\rho_\gamma^2)(\kappa+\rho_\gamma)(1+\rho_\gamma^2-\kappa\rho_\gamma)\big)\frac{(1+\rho_\gamma^2)\phi\dot\gamma-\kappa(1-\rho_\gamma^2)\mathcal{N}}{k_1\sqrt{1-\rho_\gamma^2}}.$$

We hence obtain the following.

Proposition 5.1 (c.f. [3]). *Let γ be a trajectory for a Sasakian magnetic field $\mathbb{F}_\kappa$ on a horosphere HS in $\mathbb{C}H^n(-4)$. Suppose $\rho_\gamma \neq \pm 1$, $\kappa \neq \rho_\gamma + \rho_\gamma^{-1}$ and $\kappa \neq -\rho_\gamma$.*

(1) *When $\kappa = (\rho_\gamma+\rho_\gamma^{-1})/2$, its extrinsic shape is an essential Killing helix of proper order 3. Its geodesic curvatures are*

$$k_1 = (1+\rho_\gamma^2)\sqrt{1+3\rho_\gamma^2}/(2|\rho_\gamma|), \quad k_2 = \sqrt{(1+3\rho_\gamma^2)(1-\rho_\gamma^2)}/2,$$

and its complex torsions are

$$\tau_{12} = -\operatorname{sgn}(\rho_\gamma)\frac{1+\rho_\gamma^2}{\sqrt{1+3\rho_\gamma^2}} = -\operatorname{sgn}(\rho_\gamma)\frac{k_1}{\sqrt{k_1^2+k_2^2}}, \qquad \tau_{13}=0,$$

$$\tau_{23} = -\rho_\gamma\sqrt{\frac{1-\rho_\gamma^2}{1+3\rho_\gamma^2}} = -\operatorname{sgn}(\rho_\gamma)\frac{k_2}{\sqrt{k_1^2+k_2^2}}.$$

(2) *Otherwise, its extrinsic shape is an essential Killing helix of proper order 4. Its geodesic curvatures are*

$$k_1 = \sqrt{\kappa^2(1-\rho_\gamma^2)+(1+\rho_\gamma^2)^2},$$

$$k_2 = \big|(\rho_\gamma+\kappa)(1+\rho_\gamma^2-\kappa\rho_\gamma)\big|\sqrt{1-\rho_\gamma^2}\Big/k_1,$$

$$k_3 = \big|2\kappa\rho_\gamma-1-\rho_\gamma^2\big|/k_1,$$

and its complex torsions are

$$\tau_{12} = -\frac{1}{k_1}\{\kappa(1-\rho_\gamma^2)+\rho_\gamma(1+\rho_\gamma^2)\} = -\frac{\operatorname{sgn}(\kappa+\rho_\gamma)\big(k_1+\operatorname{sgn}(\varepsilon)k_3\big)}{\sqrt{k_2^2+(k_1+\operatorname{sgn}(\varepsilon)k_3)^2}},$$

$$\tau_{14} = -\operatorname{sgn}\big(\varepsilon(\kappa+\rho_\gamma)\big)\big|\kappa\rho_\gamma-1-\rho_\gamma^2\big|\sqrt{1-\rho_\gamma^2}\Big/k_1$$

$$= -\frac{\operatorname{sgn}\big(\varepsilon(\kappa+\rho_\gamma)\big)k_2}{\sqrt{k_2^2+(k_1+\operatorname{sgn}(\varepsilon)k_3)^2}},$$

$$\tau_{34} = \operatorname{sgn}(\varepsilon)\tau_{12}, \quad \tau_{23} = \operatorname{sgn}(\varepsilon)\tau_{14}, \quad \tau_{13}=\tau_{24}=0,$$

where $\varepsilon = 2\kappa\rho_\gamma - 1 - \rho_\gamma^2$.

By use of this information on extrinsic shapes of trajectories we can get their properties.

Theorem 5.1. *Let γ be a trajectory for $\mathbb{F}_\kappa$ on HS in $\mathbb{C}H^n(-4)$. If $\kappa \neq \rho_\gamma + \rho_\gamma^{-1}$, it is unbounded in both directions, hence has a single point at infinity.*

Proof. We are enough to consider the cases that the extrinsic shape is a helix of proper order 3 and 4. For an essential Killing helix of proper order 3 on $\mathbb{C}H^n(-4)$, it is unbounded in both directions and has a single point at infinity if and only if its geodesic curvatures satisfy one of the following conditions (see [4, 8]):

i) $0 < k_2 \leq 1/2$ and $(1-4k_2^2)k_1^2 = 2\{2k_2^4 - 5k_2^2 + 1 + (1-3k_2^2)^{3/2}\}$,
ii) $1/2 < k_2 \leq 1/\sqrt{3}$ and $(4k_2^2-1)k_1^2 = 2\{-2k_2^4 + 5k_2^2 - 1 \pm (1-3k_2^2)^{3/2}\}$.

We shall check these conditions in the case $\kappa = (\rho_\gamma + \rho_\gamma^{-1})/2$. As $4k_2^2 = (1+3\rho_\gamma^2)(1-\rho_\gamma^2)$, we see $k_2^2 \leq 1/3$ holds and find that $0 < k_2^2 < 1/4$ if and only if $2/3 < \rho_\gamma^2 < 1$. By direct computation we have $1-3k_2^2 = (1-3\rho_\gamma^2)^2/4$ and hence have

$$\begin{aligned}(1-4k_2^2)k_1^2 &= (1+\rho_\gamma^2)^2(1+3\rho_\gamma^2)(3\rho_\gamma^2-2)/4 \\ &= \begin{cases} 2\{2k_2^4 - 5k_2^2 + 1 + (1-3k_2^2)^{3/2}\}, & \text{if } \rho_\gamma^2 > \frac{1}{3}, \\ 2\{2k_2^4 - 5k_2^2 + 1 - (1-3k_2^2)^{3/2}\}, & \text{if } \rho_\gamma^2 \leq \frac{1}{3}. \end{cases}\end{aligned}$$

Thus, we find that the trajectory is unbounded in both directions.

For an essential Killing helix of proper order 4 on $\mathbb{C}H^n(-4)$, it is unbounded in both directions and has a single point at infinity if and only if its geodesic curvatures satisfy one of the following conditions (see [4]):

i) When $\tau_{12} = \tau_{34}$ and $\tau_{14} = \tau_{23}$,
 i-a) $k_1^2 + k_2^2 + k_3^2 - k_1k_3 = 3$,
 i-b) $k_1^2 + k_2^2 + k_3^2 - k_1k_3 > 3$ and $\tau_H^+ = \pm 1$, where

$$\tau_H^+ = \frac{2\{k_2^2+(k_1+k_3)^2\}^2 + 9(2-k_1k_3)\{k_2^2+(k_1+k_3)^2\} - 27k_1(k_1+k_3)}{2\{k_2^2+(k_1+k_3)^2 - 3(1+k_1k_3)\}^{3/2}\sqrt{k_2^2+(k_1+k_3)^2}}.$$

ii) When $\tau_{12} = -\tau_{34}$ and $\tau_{14} = -\tau_{23}$,
 ii-a) $k_1^2 + k_2^2 + k_3^2 + k_1k_3 = 3$,
 ii-b) $k_1^2 + k_2^2 + k_3^2 + k_1k_3 > 3$ and $\tau_H^- = \pm 1$, where

$$\tau_H^- = \frac{2\{k_2^2+(k_1-k_3)^2\}^2 + 9(2+k_1k_3)\{k_2^2+(k_1-k_3)^2\} - 27k_1(k_1-k_3)}{2\{k_2^2+(k_1-k_3)^2 - 3(1-k_1k_3)\}^{3/2}\sqrt{k_2^2+(k_1-k_3)^2}}.$$

We shall check these conditions in our case. As we have

$$k_1^2+k_2^2+k_3^2-\mathrm{sgn}(\varepsilon)k_1k_3-3=(\kappa-2\rho_\gamma)^2,$$

we find that γ is unbounded in both directions with a single point at infinity when $\kappa=2\rho_\gamma$. We now compute $\tau_H^{\pm}$ in the case $\kappa\neq 2\rho_\gamma$. Since we can get

$$k_1+\mathrm{sgn}(\varepsilon)k_3=\frac{1}{k_1}(\kappa+\rho_\gamma)\{\kappa(1-\rho_\gamma^2)+\rho_\gamma(1+\rho_\gamma^2)\},$$
$$k_2^2+\{k_1+\mathrm{sgn}(\varepsilon)k_3\}^2=(\kappa+\rho_\gamma)^2,$$

we obtain $|\tau_H^{\mathrm{sgn}(\varepsilon)}|=1$, hence find that γ is unbounded in both directions also in this case. □

This result corrects Theorem 10 and Proposition 13 in [3]. The author dropped a term in the equality (8.1) in [3].

Like we do in §4, we consider the converse of Theorem 5.1.

Theorem 5.2. *Every essential Killing helix of proper order* 3 *on* $\mathbb{C}H^n$ *having a single point at infinity is the extrinsic shape of a trajectory for some Sasakian magnetic field on a horosphere. For each helix, the choices of horosphere and Sasakian magnetic field are unique.*

Proof. We take a Killing helix γ of proper order 3 on $\mathbb{C}H^n(-4)$ whose geodesic curvatures are k_1, k_2 and whose Frenet Frame is $\{\dot\gamma, Y_2, Y_3\}$.

We first consider the case $0<k_2<1/2$ and the case that $1/2\le k_2<1/\sqrt{3}$ and $(4k_2^2-1)k_1^2=-2\{2k_2^4-5k_2^2+1+(1-3k_2^2)^{3/2}\}$. In view of (4.1) we set

$$\rho=-\mathrm{sgn}(\tau_{12})\Big\{\frac{1}{3}\Big(1+2\sqrt{1-3k_2^2}\Big)\Big\}^{1/2},\qquad \kappa=\frac{1}{2}(\rho+\rho^{-1}),$$

where $\tau_{12}=\langle\dot\gamma, JY_2\rangle$, and take a tangent vector

$$v=2\big(k_1Y_2(0)-\kappa J\dot\gamma(0)\big)/(1+\rho^2).$$

In the former case we have $1>\rho^2>2/3$, and in the latter case we have $2/3\ge\rho^2>1/3$. Since the geodesic curvatures satisfy $(1-4k_2^2)k_1^2=2\{2k_2^4-5k_2^2+1+(1-3k_2^2)^{3/2}\}$ in both cases, we see $k_1^2=(1+\rho^2)^2(1+3\rho^2)/(4\rho^2)$, hence find that

$$\tau_{12}=-\mathrm{sgn}(\rho)\frac{k_1}{\sqrt{k_1^2+k_2^2}}=-\mathrm{sgn}(\rho)\frac{1+\rho^2}{\sqrt{1+3\rho^2}}.$$

Thus we get

$$\|2k_1Y_2(0)-2\kappa J\dot\gamma(0)\|^2=4k_1^2+(\rho+\rho^{-1})^2+4k_1(\rho+\rho^{-1})\tau_{12}=(1+\rho^2)^2,$$

hence find $\|v\| = 1$. Therefore, if we consider the horosphere which makes v to be the inward unit normal vector at $\gamma(0)$, then its inward unit normal $\mathcal{N}$ of this horosphere satisfies

$$\langle \dot\gamma(0), -J\mathcal{N}_{\gamma(0)}\rangle = \frac{-2(k_1\tau_{12}+\kappa)}{1+\rho^2} = \rho.$$

We take a trajectory for a Sasakian magnetic field $\mathbb{F}_\kappa$ on HS with initial vector $\dot\gamma(0)$. Then, its structure torsion is ρ and its extrinsic shape $\hat\gamma$ is a Killing helix of proper order 3 whose geodesic curvatures are k_1, k_2 and which satisfies

$$\widetilde\nabla_{\dot{\hat\gamma}}\dot{\hat\gamma}(0) = \kappa\phi\dot{\hat\gamma}(0) + (1+\rho^2)\mathcal{N}_{\gamma(0)} = \kappa J\dot{\hat\gamma}(0) + \frac{1}{2}(1+\rho^2)\mathcal{N}_{\gamma(0)} = k_1 Y_2(0).$$

This shows that the complex torsion $\tau_{12}(\hat\gamma)$ of $\hat\gamma$ satisfies $\tau_{12}(\hat\gamma) = -\mathrm{sgn}(\rho)k_1/\sqrt{k_1^2+k_2^2}$. Since Killing helices of proper order 3 are determined by their geodesic curvatures, signature of complex torsion τ_{12} and their initial velocity and acceleration vectors (see [4]), we find $\hat\gamma$ coincides with γ.

Next we study the case $1/2 \le k_2 < 1/\sqrt{3}$ and $(4k_2^2-1)k_1^2 = 2\{-2k_2^4 + 5k_2^2 - 1 + (1-3k_2^2)^{3/2}\}$. In view of (4.1) we set

$$\rho = -\mathrm{sgn}(\tau_{12})\Big\{\frac{1}{3}\Big(1-2\sqrt{1-3k_2^2}\Big)\Big\}^{1/2}, \qquad \kappa = \frac{1}{2}(\rho+\rho^{-1}).$$

We then have $0 < \rho^2 < 1/3$. We take a tangent vector $v = 2\big(k_1Y_2(0) - \kappa J\dot\gamma(0)\big)/(1+\rho^2)$. The expression of k_1 by k_2 shows $k_1^2 = (1+\rho^2)^2(1+3\rho^2)/(4\rho^2)$, which leads us to $\|v\| = 1$. By taking the horosphere which makes v as the inward unit normal vector at $\gamma(0)$, we can show that the extrinsic shape of the trajectory for $\mathbb{F}_\kappa$ with initial vector $\dot\gamma(0)$ coincides with γ along the same lines as above.

As the choice of horospheres is determined by (4.1), we find it is unique. Hence, by the above argument and by Proposition 5.1, we get the conclusion. □

The author considers that every essential Killing helix of proper order 4 on $\mathbb{C}H^n(-4)$ having a single point at infinity also is the extrinsic shape of a trajectory for some Sasakian magnetic field on a horosphere. It is true when its geodesic curvatures satisfy one of the following conditions:

1) $k_1^2 + k_2^2 + k_3^2 \le 9$,
2) $k_1^2 + k_2^2 + k_3^2 > 9$ and $4(k_1^2+k_2^2+k_3^2) > 16 + k_1^2k_3^2$.

Our way to show this is to solve the following system of equations on (κ, ρ)

$$\begin{cases} \kappa^2(1-\rho^2) + (1+\rho^2)^2 = k_1^2, \\ (\kappa+\rho)^2 = k_2^2 + (k_1+\epsilon k_3)^2, \\ 2\kappa\rho - 1 - \rho^2 = \epsilon k_1 k_3, \end{cases} \tag{5.1}$$

under the assumption that $k_1^2 + k_2^2 + k_3^2 \geq 3 + \epsilon k_1 k_3$ and

$$\frac{2\{k_2^2+(k_1+\epsilon k_3)^2\}^2+9(2-\epsilon k_1 k_3)\{k_2^2+(k_1+\epsilon k_3)^2\}-27k_1(k_1+\epsilon k_3)}{2\{k_2^2+(k_1+\epsilon k_3)^2-3(1+\epsilon k_1 k_3)\}^{3/2}\sqrt{k_2^2+(k_1+\epsilon k_3)^2}} = \pm 1.$$

Here $\epsilon = \pm 1$ is defined by the relation $\tau_{12} = \epsilon\tau_{34}$. If we solve the second and the third equations in (5.1) by assuming $\kappa > 0$, we have

$$\kappa = \frac{1}{3}\Big(2\sqrt{k_2^2+(k_1^2+\epsilon k_3)^2} \pm \sqrt{k_1^2+k_2^2+k_3^2-3-\epsilon k_1 k_3}\,\Big),$$

$$\rho = \frac{1}{3}\Big(\sqrt{k_2^2+(k_1^2+\epsilon k_3)^2} \mp \sqrt{k_1^2+k_2^2+k_3^2-3-\epsilon k_1 k_3}\,\Big),$$

where double signs take the opposite signatures. The first assumption guarantees that κ, ρ are real numbers, and the second assumption shows that these κ, ρ satisfy the first equation in (5.1). As we need to show $|\rho| < 1$, we have to pose the conditions. Thus, our problem turns to check that every essential Killing helix of proper order 4 on $\mathbb{C}H^n(-4)$ having a single point at infinity satisfies those conditions.

We here make mention of the moduli space of trajectories of structure torsion $\rho \neq \pm 1$ for non-trivial Sasakian magnetic fields on a horosphere. By Lemma 4.1, it is set-theoretically identified with the open band $(0, \infty) \times (-1, 1)$ which is parameterized by strength of magnetic fields and structure torsion of trajectories. Since it is symmetric with respect to the axis of strength, we only consider the set $(0, \infty) \times [0, 1)$ and draw figures of the leaves of horocyclic circular trajectories, bounded extrinsic circular trajectories and trajectories whose extrinsic shapes are helices of proper order 3. We then get Figure 2. The author is interested in the relationship between these leaves and leaves on the sets of "congruence classes" of circular trajectories on geodesic spheres and those of all trajectories. Also he is interested in classifying trajectories on horospheres whose extrinsic shapes are helices of proper order 4 geometrically to obtain other leaves on the moduli space of trajectories.

Bibliography

1. T. Adachi, *Kähler magnetic flows on a manifold of constant holomorphic sectional curvature*, Tokyo J. Math. 18(1995), 473–483.

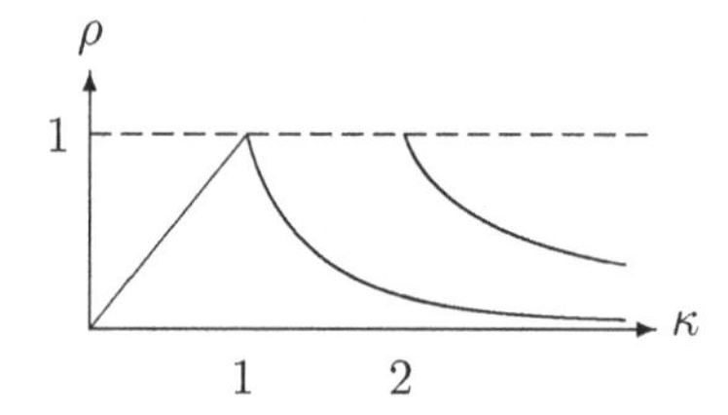

Figure 2. Leaves of extrinsic circular trajectories on *HS*

2. ______, *Lamination of the moduli space of circles and their length spectrum for a non-flat complex space form*, Osaka J. Math. 40(2003), 895–916.
3. ______, *Trajectories on geodesic spheres in a non-flat complex space form*, J. Geom. 90(2008), 1–29.
4. ______, *Essential Killing helices of order less than five on a non-flat complex space form*, J. Math. Soc. Japan 64(2012), 969–984.
5. ______, *Kähler magnetic fields on Kähler manifolds of negative curvature*, Diff. Geom. its Appl. 29(2011), S2–S8.
6. ______, *Extrinsic circular trajectories on geodesic spheres in a complex projective space*, in preparation.
7. T. Adachi & S. Maeda, *Global behaviors of circles in a complex hyperbolic space*, Tsukuba J. Math. 21(1997), 29–42.
8. ______, *Holomorphic helix of proper order 3 on a complex hyperbolic plane*, Topol. and its Appl. 146–147(2005), 201–207.
9. T. Bao & T. Adachi, *Circular trajectories on real hypersurfaces in a nonflat complex space form*, J. Geom. 96(2009), 41–55.
10. B.Y. Chen & S. Maeda, *Extrinsic characterizations of circles in a complex projective space imbedded in a Euclidean space*, Tokyo J. Math. 19(1986), 169–185.
11. S. Maeda & T. Adachi, *Sasakian curves on hypersurfaces of type (A) in a nonflat complex space form*, Results Math. 56(2009), 489–499.
12. S. Maeda & Y. Ohnita, *Helical geodesic immersion into complex space form*, Geom. Dedicata, 30(1989), 93–114.
13. K. Nomizu & K. Yano, *On circles and spheres in Riemannian geometry*, Math. Ann. 210(1974), 163–170.
14. J.S. Pak & K. Sakamoto, *Submanifolds with d-planar geodesic immersed in complex projective spaces*, Tôhoku Math. J. 38(1986), 297–311.
15. K. Suizu, S. Maeda & T. Adachi, *A characterization of the Veronese imbedding into a complex projective space*, C. R. Math. Rep. Acad. Sci. Canada 24(2002), 61–66.

Received December 24, 2012
Revised April 10, 2013

Proceedings of the 3rd International
Colloquium on Differential Geometry
and its Related Fields
Veliko Tarnovo, September 3–7, 2012

VOLUME DENSITIES OF TRAJECTORY-BALLS AND TRAJECTORY-SPHERES FOR KÄHLER MAGNETIC FIELDS

Pengfei BAI *

Division of Mathematics and Mathematical Science,
Nagoya Institute of Technology,
Nagoya 466-8555, Japan
E-mail: baipf2000@yahoo.co.jp

In this paper we study balls and spheres formed by trajectories for Kähler magnetic fields on a Kähler manifold and give estimates of their volume elements under some assumptions on sectional curvatures.

Keywords: Kähler magnetic fields; Trajectories; Magnetic exponential maps; Trajectory-balls; Trajectory-spheres; Volume density functions; Magnetic Jacobi fields; Comparison theorem.

1. Introduction

When we study a Riemannian manifold with a geometric structure, the author considers that investigating curves associated with this structure is one of natural ways. In our recent papers, we intend to study Kähler manifolds from the view point of real Riemannian geometry.

Let M be a complete Kähler manifold with complex structure J and Riemannian metric $\langle\ ,\ \rangle$. On this manifold we have a natural closed 2-form $\mathbb{B}_J$, which is called the Kähler form. A constant multiple $\mathbb{B}_k = k\mathbb{B}_J$ is said to be a *Kähler magnetic field*. As a generalization of geodesics we call a smooth curve γ a *trajectory* for $\mathbb{B}_k$ if it is parameterized by its arclength and satisfies the equation $\nabla_{\dot\gamma}\dot\gamma = kJ\dot\gamma$, where ∇ is the Riemannian connection. In their papers Comtet [7] and Sunada [8] studied them on Riemann surfaces of constant sectional curvature. There are also some results due to Adachi which leads us to study them on general Kähler manifolds (see [1, 2] and papers in their references). In this paper we take trajectory-balls and trajectory-spheres. It is needless to say that geodesic balls and

*The author is partially supported by Rotary Yoneyama Memorial Foundation.

geodesic spheres are basic objects in the study of Riemannian manifolds. Trajectory-balls can be considered as deformed objects under the influence of a magnetic field. We give a report on their volume elements based on our recent articles [4, 5].

The author is grateful to Professor T. Adachi for his advice in preparing this article.

2. Trajectory-balls and trajectory-spheres

We shall start by giving some basic definitions. We take a Kähler magnetic field $\mathbb{B}_k$ on a Kähler manifold M. For a unit tangent vector $u \in UM$, we denote by γ_u a trajectory for $\mathbb{B}_k$ with initial vector $\dot\gamma_u(0) = u$. At an arbitrary point $p \in M$ we define a *magnetic exponential map* $\mathbb{B}_k\exp_p : T_pM \to M$ by

$$\mathbb{B}_k \exp_p(w) = \begin{cases} \gamma_{w/\|w\|}\big(\|w\|\big), & \text{if } w \neq 0_p, \\ p, & \text{if } w = 0_p. \end{cases}$$

When $\kappa = 0$, as the magnetic field $\mathbb{B}_0 = 0$ is the trivial 2-form, trajectories for $\mathbb{B}_0$ are geodesics. Therefore the map $\mathbb{B}_0 \exp_p$ is the ordinary exponential map on M. Making use of magnetic exponential maps, we define a *magnetic trajectory-ball* $B_r^\kappa(p)$ and a *magnetic trajectory-sphere* $S_r^k(p)$ for $\mathbb{B}_k$ of arc-radius r centered at p by

$$B_r^k(p) = \{\mathbb{B}_k\exp_p(tv) \mid 0 \le t < r,\ v \in U_pM\},$$
$$S_r^k(p) = \{\mathbb{B}_\kappa\exp_p(ru) \mid u \in U_pM\},$$

where U_pM denotes the unit tangent space at p. Clearly, for sufficiently small r, a trajectory sphere $S_r^k(p)$ is the boundary of the trajectory-ball $B_r^k(p)$.

In order to study the influence of magnetic fields we are interested in the difference between trajectory-balls and geodesic balls or between trajectory-spheres and geodesic spheres. We first study them on a complex space form, which is a simply connected Kähler manifold of constant holomorphic sectional curvature, hence is one of a complex projective space, a complex Euclidean space and a complex hyperbolic space. On a complex Euclidean space $\mathbb{C}^n$, if we take a trajectory γ for $\mathbb{B}_k$, the distance between $\gamma(0)$ and $\gamma(r)$ is $\ell_k(r;0) = (2/|k|)\sin\big(|k|r/2\big)$. On a complex projective space $\mathbb{C}P^n(c)$ of constant holomorphic sectional curvature c (> 0), it is known that the distance $\ell_k(r;c)$ between $\gamma(0)$ and $\gamma(r)$ on a trajectory γ for $\mathbb{B}_k$ is given by the relation

$$\sqrt{k^2+c}\,\sin\big(\sqrt{c}\,\ell_k(r;c)/2\big) = \sqrt{c}\,\sin\big(\sqrt{k^2+c}\,r/2\big)$$

when $0 < r < 2\pi/\sqrt{k^2+c}$ (see [1]). On a complex hyperbolic space $\mathbb{C}H^n(c)$ of constant holomorphic sectional curvature ($c < 0$), it is also known that the distance $\ell_k(r;c)$ between $\gamma(0)$ and $\gamma(r)$ is given by the following relation:

$$\begin{cases} \sqrt{|c|-k^2}\sinh\frac{1}{2}\sqrt{|c|}\,\ell_k(r;c) = \sqrt{|c|}\sinh\frac{1}{2}\sqrt{|c|-k^2}\,r, & \text{if } |k| < \sqrt{|c|}, \\ 2\sinh\frac{1}{2}\sqrt{|c|}\ell_k(r;c) = \sqrt{|c|}\,r, & \text{if } k = \pm\sqrt{|c|}, \\ \sqrt{k^2+c}\sinh\frac{1}{2}\sqrt{|c|}\ell_k(r;c) = \sqrt{|c|}\sin\frac{1}{2}\sqrt{k^2+c}\,r, & \text{if } |k| > \sqrt{|c|}. \end{cases}$$

Here, in the case $|k| > \sqrt{|c|}$ we only consider r with $0 < r < 2\pi/\sqrt{k^2+c}$. Through out of this paper we shall regard $a/\sqrt{k^2+c}$ with a constant a to be infinity when $k^2+c \leq 0$. As a consequence of the above relationships between distances and arc-length of trajectories, we have the following:

Proposition 2.1 ([3]). *On a complex space form $\mathbb{C}M^n(c)$ of constant holomorphic sectional curvature c we have the following:*

(1) *Every trajectory-ball coincides with some geodesic ball:*
$B_r^k(p) = B^0_{\ell_k(r;c)}(p)$ *for* $0 < r < 2\pi/\sqrt{k^2+c}$.
(2) *Every trajectory-sphere coincides with some geodesic sphere:*
$S_r^k(p) = S^0_{\ell_k(r;c)}(p)$ *for* $0 < r < 2\pi/\sqrt{k^2+c}$.

3. Volume elements of trajectory-balls

We now study trajectory-balls and trajectory-spheres on a general Kähler manifold. We first study trajectory-balls. In order to study trajectory-balls, we here consider their volume elements. By estimating their volumes we would like to show the difference from geodesic balls. We define a map $\Phi_p^k : (0,r) \times U_pM \to M$ by $\Phi_p^k(t,u) = \mathbb{B}_k\exp_p(tu)$. We consider the pull back of the volume element vol_M of M by this map and define a function $\theta_k(t,u)$ by $(\Phi_p^k)^* vol_M = \theta_k(t,u)\, dt\, d\omega$, where $d\omega$ is the volume element of $S^{2n-1} = U_pM$. We shall call $\theta_k(t,u)$ the *volume density function* of a trajectory ball $B_r^k(p)$.

In order to compute the volume density function, we need to study the differential of $\mathbb{B}_k\exp_p$. A vector field Y along a trajectory γ for $\mathbb{B}_k$ is said to be a *normal magnetic Jacobi field* if it satisfies

$$\begin{cases} \nabla_{\dot\gamma}\nabla_{\dot\gamma}Y - \kappa J\nabla_{\dot\gamma}Y + R(Y,\dot\gamma)\dot\gamma = 0, \\ \langle\nabla_{\dot\gamma}Y,\dot\gamma\rangle = 0. \end{cases} \tag{3.1}$$

For a vector field X along a trajectory γ for $\mathbb{B}_k$ we decompose it into three components and denote as $X = f_X\dot\gamma + g_XJ\dot\gamma + X^\perp$ with functions f_X, g_X and a vector field $X^\perp$ which is orthogonal to both $\dot\gamma$ and $J\dot\gamma$ at each point.

We set $X^\sharp = g_X J\dot\gamma + X^\perp$ and $X^\top = f_X\dot\gamma + g_X J\dot\gamma$. A vector field Y along γ is a magnetic Jacobi field if and only if it satisfies

$$\begin{cases} f_Y' = kg_Y, \\ (g_Y'' + k^2 g_Y)J\dot\gamma + \nabla_{\dot\gamma}\nabla_{\dot\gamma}Y^\perp - kJ(\nabla\dot\gamma Y^\perp) + R(Y^\sharp, \dot\gamma)\dot\gamma = 0. \end{cases} \tag{3.2}$$

We here consider singularities of the map $\Phi_p^k : (0, r) \times U_pM \to M$. We take a trajectory γ for $\mathbb{B}_k$ with $\gamma(0) = p$. We call t_0 a *magnetic conjugate value* of p along γ if there is a non-trivial normal magnetic Jacobi field Y along γ which satisfies $Y^\sharp(0) = 0$ and $Y^\sharp(t_0) = 0$. The point $\gamma(t_0)$ is said to be a magnetic conjugate point of p along γ. We denote by $c_\gamma(p)$ the minimum magnetic conjugate value of p along γ. We set $c_\gamma(p) = \infty$ if there are no magnetic conjugate points of p along γ. This value $c_\gamma(p)$ plays the similar role as of first conjugate values along geodesics. It is clear that the differential $d\big(\mathbb{B}_k \exp_p(t_0 u)\big) : T_{t_0 u}(T_pM) \to T_{\gamma_u(t_0)}M$ of a magnetic exponential map is singular if and only if t_0 is a magnetic conjugate value of p along γ_u.

Lemma 3.1. *Given a unit tangent vector $u \in T_pM$ at an arbitrary point $p \in M$ we take normal magnetic Jacobi fields Y_j $(j = 2, 3, \ldots, 2n)$ along a trajectory γ_u for $\mathbb{B}_k$ which satisfy the following conditions*:

i) $Y_j(0) = 0,\ j = 2, \ldots, 2n,$
ii) *the vectors $\{u, \nabla_{\dot\gamma_u} Y_2(0), \ldots, \nabla_{\dot\gamma_u} Y_{2n}(0)\}$ form an orthonormal basis of T_pM.*

We then find that the volume density function of a trajectory-ball is expressed as

$$\theta_k(t, u) = \begin{vmatrix} \langle Y_2^\sharp, Y_2^\sharp\rangle & \langle Y_2^\sharp, Y_3^\sharp\rangle & \cdots & \langle Y_2^\sharp, Y_{2n}^\sharp\rangle \\ \langle Y_3^\sharp, Y_2^\sharp\rangle & \langle Y_3^\sharp, Y_3^\sharp\rangle & \cdots & \langle Y_3^\sharp, Y_{2n}^\sharp\rangle \\ \vdots & \vdots & & \vdots \\ \langle Y_{2n}^\sharp, Y_2^\sharp\rangle & \langle Y_{2n}^\sharp, Y_3^\sharp\rangle & \cdots & \langle Y_{2n}^\sharp, Y_{2n}^\sharp\rangle \end{vmatrix}^{1/2}. \tag{3.3}$$

We here give expressions of volume density functions of trajectory-balls on a complex space form. Since we know the expressions of volume density functions of geodesic balls, we have the following by Proposition 2.1:

Proposition 3.1 ([4]). *On a complex space form $\mathbb{C}M^n(c)$, the volume* *density function* $\theta_n(t;c,n)$ *of a trajectory-ball of arc-radius* t *is given as*

$$\theta_k(t;c,n) = \begin{cases} \left(\dfrac{2}{\sqrt{|c|-k^2}}\sinh\dfrac{1}{2}\sqrt{|c|-k^2}\,t\right)^{2n-1}\cosh\dfrac{1}{2}\sqrt{|c|-k^2}\,t, & \textit{if } |k|<\sqrt{|c|}, \\ t^{2n-1}, & \textit{if } k=\pm\sqrt{|c|}, \\ \left(\dfrac{2}{\sqrt{k^2+c}}\sin\dfrac{1}{2}\sqrt{k^2+ct}\right)^{2n-1}\cos\dfrac{1}{2}\sqrt{k^2+c}t, & \textit{if } |k|>\sqrt{|c|}, \end{cases}$$

for $0<t<\pi/\sqrt{k^2+c}$.

We can also get the above expressions directly by use of (3.3). Solving the equations (3.2), we find that a normal magnetic Jacobi field Y along a trajectory γ for $\mathbb{B}_k$ which satisfies $Y(0)=0$ is of the following form:

$$\begin{cases} f_Y(t)=ak\big(\cosh\sqrt{|c|-k^2}\,t-1\big), \\ g_Y(t)=a\sqrt{|c|-k^2}\sinh\sqrt{|c|-k^2}\,t, & \text{if } |k|<\sqrt{|c|}, \\ Y^{\perp}(t)=e^{\sqrt{-1}kt/2}\sinh\dfrac{1}{2}\sqrt{|c|-k^2}\,t\,E(t), \end{cases}$$

$$\begin{cases} f_Y(t)=a|c|t^2/2, \\ g_Y(t)=akt, & \text{if } k=\pm\sqrt{|c|}, \\ Y^{\perp}(t)=te^{\sqrt{-1}kt}\,E(t), \end{cases}$$

$$\begin{cases} f_Y(t)=a\big(1-\cos\sqrt{k^2+ct}\big), \\ g_Y(t)=a\sqrt{k^2+c}\sin\sqrt{k^2+ct}, & \text{if } |k|>\sqrt{|c|}. \\ Y^{\perp}(t)=e^{\sqrt{-1}kt/2}\sin\dfrac{1}{2}\sqrt{k^2+ct}\,E(t), \end{cases}$$

Here, a is a constant and E is a parallel vector field along γ whose initial $E(0)$ is orthogonal to $\dot\gamma(0), J\dot\gamma(0)$. By taking $Y_2=f_Y\dot\gamma+g_YJ\dot\gamma$ with $g_Y'(0)=1$ and by taking $Y_j=Y_j^{\perp}$ $(j=3,\dots,2n)$ so that $\{E_3(0),\dots,E_{2n}(0)\}$ is orthonormal, we can get Proposition 3.1.

4. Estimates of volume density functions of trajectory-balls

On complex space forms trajectory-balls coincide with some geodesic balls. Since complex space forms are model spaces of Kähler manifolds, it is natural to consider that the difference between trajectory-balls and geodesic balls should show some "difference" of a base Kähler manifold from model spaces. For the sake of getting information on trajectory-balls, we estimate their volume density functions on general Kähler manifolds by functions closely related with density functions on complex space forms. Our results correspond to Bishop's comparison theorem on volumes of geodesic balls (c.f. [6]).

We define a function $\mathfrak{s}_k(t;c)$ by

$$\mathfrak{s}_k(t;c) = \begin{cases} (2/\sqrt{k^2+c}\,)\sin(\sqrt{k^2+c}\,t/2), & \text{if } k^2+c>0,\\ t, & \text{if } k^2+c=0,\\ (2/\sqrt{|c|-k^2}\,)\sinh(\sqrt{|c|-k^2}\,t/2), & \text{if } k^2+c<0. \end{cases}$$

Here, we regard $\pi/\sqrt{k^2+c}$ as infinity when $k^2+c \le 0$. We set $\mathfrak{c}_k(t;c)$ to be the differential of $\mathfrak{s}_k(t;c)$. That is, the function $\mathfrak{c}_k(t;c)$ is given as

$$\mathfrak{c}_k(t;c) = \begin{cases} \cos(\sqrt{k^2+c}\,t/2), & \text{if } k^2+c>0,\\ 1, & \text{if } k^2+c=0,\\ \cosh(\sqrt{|c|-k^2}\,t/2), & \text{if } k^2+c<0. \end{cases}$$

Volume density functions of trajectory-balls on complex space forms are then expressed as $\theta_k(t;c,n) = \big(\mathfrak{s}_k(t;c)\big)^{2n-1}\mathfrak{c}_k(t;c)$ by Proposition 3.1.

We prepare a function of estimation. We define $\alpha_k(t;c,n)$ by $\alpha_k(t;c,n) = \mathfrak{s}_{2k}(t;4c)\ \big(\mathfrak{s}_k(t;4c)\big)^{2(n-1)}$. Since the function $\mathfrak{s}_k(t;c)$ satisfies $\mathfrak{s}_k(t;c_1) > \mathfrak{s}_k(t;c_2)$ if $c_1 < c_2$, the function of estimation satisfies the following:

$$\begin{cases} \alpha_k(t;c/4,n) > \theta_k(t;c,n) > \alpha_k(t;c,n) > \theta_k(t;4c,n), & \text{if } c>0\ ,\\ \alpha_k(t;0,n) = \theta_k(t;0,n), & \text{if } c=0\ ,\\ \alpha_k(t;c/4,n) < \theta_k(t;c,n) < \alpha_k(t;c,n) < \theta_k(t;4c,n), & \text{if } c<0. \end{cases}$$

By use of this function, we can estimate volume density functions of trajectory-balls under an assumption on sectional curvatures from below and an assumption from above.

Theorem 4.1 ([4]). *Let M be a Kähler manifold of complex dimension n and k be a nonzero constant. We take an arbitrary $u \in UM$. If sectional curvatures along a trajectory γ_u for $\mathbb{B}_k$ satisfy* $\max\{Riem(v,\dot\gamma_u(t)) \mid v \in U_{\gamma_u(t)}M, v \perp \dot\gamma_u(t)\} \le c$ *with some constant c for $0 \le t < \pi/\sqrt{k^2+c}$, then we have the following properties:*

(1) *The function $t \mapsto \theta_k(t,u)/\alpha_k(t;c,n)$ is strictly monotone increasing for $0 \le t \le \pi/\sqrt{k^2+c}$;*
(2) *$\theta_k(t,u) \ge \alpha_k(t;c,n)$ for $0 < t \le \pi/\sqrt{k^2+c}$.*

Theorem 4.2 ([4]). *Let M be a Kähler manifold of complex dimension n and k be a nonzero constant. If sectional curvatures along a trajectory γ_u for $\mathbb{B}_k$ satisfy* $\min\{Riem(v,\dot\gamma(t)) \mid v \in U_{\gamma_u(t)}M, v \perp \dot\gamma_u(t)\} \geq c$ *with some constant c for $0 \leq t < \mathfrak{c}_{\gamma_u}(\gamma(0))$ $(\leq \pi/\sqrt{k^2+c})$, then we have the following properties*:

(1) *The function $t \mapsto \theta_k(t,u)/\alpha_k(t;c,n)$ is strictly monotone decreasing for $0 \leq t \leq \mathfrak{c}_{\gamma_u}(\gamma(0))$;*
(2) *$\theta_k(t,u) \leq \alpha_k(t;c,n)$ for $0 < t \leq \mathfrak{c}_{\gamma_u}(\gamma(0))$.*

In order to show these theorems we first compute the differential of the volume density function. We suppose there are no magnetic conjugate points of $\gamma_u(0)$ on the interval $[0,r]$ along a trajectory γ_u for $\mathbb{B}_k$. We take normal magnetic Jacobi fields $W_2,\ldots,W_{2n}$ along γ_u for $\mathbb{B}_k$ so that they satisfy

i) $W_j(0)=0$,
ii) $\{W_2^\sharp(r),\ldots,W_{2n}^\sharp(r)\}$ is an orthonormal basis of $\left(T_{\gamma_u(r)}M\right)^\sharp$,
iii) $\left(\nabla_{\dot\gamma_u}W_2\right)(0)$ is parallel to Ju.

By direct computation one can find with the aid of Lemma 3.1 that the logarithmic derivative of the volume density function is given by derivatives of norms of magnetic Jacobi fields:

$$\frac{1}{\theta_k(r,u)}\frac{\partial}{\partial t}\theta_k(t,u)\bigg|_{t=r} = \sum_{j=2}^{2n}\left\langle W_j^\sharp(r), \left(\nabla_{\dot\gamma_u}W_j^\sharp\right)(r)\right\rangle. \tag{4.1}$$

We hence need results on derivatives of norms of magnetic Jacobi fields. We define a function $\mathfrak{t}_k(t;c)$ by

$$\mathfrak{t}_k(t;c) = \frac{\mathfrak{c}_k(t;c)}{\mathfrak{s}_k(t;c)} = \begin{cases} \left(\sqrt{k^2+c}/2\right)\cot\left(\sqrt{k^2+c}\,t/2\right), & \text{if } k^2+c>0\,, \\ 1/t, & \text{if } k^2+c=0, \\ \left(\sqrt{|c|-k^2}/2\right)\coth\left(\sqrt{|c|-k^2}\,t/2\right), & \text{if } k^2+c<0. \end{cases}$$

A comparison theorem on magnetic Jacobi fields which corresponds to Rauch's comparison theorem is as follows.

Proposition 4.1 ([2]). *Suppose sectional curvatures satisfy*
$\max\{\mathrm{Riem}(v,\dot\gamma_u(t)) \mid v \in U_{\dot\gamma(t)}M, v\perp\dot\gamma(t)\} \leq c$
with some constant c for $(0 \leq t < \pi/\sqrt{k^2+c})$ along a trajectory γ_u for $\mathbb{B}_k$. We then have the following:

(1) $\mathfrak{c}_\gamma(\gamma(0)) \geq \pi/\sqrt{k^2+c}$.

(2) *If Y is a normal magnetic Jacobi field along γ with $Y(0)=0$, then for* $0 \le t < \pi/\sqrt{k^2+c}$ *we have*

$$\langle \nabla_{\dot\gamma} Y^\sharp(t), Y^\sharp(t)\rangle \ge g_Y(t)^2 \mathfrak{t}_{2k}(t;4c) + \| Y^\perp(t) \|^2 \mathfrak{t}_k(t;4c)$$

for $0 < t < \pi/\sqrt{k^2+c}$. *If the equality holds at some* t_0 *with* $0 < t_0 < \pi/\sqrt{k^2+c}$, *then the equality holds at every* t *with* $0 \le t \le t_0$. *In this case, the magnetic Jacobi field* Y *is of the form*

$$\begin{aligned} Y^\sharp(t) &= g_Y'(0)\mathfrak{s}_{2k}(t;4c)J\dot\gamma(t) \\ &\quad + \|\nabla_{\dot\gamma} Y^\perp(0)\|\, \mathfrak{s}_k(t;4c)\big\{\cos(kt/2)E(t) + \sin(kt/2)JE(t)\big\} \end{aligned}$$

with a parallel vector field E *satisfying* $E(0) = \nabla_{\dot\gamma}Y^\perp(0) \,/\, \|\nabla_{\dot\gamma}Y^\perp(0)\|$, *and the curvature tensor satisfies* $R(Y^\sharp, \dot\gamma)\dot\gamma = cY^\sharp$ *for* $0 \le t \le t_0$. *Here, in the case* $\nabla_{\dot\gamma}Y^\perp(0) = 0$, *we take* E *as a null vector field.*

Being different from ordinary Jacobi fields, for magnetic Jacobi fields components parallel to the velocity vector of the trajectory and components orthogonal to it are interacted each other. We therefore need to consider only components orthogonal to the velocity vector.

We have the similar result which gives an estimate from above under an assumption on sectional curvatures from below.

By use of (4.1) and Proposition 4.1 we have

$$\frac{\partial}{\partial t} \log \theta_k(t,u)\Big|_{t=r} \ge \sum_{j=2}^{2n} g_j(r)^2\, \mathfrak{t}_{2k}(r;4c) + \sum_{j=2}^{2n} \|W_j^\perp(r)\|^2\, \mathfrak{t}_k(r;4c).$$

As $\{W_2^\sharp(r), \ldots, W_{2n}^\sharp(r)\}$ is an orthonormal basis of $\big(T_{\gamma(r)}M\big)^\sharp$, we see

$$\sum_{j=2}^{2n} g_j(r)^2 = \sum_{j=2}^{2n} \langle W_j^\sharp(r), J\dot\gamma(r)\rangle = \|J\dot\gamma(r)\| = 1.$$

We therefore obtain

$$\frac{\partial}{\partial t} \log \theta_k(t,u)\Big|_{t=r} \ge \mathfrak{t}_{2k}(r;4c) + 2(n-1)\mathfrak{t}_k(r;4c) = \frac{\partial}{\partial t} \log \alpha_k(t;c,n)\Big|_{t=r},$$

and find that the function $\theta_k(t,u)/\alpha_k(t;c,n)$ is monotone increasing. Since we have $\lim_{t\downarrow 0} \theta_k(t,u)/\alpha_k(t;c,n) = 1$, we get the second assertion of Theorem 4.1. If we use an estimate of differentials of norms of magnetic Jacobi fields from above which corresponds to Proposition 4.1, we can obtain Theorem 4.2 by the same argument (see [4] for more detail).

When we study volumes of geodesic balls, we have an important result which is called Bishop's comparison theorem. It gives an estimate from above under an assumption on Ricci curvatures bounded from below. We here give a result on volume density functions of trajectory-balls which corresponds to Bishop's comparison theorem. We set $\beta(t,c;n) = \big(\mathfrak{s}_0(t;4c)\big)^{n-1}$, which is the volume density of a geodesic ball of radius t in a real space form $\mathbb{R}M^n(c)$ of constant sectional curvature c, that is either a standard sphere $S^n(c)$, a Euclidean space $\mathbb{R}^n$ or a real hyperbolic space $H^n(c)$ according as c is positive, zero or negative.

Theorem 4.3 ([4]). *Let M be a Kähler manifold of complex dimension n. We take an arbitrary $u \in UM$ and consider a trajectory γ_u for $\mathbb{B}_k$ with $\dot\gamma_u(0) = u$. If Ricci curvatures satisfy* $\mathrm{Ric}\big(\dot\gamma(t)\big) \geq (2n-1)c$ *for some constant c for $0 \leq t < c_\gamma(\gamma(0))$, then we have the following properties.*

(1) $c_{\gamma_u}(\gamma_u(0)) \leq \pi/\sqrt{c + \frac{(n+1)k^2}{2(2n-1)}}$.
(2) *The function $t \mapsto \theta_k(t,u)/\beta\big(t; c + \frac{(n+1)k^2}{2(2n-1)}, 2n\big)$ is monotone decreasing for $0 \leq t \leq c_{\gamma_u}(\gamma_u(0))$.*
(3) $\theta_k(t,u) \leq \beta\big(t; c + \frac{(n+1)k^2}{2(2n-1)}, 2n\big)$ *for $0 \leq t \leq c_{\gamma_u}(\gamma_u(0))$.*

In the second assertion of the above theorem the function may have stationary points and in the third assertion equality may hold at some t_0 with $0 < t_0 \leq c_{\gamma_u}(\gamma_u(0))$. We can say a condition when such things occur. For example, if such a thing occurs at $t_0$with $0 < t_0 \leq c_{\gamma_u}(\gamma_u(0))$, then the curvature tensor satisfies

$$R(J\dot\gamma_u,\dot\gamma_u)\dot\gamma_u = \Big\{c - \frac{3(n-1)}{2(2n-1)}k^2\Big\}J\dot\gamma_u,\quad R(E_j,\dot\gamma_u)\dot\gamma_u = \Big\{c + \frac{3}{4(2n-1)}\kappa^2\Big\}E_j$$

with orthonormal parallel vector fields $E_3,\ldots,E_{2n}$ along γ_u satisfying that $E_j \perp \dot\gamma_u$, $E_j \perp J\dot\gamma_u$ and $E_{2m} = JE_{2m-1}$ for $0 < t \leq t_0$.

5. Volume densities of trajectory-spheres

Next we study trajectory-spheres by considering their volume elements. For a geodesic sphere $S_r^0(p)$, its volume form is given as $\theta_0(r;u)d\omega$ through the exponential map at p because geodesics emanating from p cross $S_r^0(p)$ orthogonally. But for trajectory-spheres, the situation is not the same. Trajectories emanating from the center of a trajectory-sphere do not cross it orthogonally. We therefore have to consider volume elements of trajectory-spheres in another way. We define a map $\Phi_{p,r}^k : U_pM \to M$ by $\Phi_{p,r}^k(u) = \mathbb{B}_k \exp_p(ru)$. Its image is the trajectory-sphere of radius r centered at p. For a trajectory γ for $\mathbb{B}_k$ with $\gamma(0) = p$, we call t_0 a *magnetic spherical conjugate value* of p along γ if there is a non-trivial normal magnetic Jacobi field Y along γ which satisfies $Y(0) = 0$ and $Y(t_0) = 0$. The point $\gamma(t_0)$ is said to be a magnetic spherical conjugate point of p along γ. We denote by $c^s_\gamma(p)$ the minimum magnetic spherical conjugate value of p along γ. It is clear that $c^s_\gamma(p) \le c_\gamma(p)$. We set $c^s_\gamma(p) = \infty$ if there are no magnetic spherical conjugate points of p along γ. The differential $\left(d\Phi^{\,k}_{p,r}\right)_u : T_u(T_pM) \to T_{\gamma_u(r)}M$ of a magnetic exponential map is singular if and only if r is a magnetic spherical conjugate value of p along γ_u.

When $0 < r < c^s_{\gamma_u}(\gamma_u(0))$, the trajectory-sphere $S_r^k(p)$ is locally an embedded real hypersurface near $\Phi^{\,k}_{p,r}(u)$. Therefore, we can define the volume element of this real hypersurface near this point. By pulling back this volume element by $\Phi^{\,k}_{p,r}$ we denote it by $\sigma_k^r(u)d\omega$ by using the volume element $d\omega$ of $S^{2n-1} = U_pM$, and get a density function $\sigma_k^r(u)$ which is defined near u. In view of volume forms of trajectory-balls, when $0 < r < c^s_\gamma(\gamma(0))$ we find that density function $\sigma_k^r(u)$ of $S_r^k(p)$ is given as

$$\sigma_k^r(u) = \begin{vmatrix} \langle Y_2(r), Y_2(r)\rangle & \cdots & \langle Y_2(r), Y_{2n}(r)\rangle \\ \vdots & & \vdots \\ \langle Y_{2n}(r), Y_2(r)\rangle & \cdots & \langle Y_{2n}(r), Y_{2n}(r)\rangle \end{vmatrix}^{1/2} \tag{5.1}$$

by use of normal magnetic Jacobi fields $Y_2, \dots, Y_{2n}$ along γ_u for $\mathbb{B}_k$ which satisfy $Y_j(0) = 0$ and that $\{u, (\nabla_{\dot\gamma_u} Y_2)(0), \dots, (\nabla_{\dot\gamma_u} Y_{2n})(0)\}$ is an orthonormal basis of T_pM. We shall call $\sigma_k^r(u)$ the *volume density function* of $S_r^k(p)$.

We first consider volume density functions of trajectory-spheres in a complex space form $\mathbb{C}M^n(c)$. Since a trajectory sphere $S_r^k(p)$ coincides with the geodesic sphere $S^0_{\ell_k(r;c)}(p)$ when $0 < r < 2\pi/\sqrt{k^2+c}$ by Proposition 2.1, we find that its volume density function $\sigma_k^r(c,n)$, which is a constant function, satisfies $\sigma_k^r(c,n) = \theta_k(r;c,n) = \theta_0(\ell_k(r;c);c,n)$. We can also get this expression by use of the expression of magnetic Jacobi fields on $\mathbb{C}M^n(c)$ in §3 and (5.1). Hence we may say that our definition of volume density functions of trajectory-spheres is consistent. If we set

$$\begin{aligned} \mu_k(t;c) &= \{(k^2/4)\mathfrak{s}_k(t;c)^2 + \mathfrak{c}_k(t;c)^2\}^{1/2} \\ &= \begin{cases} \sqrt{\dfrac{|c|\cosh^2\left(\sqrt{|c|-k^2}\,t/2\right) - k^2}{|c|-k^2}}, & \text{if } k^2+c<0, \\ \sqrt{|c|t^2+4}/2, & \text{if } k^2+c=0, \\ \sqrt{\dfrac{k^2+c\cos^2\left(\sqrt{k^2+c}\,t/2\right)}{k^2+c}}, & \text{if } k^2+c>0, \end{cases} \end{aligned}$$

we have the following:

Proposition 5.1 ([5]). *In a complex space form $\mathbb{C}M^n(c)$, the volume density function of a trajectory-sphere of arc-radius r is given as $\sigma_k^r(c,n) = \mu_k(r;c)\big(\mathfrak{s}_k(r;c)\big)^{2n-1}$.*

We next consider to estimate volume density functions of trajectory-spheres on a general Kähler manifold. We prepare functions of estimation. We set $\delta_k(t;c,n)$ and $\epsilon_k(t;c,n)$ by

$$\delta_k(t;c,n) = \mathfrak{s}_k(t;c)\,\mu_k(t;c)\,\big\{\min\{\mathfrak{s}_k(t;c)\,\mu_k(t;c), \mathfrak{s}_k(t;4c)\}\big\}^{2(n-1)},$$

$$\epsilon_k(r;c,n) = \mathfrak{s}_k(r;c)\,\mu_k(r;c)\,\big\{\max\{\mathfrak{s}_k(r;c)\,\mu_k(r;c), \mathfrak{s}_k(r;4c)\}\big\}^{2(n-1)}.$$

Then they satisfy $\delta_k(t;c,n) > \mathfrak{s}_k(t;c)\,\mu_k(t;c)\,\big(\mathfrak{s}_{2k}(t;4c)\big)^{2(n-2)}$,

$$\begin{cases} \sigma_k^r(c,n) > \delta_k(r;c,n), & \text{when } c > 0,\\ \sigma_k^r(0,n) = \delta_k(r;0,n), & \text{when } c = 0,\\ \sigma_k^r(4c,n) > \delta_k(r;c,n) > \sigma_k^r(c,n), & \text{when } c < 0, \end{cases}$$

and

$$\begin{cases} \sigma_k^r(c,n) > \epsilon_k(r;c,n) > \sigma_k^r(4c,n), & \text{when } c > 0,\\ \sigma_k^r(0,n) = \epsilon_k(r;0,n), & \text{when } c = 0,\\ \epsilon_k(r;c,n) > \sigma_k^r(c,n), & \text{when } c < 0, \end{cases}$$

for $0 < r \le \pi/\sqrt{k^2+c}$. We shall give estimations of volume density functions of trajectory-spheres by making vary their arc-radius. We put $\sigma_k(t;u) = \sigma_k^t(u)$ and consider a function $t \mapsto \sigma_k(t;u)$.

Theorem 5.1 ([5]). *Let M be a Kähler manifold of complex dimension n, and $u \in U_pM$ be an arbitrary unit tangent vector at an arbitrary point $p \in M$. If sectional curvatures satisfy $\max\big\{\mathrm{Riem}\big(v, \dot\gamma_u(t)\big) \,\big|\, v \in U_{\gamma_u(t)}M,\ v \perp \dot\gamma_u(t)\big\} \le c$ with some constant c for $0 \le t < \pi/\sqrt{k^2+c}$, then we have the following properties.*

1. *The function $t \mapsto \sigma_k(t,u)/\delta_k(t;c,n)$ is monotone increasing for $0 \le t \le \pi/\sqrt{k^2+c}$.*
2. *$\sigma_k(t,u) \ge \delta_k(t;c,n)$ for $0 < t \le \pi/\sqrt{k^2+c}$. If $\sigma_k(t_0,u) = \delta_k(t_0;c,n)$ holds at some t_0 with $0 < t_0 < \pi/\sqrt{k^2+c}$, then $c \ge 0$ and on the interval $[0,t_0]$ we have $\sigma_k(t,u) \equiv \delta_k(t;c,n)$ and $R\big(v, \dot\gamma_u(t)\big)\dot\gamma_u(t) = cv$ for every $v \in T_{\gamma_u(t)}M$ which is orthogonal to $\dot\gamma_u(t)$.*

Theorem 5.2 ([5]). *Let M be a Kähler manifold of complex dimension n, and $u \in U_pM$ be an arbitrary unit tangent vector at an arbitrary point $p \in$*

M. If sectional curvatures satisfy $\min\{\mathrm{Riem}(v, \dot\gamma_u(t)) \mid v \in U_{\gamma_u(t)}M,\ v \perp \dot\gamma_u(t)\} \geq c$ *with some constant c for $0 \leq t < c_{\gamma_u}(\gamma_u(0))$, then we have the following properties.*

(1) *The function $t \mapsto \sigma_k(t,u)/\epsilon_k(t;c,n)$ is monotone decreasing for $0 \leq t \leq c_{\gamma_u}(\gamma_u(0))$.*
(2) *$\sigma_k(t,u) \leq \epsilon_k(t;c,n)$ for $0 < t \leq c_{\gamma_u}(\gamma_u(0))$. If $\sigma_k(t_0,u) = \epsilon_k(t_0;c,n)$ holds at some t_0 with $0 < t_0 < c_{\gamma_u}(\gamma_u(0))$, then $c \leq 0$ and on the interval $[0,t_0]$ we have $\sigma_k(t,u) \equiv \epsilon_k(t;c,n)$ and $R(v,\dot\gamma(t))\dot\gamma(t) = cv$ for all $v \in T_{\gamma(t)}M$ which are orthogonal to $\dot\gamma(t)$.*

We can show these results along the almost same way as for Theorems 4.1, 4.2. If we take normal magnetic Jacobi fields $W_2, \ldots, W_{2n}$ along a trajectory γ_u for $\mathbb{B}_k$ satisfying

i) $W_j(0) = 0$ for $j = 2, \ldots, 2n$,
ii) $\{W_2(r), \ldots, W_{2n}(r)\}$ is an orthonormal basis of $T_{\gamma(r)}S_r^{\ k}(p)$,

then we have

$$\frac{1}{\sigma_k(t,u)}\,\frac{\partial}{\partial t}\sigma_k(t,u)\bigg|_{t=r} = \sum_{j=2}^{2n}\langle(\nabla_{\dot\gamma_u}W_j)(r), W_j(r)\rangle \tag{5.2}$$

for $0 < t < c^{\,s}_{\gamma_u}(p)$. We hence need comparison theorems on magnetic Jacobi fields of hole components. More practically, as we have $\langle\nabla_{\dot\gamma_u}Y, Y\rangle = kf_Y g_Y + \langle\nabla_{\dot\gamma_u}Y^\sharp, Y^\sharp\rangle$ for a normal magnetic Jacobi field Y, we need to take in account of the term $kf_Y g_Y$. When $k = 0$, which is the case of ordinary Jacobi fields, this component is zero. We therefore need to treat magnetic Jacobi fields a bit more carefully than treating Jacobi fields in studying Rauch's comparison theorem. We define a function $\nu_k(\cdot\,;c) : \mathbb{R} \to \mathbb{R}$ by

$$\nu_k(t;c) = \begin{cases} \dfrac{|c|\cosh\sqrt{|c|-k^2}\,t - k^2}{|c|\cosh^2\big(\sqrt{|c|-k^2}\,t/2\big) - k^2}, & \text{if } k^2+c < 0, \\ (2|c|t^2+4)/(|c|t^2+4), & \text{if } k^2+c = 0, \\ \dfrac{k^2 + c\cos\sqrt{k^2+c}\,t}{k^2 + c\cos^2\big(\sqrt{k^2+c}\,t/2\big)}, & \text{if } k^2+c > 0. \end{cases}$$

We then have the following:

Proposition 5.2 ([5]). *Let γ be a trajectory for a non-trivial Kähler magnetic field $\mathbb{B}_k$ on a Kähler manifold M. Suppose sectional curvatures satisfy* $\max\{\mathrm{Riem}(v,\dot\gamma(t)) \mid v \in U_{\gamma(t)}M,\ v \perp \dot\gamma(t)\} \leq c$ *for some constant c for*

$0 \le t \le \pi/\sqrt{k^2+c}$. *Then, for a normal magnetic Jacobi field* Y *along* γ *satisfying* $Y(0) = 0$, *we have*

$$\langle \nabla_{\dot\gamma} Y(t), Y(t)\rangle \ge \|Y^{\top}(t)\|^2\, \mathfrak{t}_k(t;c)\nu_k(t;c) + \|Y^{\perp}(t)\|^2\, \mathfrak{t}_k(t;4c)$$

for $0 < t < \pi/\sqrt{k^2+c}$. *If the equality holds at some* t_0 *with* $0 < t_0 < \pi/\sqrt{k^2+c}$, *then it holds at every* t *with* $0 \le t \le t_0$. *In this case, the normal magnetic Jacobi field* Y *is of the form*

$$\begin{aligned} Y(t) = \|\nabla_{\dot\gamma} Y^{\top}(0)\|\, \mathfrak{s}_k(t;c)\bigl\{(k/2)\mathfrak{s}_k(t;c)\dot\gamma(t) + \mathfrak{c}_k(t;c)J\dot\gamma(t)\bigr\} \\ + \|\nabla_{\dot\gamma} Y^{\perp}(0)\|\, \mathfrak{s}_k(t;4c)\bigl\{\cos(kt/2)E(t) + \sin(kt/2)JE(t)\bigr\} \end{aligned}$$

with a parallel vector field E *along* γ *satisfying*
$E(0) = \nabla_{\dot\gamma} Y^{\perp}(0)/\|\nabla_{\dot\gamma} Y^{\perp}(0)\|$,
and the curvature tensor satisfies $R(Y^{\sharp}, \dot\gamma)\dot\gamma \equiv cY^{\sharp}$ *for* $0 \le t \le t_0$. *Here, when* $\nabla_{\dot\gamma} Y^{\perp}(0) = 0$ *we take* E *as a null vector field.*

When we use (5.2) we may suppose $W_2(r)^{\perp} = W_2(r)$. With the aid of Proposition 5.2, we have

$$\begin{aligned} &\frac{1}{\sigma_k(t,u)}\frac{\partial}{\partial t}\sigma_k(t,u)\Big|_{t=r} \\ &\ge \mathfrak{t}_k(t;c)\nu_k(t;c) + \sum_{j=3}^{2n}\bigl\{\|Y_j^{\top}(t)\|^2\, \mathfrak{t}_k(t;c)\nu_k(t;c) + \|Y^{\perp}(t)\|^2\, \mathfrak{t}_k(t;4c)\bigr\} \\ &\ge \min\biggl\{\frac{\partial}{\partial t}\log\bigl(\mathfrak{s}_k(t;c)\mu_k(t:c)\bigr)^{2n-1}\Big|_{t=r}, \\ &\qquad \frac{\partial}{\partial t}\log\Bigl\{\mathfrak{s}_k(t;c)\mu_k(t:c)\bigl(\mathfrak{s}_k(t;4c)\bigr)^{2(n-1)}\Bigr\}\Big|_{t=r}\biggr\}, \end{aligned}$$

and get the estimate in Theorem 5.1.

In order to show Theorem 5.2, we use the following estimate on magnetic Jacobi fields from above. We can obtain Theorem 5.2 by the same argument by make use of the following proposition (see [5] for more detail).

Proposition 5.3 ([5]). *Let* γ *be a trajectory for a non-trivial Kähler magnetic field* $\mathbb{B}_k$ *on a Kähler manifold* M. *Suppose sectional curvatures satisfy* $\min\bigl\{\mathrm{Riem}(v,\dot\gamma(t)) \bigm| v \in U_{\gamma(t)}M,\ v \perp \dot\gamma(t)\bigr\} \ge c$ *for some constant* c *for* $0 \le t \le c_{\gamma_u}(\gamma_u(0))$. *Then, for a normal magnetic Jacobi field* Y *along* γ *satisfying* $Y(0) = 0$, *we have*

$$\langle \nabla_{\dot\gamma} Y(t), Y(t)\rangle \le \|Y^{\top}(t)\|^2\, \mathfrak{t}_k(t;c)\nu_k(t;c) + \|Y^{\perp}(t)\|^2\, \mathfrak{t}_k(t;4c)$$

for $0 < t < c_{\gamma_u}(\gamma_u(0))$. *If the equality holds at some* t_0 *with* $0 < t_0 < c_{\gamma_u}(\gamma_u(0))$, *then it holds at every* t *with* $0 \le t \le t_0$. *In this case, the normal magnetic Jacobi field Y is of the form*

$$Y(t) = \|\nabla_{\dot\gamma} Y^{\top}(0)\|\, \mathfrak{s}_k(t;c)\big\{(k/2)\mathfrak{s}_k(t;c)\dot\gamma(t) + \mathfrak{c}_k(t;c)J\dot\gamma(t)\big\} + \|\nabla_{\dot\gamma} Y^{\perp}(0)\|\, \mathfrak{s}_k(t;4c)\big\{\cos(kt/2)E(t) + \sin(kt/2)JE(t)\big\}$$

with a parallel vector field E along γ satisfying
$E(0) = \nabla_{\dot\gamma}Y^{\perp}(0)/\|\nabla_{\dot\gamma}Y^{\perp}(0)\|$,
and the curvature tensor satisfies $R(Y^{\sharp}, \dot\gamma)\dot\gamma \equiv cY^{\sharp}$ *for* $0 \le t \le t_0$. *Here, when* $\nabla_{\dot\gamma}Y^{\perp}(0) = 0$ *we take E as a null vector field.*

Acknowledgements

The author would like to express his hearty thanks to the member of The Rotary Club of OKAZAKI, particularly to the counselor Mr. Hitoshi DO-MAE and the scholarship committee-person Mr. Tomoharu SUGIMOTO, for their hospitality as Rotarians and support for his stay in Japan.

Bibliography

1. T. Adachi, *Kähler magnetic flows on a manifold of constant holomorphic sectional curvature*, Tokyo J. Math. 18(1995), 473-483
2. ———, *Magnetic Jacobi fields for Kähler magnetic fields*, Recent Progress in Differential Geometry and its Related Fields, T. Adachi, H. Hashimoto, M. Hristov eds, World Scientific (2011), 41–53.
3. ———, *A theorem of Cartan-Hadamard type for Kähler magnetic fields*, J. Math. Soc. Japan 64(2012), 969–984.
4. P. Bai & T. Adachi, *Volumes of trajectory-balls for Kähler magnetic fields*, preprint 2011.
5. ———, *An estimate of the spread of trajectories for Kähler magnetic fields*, to appear in Hokkaido Math. J.
6. I. Chavel, *Riemannian Geometry: A Modern Introduction*, Cambrdge Tracts in Math. 108, Cambridge Univ. Press 1993.
7. A. Comtet, *On the Landau levels on the hyperbolic plane*, Ann. Phys. 173(1987), 185–209.
8. T. Sunada, *Magnetic flows on a Riemann surface*, Proc. KAIST Math. Workshop 8(1993), 93–108.

Received December 12, 2012
Revised January 11, 2013

Proceedings of the 3rd International
Colloquium on Differential Geometry
and its Related Fields
Veliko Tarnovo, September 3–7, 2012

GEOMETRIC GROUP STRUCTURES AND TRAJECTORIES OF RATIONAL BÉZIER CURVES

Milen J. HRISTOV *

Department of Algebra and Geometry,
Faculty Mathematics and Informatics,
"St. Cyril and St. Methodius" University,
5000 Veliko Tarnovo, Bulgaria
E-mail: m.hristov@uni-vt.bg

On the projectively extended euclidean plane of a fixed triangle, point-wise binary operations and transformations described in barycentric coordinates arise in a geometric way. We apply these operations over the points of a standard Bézier parabola and a weigh point, and get families of associated curves. There exist two geometric multiplicative group structures. We describe the trajectories of a rational Bézier curve by means of group orbits of its points. We calculate the curvature of a smooth curve in barycentrics and apply the formula to a rational Bézier curve.

Keywords: Barycentric coordinates; Multiplicative group; Rational Bézier curve; Orbits and trajectories; Curvature of a curve in barycentrics.

1. On the geometry of the triangle

Let $\mathcal{K} = \{O, \vec{a}_0, \vec{a}_1, \vec{a}_2,\}$ be an orthonormal coordinate system in the euclidean space $\mathbb{R}^3$. We consider the plane $\alpha : x+y+z = 1$ which passes trough the points $A_i : \overrightarrow{OA_i} = \vec{a}_i$, $i = 0, 1, 2$. The triangle $A_0A_1A_2$ is called a basic triangle. Let $\mathbb{E}_2^*$ be the real projective plane associated with α which consists of the set Σ^* of all the points $P(x, y, z)$ (finite (infinite) if $x+y+z = 1$ $(x + y + z = 0)$) and the set Λ^* of all the lines (finite and the infinite line $\omega : x + y + z = 0$), related with the natural incidence (z). An arbitrary line g is identified with the plane $\beta : ax + by + cz = 0$ such that $g = \beta \cap \alpha$. The ordered triplet of reals (a, b, c), i.e. a normal vector to β which is determined uniquely up to a nonzero real multiplier, is said to be barycentric coordinates of the line g and is written $g[a, b, c]$. Thus the infinite line is

*Supported by Scientific researches fund of "St. Cyril and St. Methodius" University of Veliko Tarnovo under contract RD-642-01/26.07.2010.

$\omega[1,1,1]$. Each finite point $P(p_0,p_1,p_2)$ satisfies the vector-algebraic linear system

$$\begin{cases} \overrightarrow{OP} = \sum_{\ell=0}^{2} p_\ell \overrightarrow{OA_\ell} \\ \sum_{\ell=0}^{2} p_\ell = 1. \end{cases} \tag{1}$$

The ordered triplet of reals (p_0,p_1,p_2) (instead the usual (x,y,z)) is known as barycentric coordinates (b-coordinates) of P with respect to the basic $\triangle A_0A_1A_2$ [1, pp.178–179]. Thus we see $A_0(1,0,0)$, $A_1(0,1,0)$, $A_2(0,0,1)$. If $A_0P \cap A_1A_2 = A_0'$ we set

$$\lambda = (A_1A_2A_0') = (A_2A_1A_0')^{-1} = \frac{\overline{A_1A_0'}}{\overline{A_2A_0'}} \quad \text{and} \quad \mu = (A_0A_0'P).$$

The reals λ, μ are known as affine ratios of the point P. By use of these one finds that (p_0,p_1,p_2) is the solution of the system of linear equations which is represented by the matrix equation

$$\boxed{AX_1 = B} : \; A{=}A(\lambda,\mu){=}\begin{pmatrix} 1 & \lambda & 0 \\ 0 & 1{-}\lambda & \mu \\ 0 & 0 & 1{-}\mu \end{pmatrix}, B{=}\begin{pmatrix} 0 \\ 0 \\ 1 \end{pmatrix}, X_1{=}\begin{pmatrix} p_2 \\ p_1 \\ p_0 \end{pmatrix}. \tag{2}$$

Here the reverse order for variables in X_1 is used for technical reasons. Since $\det A = (1-\lambda)(1-\mu) \neq 0$ for any finite point, the inverse matrix A^{-1} is

$$A^{-1} = \frac{1}{\det A} \begin{pmatrix} \det A & \lambda(\mu{-}1) & \lambda\mu \\ 0 & 1{-}\mu & -\mu \\ 0 & 0 & 1{-}\lambda \end{pmatrix},$$

and the triplet

$$p_0 = \frac{1-\lambda}{\det A}, \quad p_1 = -\frac{\mu}{\det A}, \quad p_2 = \frac{\lambda\mu}{\det A} \tag{3}$$

is the unique solution of (2). The expressions for the affine ratios and for $\det A$ in terms of b-coordinates are as follows

$$\lambda = -\frac{p_2}{p_1}, \quad \mu = -\frac{p_1+p_2}{p_0} \quad \text{and} \quad \det A = \frac{\lambda\mu}{p_2} = \frac{p_1+p_2}{p_0p_1} = \frac{1-p_0}{p_0p_1}. \tag{4}$$

Here, $p_0p_1p_2 \neq 0$ for all the points P which are not on the edges of $\triangle A_0A_1A_2$. For such a point P the points A_0', A_1', A_2' which are given by $A_i' = A_iP \cap A_jA_k$ are known as Ceva traces of P. Moreover, since the set $\mathcal{A}$ of all the matrices A in (2) is a Lie group [2, pp.110–111], the b-orbit of the point P is formed by all the points $P_n(x_n,y_n,z_n)$ whose b-coordinates satisfy $A^n(z_n, y_n, x_n)^T = (0,0,1)^T$. Basic facts of the barycentric analytic geometry for the real projective plane are summarized in [2, Theorem 1.1., Theorem 1.2., pp.111–113]. The results do not depend on the choice of basic triangle. From now on, we fix an arbitrary triangle $\mathcal{A}_0\mathcal{A}_1\mathcal{A}_2$ and denote it by $\triangle$. We set $a_i = |\overrightarrow{\mathcal{A}_j\mathcal{A}_k}|$ for an arbitrary cyclic permutation (i,j,k) of $(0,1,2)$ which is the length of edge $\mathcal{A}_j\mathcal{A}_k$.

Some binary operations. We define the notion of *point-matrix multiplication* in a following way. Given finite points $P_1(p_0, p_1, p_2)$ and $P_2(q_0, q_1, q_2)$ with respect to $\triangle$, we denote by λ_ℓ, μ_ℓ and A_ℓ the affine ratios and matrices in the sense of (2) corresponding to the points P_ℓ, $\ell = 1, 2$. We give the following:

Definition 1.1. ***Point-matrix multiplication*** ($\triangle$) of P_1 and P_2 is the point $P = P_1 \triangle P_2$ whose barycentric coordinates (x_0, x_1, x_2) satisfy the matrix equation (2) with $A = A_1A_2$. We call this point the *point-matrix product* (or $\triangle$-product for short) of P_1 and P_2.

Because of $[A_1, A_2] = A_1A_2 - A_2A_1 \neq 0$ [2, p.110, (2)] we find $P_1 \triangle P_2 \neq P_2 \triangle P_1$ in general. The point $A_0(1,0,0)$ is the $\triangle$-unit and the $\triangle$-multiplication is associative. We state the following:

Lemma 1.1. *Suppose finite points $P_1(p_0, p_1, p_2)$ and $P_2(q_0, q_1, q_2)$ with respect to $\triangle$ are given. Let λ_ℓ, μ_ℓ and $A_\ell(\lambda_\ell, \mu_\ell)$ be the affine ratios and matrices corresponding to these points P_1 and P_2.*

(i) *The point $P_1 \triangle P_2$ has the following barycentric coordinates*

$$x_0 = \frac{p_0(1-\lambda_2)}{\det A_2}, \quad x_1 = \frac{p_1(1-\mu_2) - p_0\mu_2}{\det A_2},$$
$$x_2 = p_2 + \frac{p_0\lambda_2\mu_2 - p_1\lambda_2(1-\mu_2)}{\det A_2}.$$

Its expressions by the affine ratios λ_1, μ_1 of P_1 and λ_2, μ_2 of P_2 are

$$x_0 = \frac{(1-\lambda_1)(1-\lambda_2)}{\det(A_1A_2)}, \quad x_1 = \frac{1}{\det(A_1A_2)} \begin{vmatrix} -\mu_1 & 1-\lambda_1 \\ \mu_2 & 1-\mu_2 \end{vmatrix},$$
$$x_2 = \frac{\lambda_1\mu_1}{\det A_1} - \lambda_2 x_1 .$$

(ii) *The set of all the points $P_2(q_0, q_1, q_2)$ which are $\triangle$-commutative with the fixed point $P_1(p_0, p_1, p_2)$ (i.e. $P_1 \triangle P_2 = P_2 \triangle P_1$) is the curve satisfying the following b-parametric equations*

$$\mathcal{C}_{P_1}: \begin{cases} q_0 = \dfrac{p_0}{k+(1-k)p_0} \\ q_1 = \dfrac{kp_1(1-p_0)}{\big[k+(1-k)p_0\big]\,(kp_2+p_1)}\,, \qquad k\in\mathbb{R}\backslash\{\lambda_1{}^{-1},\mu_1{}^{-1}\}. \\ q_2 = \dfrac{k^2p_2(1-p_0)}{\big[k+(1-k)p_0\big]\,(kp_2+p_1)} \end{cases}$$

This curve passes through the points $A_0(k=0)$, $P_1(k=1)$ and its type is parabola with a point at infinity $U(0,-1,1)$ exactly when $\lambda_1=\mu_1$ and hyperbola with points at infinity $U(0,-1,1)$ and $U'(\mu_1-\lambda_1,-\mu_1,\lambda_1)$ otherwise.

Proof. (i) Since $(A_1A_2)X_1 = B$ we have $(x_2,x_1,x_0)^T = A_2^{-1}A_1^{-1}B = A_2^{-1}(p_2,p_1,p_0)^T$ and get the first expressions. We find the second expressions by replacing p_2,p_1,p_0 by (3) in the first ones and counting $x_2=1-x_0-x_1$.

(ii) Let λ_ℓ and μ_ℓ be the affine ratios of the points P_ℓ, $\ell=1,2$. Let $P_1{\scriptstyle\triangle}P_2(x_0,x_1,x_2)$ and $P_2{\scriptstyle\triangle}P_1(y_0,y_1,y_2)$. Since the expression for x_0 is symmetric in (λ_1,λ_2) and (μ_1,μ_2), we find that $x_0=y_0$ always holds. From the equalities $x_2=1-x_0-x_1$, $y_2=1-x_0-y_1$ it follows that $P_1{\scriptstyle\triangle}P_2=P_2{\scriptstyle\triangle}P_1$ if and only if $x_1=y_1$. By use of the expressions of x_1 and y_1 by the affine ratios, we get

$$x_1=y_1 \iff \begin{vmatrix}-\mu_1 & 1-\lambda_1\\ \mu_2 & 1-\mu_2\end{vmatrix} = \begin{vmatrix}-\mu_2 & 1-\lambda_2\\ \mu_1 & 1-\mu_1\end{vmatrix} \iff \begin{vmatrix}\mu_1 & \lambda_1\\ \mu_2 & \lambda_2\end{vmatrix} = 0\,.$$

We hence have $x_1=y_1$ if and only if $(\lambda_2,\mu_2)=k(\lambda_1,\mu_1)$ with a real $k\neq\frac{1}{\lambda_1},\frac{1}{\mu_1}$. From (3) and the equality $\det A_2=(1-\lambda_2)(1-\mu_2)$, we obtain the b-coordinates of P_2 in the form

$$P_2\Big(q_0=\frac{1}{1-k\mu_1}\,,\, q_1=\frac{-k\mu_1}{(1-k\lambda_1)(1-k\mu_1)}\,,\, q_2=\frac{k^2\lambda_1\mu_1}{(1-k\lambda_1)(1-k\mu_1)}\Big)\,.$$

Substituting (4) for λ_1,μ_1 we get the curve $\mathcal{C}_{P_1}$ generated by the point $P_1(p_0,p_1,p_2)$. Obviously $P_2\equiv A_0(1,0,0)$ and $P_2\equiv P_1$ when $k=0$ and $k=1$, respectively. This means that $\mathcal{C}_{P_1}$ passes through A_0 and P_1. The type of the curve depends on the number of its common points with the line at infinity $\omega: x_0+x_1+x_2=0$. Thus the type depends on the number of the solutions with respect to k of the equation $\det A(k\lambda_1,k\mu_1)=0$, which is equivalent to $\lambda_1\mu_1k^2-(\lambda_1+\mu_1)k+1=0$. This equation has a unique real solution $(k_1=k_2)$ if and only if $\lambda_1=\mu_1$, and has two real solutions

$k_1 \neq k_2$ otherwise. Thus the type of the curve is obtained. After calculating k_1, k_2 and replacing them by q_0, q_1, q_2, we get the corresponding points at infinity. □

Corollary 1.1. *The commutative subgroups of $\mathcal{A}$ are defined geometrically by the points with proportional affine ratios, i.e.* $\begin{vmatrix} \lambda_1 & \lambda_2 \\ \mu_1 & \mu_2 \end{vmatrix} = 0$. *Thus, starting from a point*

$$P\left(\frac{1}{1-\mu}; \frac{-\mu}{(1-\lambda)(1-\mu)}; \frac{\lambda\mu}{(1-\lambda)(1-\mu)}\right)$$

with matrix $A(\lambda, \mu)$, we find the 1-parameter set of the points (i.e. the curve)

$$\mathcal{C}_P(k)\left(\frac{1}{1-k\mu}; \frac{-k\mu}{(1-k\lambda)(1-k\mu)}; \frac{k^2\lambda\mu}{(1-k\lambda)(1-k\mu)}\right)$$

is a commutative subgroup $\mathcal{A}_P = \left\{A(k\lambda, k\mu) \,:\, k \in \mathbb{R}\backslash\{\frac{1}{\lambda}; \frac{1}{\mu}\}\right\}$ *in* $\mathcal{A}$.

We call $\mathcal{C}_P$ *the curve of P-$\triangle$-commutability*, or *the* $[P\text{-}\triangle]$*-commutative curve*.

We recall the notion of *barycentric (b)-product of points* which was given in [3, p. 573] and [4, p.18] by omitting its geometric construction. Its interpretation as a collineation is shown in [4, p.125]. If $P_1(p_0, p_1, p_2)$ and $P_2(q_0, q_1, q_2)$ are two points expressed by b-coordinates with respect to $\triangle$, then the point $M\left(\frac{p_0q_0}{\Pi}, \frac{p_1q_1}{\Pi}, \frac{p_2q_2}{\Pi}\right)$ with $\Pi = \sum_{s=0}^{2} p_sq_s$ is called *barycentric (b)-product* of P_1 and P_2, and is denoted as $M = P_1.P_2$. Obviously, $P_1.P_2 = P_2.P_1$ and the medicenter $G(\frac{1}{3}, \frac{1}{3}, \frac{1}{3})$ is the "b-unit" (i.e. $G.P = P$). The *n-th b-power* $P^n(p_0^n, p_1^n, p_2^n)$ of a point $P(p_0, p_1, p_2)$ is defined in a standard way. The b-inverse P^{-1} of P is the point satisfying $P^{-1}.P = G$. It exists if and only if $P(p_0, p_1, p_2)$ satisfies $p_0p_1p_2 \neq 0$ (i.e. P does not lies on the edges of the basic $\triangle$). So

$$P^{-1}\left(\frac{p_1p_2}{p_0p_1 + p_1p_2 + p_2p_0}, \frac{p_2p_0}{p_0p_1 + p_1p_2 + p_2p_0}, \frac{p_0p_1}{p_0p_1 + p_1p_2 + p_2p_0}\right).$$

The notion *b-cross product* $(\times)$ is also well known. To different points $P_1(p_0, p_1, p_2)$ and $P_2(q_0, q_1, q_2)$ one corresponds the point

$$P_1 \times P_2\left(\delta\begin{vmatrix} p_1 & p_2 \\ q_1 & q_2 \end{vmatrix}, \delta\begin{vmatrix} p_2 & p_0 \\ q_2 & q_0 \end{vmatrix}, \delta\begin{vmatrix} p_0 & p_1 \\ q_0 & q_1 \end{vmatrix}\right), \text{ where } \delta^{-1} = \begin{vmatrix} 1 & 1 & 1 \\ p_0 & p_1 & p_2 \\ q_0 & q_1 & q_2 \end{vmatrix}.$$

This operation is commutative (i.e. $P_1 \times P_2 = P_2 \times P_1$) and does not furnish with a unit.

The associativity of $\vartriangle$-multiplication and of the b-product is clear. By combining the above operations one gets the points

$$P_1 \vartriangle (P_2.P_3), \quad P_1 \vartriangle (P_2 \times P_3), \quad P_1.(P_2 \times P_3).$$

The following **transformations with respect to** $\triangle$ are well known.

- The isogonal point-conjugation $\varphi_\triangle : \Sigma^* \longrightarrow \Sigma^*$ for $P(p_0, p_1, p_2)$ with $p_0 p_1 p_2 \neq 0$ is defined by

$$\varphi_\triangle : P(p_0, p_1, p_2) \longmapsto P'\Big(\frac{ka_0^2}{p_0}, \frac{ka_1^2}{p_1}, \frac{ka_2^2}{p_2}\Big),$$

  where $k = \Big(\sum_{i=0}^{2} \frac{a_i^2}{p_i}\Big)^{-1}$ is a normalizing multiplier.
- The isotomic point-conjugation $\psi_\triangle : \Sigma^* \longrightarrow \Sigma^*$ for $P(p_0, p_1, p_2)$ with $p_0 p_1 p_2 \neq 0$ is defined by

$$\psi_\triangle : P(p_0, p_1, p_2) \longmapsto P''\Big(\frac{\ell}{p_0}, \frac{\ell}{p_1}, \frac{\ell}{p_2}\Big), \quad \ell = \Big(\sum_{i=0}^{2} \frac{1}{p_i}\Big)^{-1}.$$

  It is easy to see that $P'' = \psi_\vartriangle(P) = P^{-1}$.
- The isocircular transformation $f_\triangle : \Sigma^* \longrightarrow \Sigma^*$ is defined by

$$f_\triangle : P(p_0, p_1, p_2) \longmapsto P^*\Big(\frac{sp_0}{a_0}, \frac{sp_1}{a_1}, \frac{sp_2}{a_2}\Big), \quad s = \Big(\sum_{i=0}^{2} \frac{p_i}{a_i}\Big)^{-1}.$$

- The dual conjugation d_2 in the projective plane $\mathbb{E}_2^*$ is defined as follows

$$d_2 : \begin{cases} \text{point } P(p_0, p_1, p_2) & \longmapsto \quad \text{line } p[p_0, p_1, p_2] \\ \text{line } q[q_0, q_1, q_2] & \longmapsto \quad \text{point } Q(q_0, q_1, q_2) \\ P \,\mathsf{z}\, q & \longmapsto \quad p \,\mathsf{z}\, Q\,. \end{cases}$$

- The projective isotomic conjugation (or polar map) $\pi_\triangle = d_2 \circ \psi_\triangle : \Sigma^* \longleftrightarrow \Lambda^*$. It is given by

$$\pi_\triangle : \text{point } P(p_0, p_1, p_2) \text{ (pole)} \longleftrightarrow \text{line } p\Big[\frac{1}{p_0}, \frac{1}{p_1}, \frac{1}{p_2}\Big] \text{ (polar line)},$$

  that is, $\pi_\vartriangle(P) = p[p_1p_2, p_2p_0, p_0p_1]$.

By the analogy with the polar map we consider the compositions $d_2 \circ \varphi_\triangle$ and $d_2 \circ f_\triangle$ and call them the projective isogonal conjugation and the projective isocircular conjugation, respectively.

By means of the dual conjugation d_2, each binary operation is naturally extendable over the set of lines (Λ^*) by the same analytic way. For arbitrary g_1 and $g_2 \in \Lambda^*$, we define

- their $\triangle$-product to be the line $g_1 \triangle g_2 \stackrel{\text{def}}{=} d_2(d_2g_1 \triangle d_2g_2)$,
- their b-product to be the line $g_1 . g_2 \stackrel{\text{def}}{=} d_2(d_2g_1 . d_2g_2)$,
- their $\times$-product to be the line $g_1 \times g_2 \stackrel{\text{def}}{=} d_2(d_2g_1 \times d_2g_2)$.

The binary operation $d_2(P_1 \times P_2)$ is defined for arbitrary two points P_1, P_2 and is extended over the lines by $d_2(g_1 \times g_2)$. In [4, p.15] this operation is called **wedge operator**. Those $d_2(P_1 \times P_2)$ and $d_2(g_1 \times g_2)$ are denoted by $P_1 \wedge P_2$ and $g_1 \wedge g_2$, respectively. Geometrically, they are the following:

- $P_1 \wedge P_2 = d_2(P_1 \times P_2)$ is the line P_1P_2,
- $g_1 \wedge g_2 = d_2(g_1 \times g_2)$ is the point $g_1 \cap g_2$.

"Mixed" binary operations are defined likewise. For an arbitrary $P \in \Sigma^*$ and an arbitrary $g \in \Lambda^*$ whose dual conjugated are $d_2P = p \in \Lambda^*$ and $d_2g = G \in \Sigma^*$, respectively, we define their operations as follows:

$$g \triangle P \stackrel{\text{def}}{=} d_2g \triangle P = G \triangle P,$$
$$P \triangle g \stackrel{\text{def}}{=} d_2P \triangle g = p \triangle g = d_2p \triangle d_2g = P \triangle G;$$

$$g.P \stackrel{\text{def}}{=} d_2g.P = G.P,$$
$$P.g \stackrel{\text{def}}{=} d_2P.g = p.g = d_2p.d_2g = P.G;$$

$$g \times P \stackrel{\text{def}}{=} d_2g \times P = G \times P,$$
$$P \times g \stackrel{\text{def}}{=} d_2P \times g = p \times g = d_2p \times d_2g = P \times G;$$

$$g \wedge P \stackrel{\text{def}}{=} d_2g \wedge P = G \wedge P,$$
$$P \wedge g \stackrel{\text{def}}{=} d_2P \wedge g = p \wedge g = d_2p \wedge d_2g = P \wedge G.$$

These definitions show that

$$g \triangle P \neq P \triangle g, \quad g.P = P.g, \quad g \times P = P \times g, \quad \text{and} \quad g \wedge P \neq P \wedge g.$$

We summarize properties of these operations in the following corollary.

Corollary 1.2. *We have the following multiplicative group-structures.*

i) $(\Sigma^*, \triangle)$ *is a non-abelian group whose* $\triangle$*-unit is the vertex* $A_0(1,0,0)$, *and* $(\Lambda^*, \triangle)$ *is a non-abelian group whose* $\triangle$*-unit is the line* $d_2A_0 = A_1A_2[1,0,0]$ *of* $\triangle A_0A_1A_2$.

ii) *If one puts* $\Sigma_0^* = \{P(p_0, p_1, p_2) \in \Sigma^* : p_0p_1p_2 \neq 0\}$, *then* $(\Sigma_0^*, \cdot)$ *is an abelian group whose unit is the medicenter* G *of* $\triangle A_0A_1A_2$ *[3, p.573]. The dual version similarly holds:* $(\Lambda_0^* = d_2\Sigma_0^*\, ,\ \cdot\ )$ *is an abelian group whose unit is the infinity line* $d_2G = \omega[1,1,1]$.

Let **id** be the identity transformation. One can verify the following by direct computation.

Lemma 1.2. *The following statements are valid:*

i) $\varphi_\triangle$, $\psi_\triangle$ *and* $\pi_\triangle$ *are involutive transformations, that is*

$$\varphi_\triangle \neq \mathrm{id},\ \psi_\triangle \neq \mathrm{id},\ \pi_\triangle \neq \mathrm{id} \quad \textit{and} \quad \varphi_\triangle^2 = \mathrm{id}, \quad \psi_\triangle^2 = \mathrm{id}, \quad \pi_\triangle^2 = \mathrm{id};$$

ii) $\varphi_\triangle$ *and* $\psi_\triangle$ *are rescaling equivalent, that is*

$$\ell\, \varphi_\triangle(P) = k\, \psi_\triangle(P)\mathrm{diag}(a_0^2, a_1^2, a_2^2);$$

iii) *The composition* $f_\triangle^2 = f_\triangle \circ f_\triangle$ *satisfies* $f_\triangle^2 = \psi_\triangle \circ \varphi_\triangle$.

Remark 1.1. Each of the above analytically described transformations have geometric constructions, concerned with the well known Ceva and Menelaus theorems. Many geometric properties of special points and lines of a triangle in terms of this transformations are studied by many authors.

Remark 1.2. When Δ is equilateral, then $\psi_\triangle \equiv \varphi_\triangle$ and $f_\triangle \equiv \mathrm{id}$.

Remark 1.3. All the triangles in a projectively extended euclidean plane ($\mathbb{RP}_2$) are affinely equivalent to an equilateral triangle. This means the following. For an arbitrary $\triangle B_0B_1B_2$ with vertices $B_k(d_k, e_k, f_k)$ in b-coordinates, there is a unique affine transformation σ (an invertable linear transformation which preserves the infinity line) which maps $\triangle B_0B_1B_2$ onto $\triangle A_0A_1A_2$ so that $B_0 \mapsto A_0(1,0,0)$, $B_1 \mapsto A_1(0,1,0)$, $B_2 \mapsto A_2(0,0,1)$. Its matrix C_σ has the inverse $C_\sigma^{-1} = \begin{pmatrix} d_0 & d_1 & d_2 \\ e_0 & e_1 & e_2 \\ f_0 & f_1 & f_2 \end{pmatrix}$, and both C_σ, C_σ^{-1} are of left stochastic-like type (stochastic-like by columns). The converse is also true. The affine ratio is a basic invariant for an affine transformation. From algebraic point of view, C_σ is the transport matrix between the basis $\{\vec{b}_0, \vec{b}_1, \vec{b}_2\}$ and the orthonormal basis $\{\vec{a}_0, \vec{a}_1, \vec{a}_2\}$, where $\vec{b}_k = \overrightarrow{OB}_k$, $\vec{a}_k = \overrightarrow{OA}_k$, k=0,1,2.

Thus by means of Remarks 1.2 and 1.3, jointly with the affine invariance of the Bézier curves, we only consider the basic $\triangle A_0A_1A_2$ in the following sections.

2. Barycentric orbits

We take into account the group structures in Corollary 1.2. The matrix A in (2) generates an affine transformation and the set $\mathcal{A}$ of the such matrices is a Lie group. As was noted, the solutions of the systems $A^n X = B$, $n \in \mathbb{Z}$ generate the following sequence of points, which is the sequence of $\triangle$-powers of a point P with matrix A, in general:

$$\overset{b}{\mathcal{O}}(P) = \{\ldots, P_{-n}, \ldots, P_{-2}, P_{-1}, \boxed{P \equiv P_1}, P_2, \ldots, P_n, \ldots\}. \tag{5}$$

We call (5) a *b-$\triangle$-power-orbit*, shortly a *b-orbit of* P. Applying the dual conjugation d_2 to (5), we get the b-orbit of lines in general:

$$\overset{b}{\mathcal{L}}(p) = d_2\Big(\overset{b}{\mathcal{O}}(P)\Big) = \{p_m = d_2(P_m) : \ m \in \mathbb{Z} \setminus \{0\}\}.$$

We have the isotomic b-orbit of points

$$\overset{b}{\mathcal{T}}(P) = \psi\Big(\overset{b}{\mathcal{O}}(P)\Big) = \{T_m = \psi(P_m) : \ m \in \mathbb{Z} \setminus \{0\}\}$$

and its dual b-orbit of polar lines

$$\overset{b}{\mathcal{P}}(p) = d_2\Big(\overset{b}{\mathcal{T}}(P)\Big) = \pi\Big(\overset{b}{\mathcal{O}}(P)\Big) = \{t_m = \pi(P_m) : \ m \in \mathbb{Z} \setminus \{0\}\}.$$

We note that the orbit $\overset{b}{\mathcal{O}}(\psi(P))$ is formed by the solutions of the system $A^n Y = B$, where Y is the column of the b-coordinates of $P^{-1} = \psi(P)$ in reverse order.

By means of the b-product, there is the *b-power orbit* (*b-p-orbit*) *of* P in general which is formed by the sequence of b-power points:

$$\overset{p}{\mathcal{O}}(P) = \{\ldots, P^{-n}, \ldots, P^{-2}, P^{-1}, \boxed{P}, P^2, \ldots, P^n, \ldots\}. \tag{6}$$

Clearly (6) is isotomic invariant, that is, $\overset{p}{\mathcal{O}}(P) \equiv \overset{p}{\mathcal{O}}(\psi(P)) \equiv \psi\Big(\overset{p}{\mathcal{O}}(P)\Big)$ (Lemma 1.2 (i)). And there exists its associated b-p-orbit of lines

$$\overset{p}{\mathcal{L}}(p) = d_2\Big(\overset{p}{\mathcal{O}}(P)\Big) \equiv \pi\Big(\overset{p}{\mathcal{O}}(P)\Big).$$

We get the following

Theorem 2.1. *Given a point $P(p_0, p_1, p_2)$ with respect to the basic $\triangle A_0 A_1 A_2$, we denote by λ and μ its affine ratios. Let A be the matrix* *satisfying (2). Then we have*

(i) $\overset{b}{\mathcal{O}}(P) = \{P_n(x_n, y_n, 1 - x_n - y_n) \, : \, n \in \mathbb{Z}\}$

$$= \left\{ P_n\left(\frac{(1-\lambda)^n}{(\det A)^n}, \frac{\mu[(1-\lambda)^n - (1-\mu)^n]}{(\lambda - \mu)(\det A)^n}, 1 + \frac{\mu(1-\mu)^n - \lambda(1-\lambda)^n}{(\lambda - \mu)(\det A)^n} \right) : n \in \mathbb{Z} \right\},$$

where $P_0(1, 0, 0) \equiv A_0$;

(ii) $\overset{p}{\mathcal{O}}(P) = \left\{ P^n\left(\frac{(1-\lambda)^n}{k_n}, \frac{(-1)^n \mu^n}{k_n}, \frac{\lambda^n \mu^n}{k_n} \right) : n \in \mathbb{Z} \right\},$

where $k_n = (1-\lambda)^n + \mu^n\{\lambda^n + (-1)^n\}$ *and* $P^0 \equiv G$, *which is the medicenter of* $\triangle A_0 A_1 A_2$.

By means of Lemma 1.1 (ii) and Corollary 1.1, each point of the orbits $\overset{b}{\mathcal{O}}(P)$, $\overset{b}{\mathcal{O}}(\psi(P))$, $\psi\left(\overset{b}{\mathcal{O}}(P)\right)$ and $\overset{p}{\mathcal{O}}(P)$ generates a curve of $\triangle$-commutability $\mathcal{C}_{P_n}$, $\mathcal{C}_{\psi(P_n)}$, $\mathcal{C}_{(\psi(P))_n}$ and $\mathcal{C}_{P^n}$, respectively.

3. Rational Bézier curves and its barycentric trajectories

We consider the basic triangle $A_0A_1A_2$, which further will be denoted by Δ. We are going to introduce Bézier curves in the following manner. Over the median through A_1, we take a point

$$W_u\left(\frac{1}{2+u}, \frac{u}{2+u}, \frac{1}{2+u}\right), \quad u = \text{const} > 0.$$

Let

$$\mathcal{C}_\Delta = \left\{ P_t\Big(p_{2,0}(t) = (1-t)^2,\ p_{2,1}(t) = 2t(1-t),\ p_{2,2}(t) = t^2\Big) \ : \ t \in [0,1] \right\}$$

be the curve formed by the 1-parameter locus of points whose b-coordinates $p_{2,\ell}(t)$, $\ell = 0, 1, 2$ are the three polynomials of Bernstein of second degree. Clearly W_0, W_1 and $W_\infty = \lim_{u \to \infty} W_u$ are the midpoint M_1 of A_0A_2, the medicenter G and the vertex A_1 of Δ, respectively. Likewise $P_0 \equiv A_0$, $P_1 \equiv A_2$, and the point $P_{\frac{1}{2}} = W_2$ is the midpoint of the median through A_1. By means of b-product, cross product and point-matrix ($\triangle$) product, we get the following curves:

$$\overset{(u)}{\mathcal{C}} = \{\mathcal{P} = W_u . P_t\ , \quad t \in [0,1]\ \},$$
$$\overset{(u)}{\mathcal{C}^*} = \{\mathcal{P}^* = W_u \times P_t\ , \quad t \in [0,1]\ \},$$
$$\overset{(u)}{\mathcal{C}^\triangle} = \{\mathcal{P}^\triangle = W_u {\scriptstyle\triangle} P_t\ , \quad t \in [0,1]\ \},$$
$$\overset{(u)}{{}^\triangle\mathcal{C}} = \{{}^\triangle\mathcal{P} = P_t {\scriptstyle\triangle} W_u\ , \quad t \in [0,1]\ \}.$$

Thus we have

$$\overset{(u)}{\mathcal{C}} = \left\{\mathcal{P}\left(\frac{(1-t)^2}{B_2(u,t)}, \frac{2ut(1-t)}{B_2(u,t)}, \frac{t^2}{B_2(u,t)}\right) : t \in [0,1]\right\}, \tag{7}$$

for $u > 0$, where $B_2(u,t) = 2(1-u)t^2 - 2(1-u)t + 1$, and have

$$\overset{(u)}{\mathcal{C}^*} = \left\{\mathcal{P}^*\left(\frac{2t-(u+2)t^2}{\mathcal{B}_2(u,t)}, \frac{2t-1}{\mathcal{B}_2(u,t)}, \frac{2t(t-1)+u(1-t)^2}{\mathcal{B}_2(u,t)}\right) : t \in [0,1]\right\}, \tag{8}$$

for $u \in (0,\infty) \setminus \{1\}$, where $\mathcal{B}_2(u,t) = (2t-1)(1-u)$. We note that when one puts $t = \frac{1}{2}$ in (8), then $\mathcal{P}^*$ is the infinity point

$$(W_u \times P_{\frac{1}{2}}) = (W_u \times W_2)(-1,0,1)$$

of the edge A_0A_2. Calculating the affine ratios we find

$$\lambda_u = -\frac{1}{u}, \qquad \mu_u = -(u+1) \quad \text{for} \quad W_u,$$
$$\lambda_t = \frac{t}{2(t-1)}, \quad \mu_t = \frac{t^2-2t}{(1-t)^2} \quad \text{for} \quad P_t \in C_\Delta.$$

Hence, by using Lemma 1.1 (i) we get

$$\overset{(u)}{\mathcal{C}^\triangle} = \left\{\mathcal{P}^\triangle\left(\frac{(t-1)^2}{u+2}, \frac{2(t-1)(u-t^2+2t)}{(u+2)(t-2)}, \frac{(t-2)(t^2+1)-tu}{(u+2)(t-2)}\right) : t \in [0,1]\right\}, \tag{9}$$

$$\overset{(u)}{{}^\triangle\mathcal{C}} = \left\{{}^\triangle\mathcal{P}\left(\frac{(t-1)^2}{u+2}, \frac{u(1-t)[(u+3)t^2-(u+5)t-2(u+1)]}{(u+1)(u+2)(t-2)}, t^2 + \frac{(1-t)[(u+3)t^2-(u+5)t-2(u+1)]}{(u+1)(u+2)(t-2)}\right) : t \in [0,1]\right\}. \tag{10}$$

By (1) one gets the standard vector equations of the curves (7), (8), (9) and (10).

Remark 3.1.

(i) The curve (7) is the well known **rational Bézier curve of second order with normal parameterization**. The point W_u is called **weigh point**. The curve $\overset{(1)}{\mathcal{C}} = \mathcal{C}_\Delta$ is the standard Bézier parabola, the curves $\overset{(1-)}{\mathcal{C}}$ are Bézier ellipses and $\overset{(1+)}{\mathcal{C}}$ are Bézier hyperbolas.

(ii) We call the curve (8) a **cross ($\times$-)product Bézier curve** (see [1, p.184]).

(iii) We shall call the curve (9) a **left-weighted ($\vartriangle$-)product Bézier curve**, and the curve (10) a **right-weighted ($\vartriangle$-)product Bézier curve**.

(iv) The curves of $\vartriangle$-commutability which are associated to any point of (7), (8), (9) and (10) are well defined by means of Lemma 1.1 (ii).

Definition 3.1. We take arbitrary points $\mathcal{P}$, $\mathcal{P}^*$, $\mathcal{P}^\vartriangle$, ${}^\vartriangle\mathcal{P}$ over the curves $\overset{(u)}{\mathcal{C}}$, $\overset{(u)}{\mathcal{C}^*}$, $\overset{(u)}{\mathcal{C}^\vartriangle}$, $\overset{(u)}{{}^\vartriangle\mathcal{C}}$, respectively. We denote by $\mathcal{P}_n$, $\mathcal{P}^*_n$, $\mathcal{P}^\vartriangle_n$, ${}^\vartriangle\mathcal{P}_n$ the n-th points from the b-orbit (5) of these points, respectively. We call the curves

$$b\overset{(u)}{\mathcal{C}_n} = \{\mathcal{P}_n : \mathcal{P} \in \overset{(u)}{\mathcal{C}}\}, \qquad b\overset{(u)}{\mathcal{C}^*_n} = \{\mathcal{P}^*_n : \mathcal{P}^* \in \overset{(u)}{\mathcal{C}^*}\},$$

$$b\overset{(u)}{\mathcal{C}^\vartriangle_n} = \{\mathcal{P}^\vartriangle_n : \mathcal{P}^\vartriangle \in \overset{(u)}{\mathcal{C}^\vartriangle}\}, \qquad b\overset{(u)}{{}^\vartriangle\mathcal{C}_n} = \{{}^\vartriangle\mathcal{P}_n : {}^\vartriangle\mathcal{P} \in \overset{(u)}{{}^\vartriangle\mathcal{C}}\}$$

the n-th b-trajectories of $\overset{(u)}{\mathcal{C}}$, $\overset{(u)}{\mathcal{C}^*}$, $\overset{(u)}{\mathcal{C}^\vartriangle}$ and $\overset{(u)}{{}^\vartriangle\mathcal{C}}$, respectively.

For n-th power points $\mathcal{P}^n$, $(\mathcal{P}^*)^n$, $(\mathcal{P}^\vartriangle)^n$, $({}^\vartriangle\mathcal{P})^n$ from the b-p-orbits (6) of those points $\mathcal{P}$, $\mathcal{P}^*$, $\mathcal{P}^\vartriangle$, ${}^\vartriangle\mathcal{P}$, we call the curves

$$p\overset{(u)}{\mathcal{C}_n} = \{\mathcal{P}^n : \mathcal{P} \in \overset{(u)}{\mathcal{C}}\}, \qquad p\overset{(u)}{\mathcal{C}^*_n} = \{(\mathcal{P}^*)^n : \mathcal{P}^* \in \overset{(u)}{\mathcal{C}^*}\},$$

$$p\overset{(u)}{\mathcal{C}^\vartriangle_n} = \{(\mathcal{P}^\vartriangle)^n : \mathcal{P}^\vartriangle \in \overset{(u)}{\mathcal{C}^\vartriangle}\}, \qquad p\overset{(u)}{{}^\vartriangle\mathcal{C}_n} = \{({}^\vartriangle\mathcal{P})^n : {}^\vartriangle\mathcal{P} \in \overset{(u)}{{}^\vartriangle\mathcal{C}}\}$$

the n-th b-p-trajectories of $\overset{(u)}{\mathcal{C}}$, $\overset{(u)}{\mathcal{C}^*}$, $\overset{(u)}{\mathcal{C}^\vartriangle}$, $\overset{(u)}{{}^\vartriangle\mathcal{C}}$, respectively.

The polar (π-) images of all the points from some n-th trajectory defines a family of polar lines. The envelope of this family will be denoted as the name of the trajectory, by replacing $\mathcal{C}$ with $\mathcal{E}$. As an example, the trajectories $b\overset{(1)}{\mathcal{C}_n}$ of the standard Bézier parabola and its corresponding envelops $b\overset{(1)}{\mathcal{E}_n}$ are discussed in [2, p.117, Theorem 2.5]. Through any point M of some n-th trajectory, the line of $\vartriangle$-commutability $\mathcal{C}_M$ passes. The envelope of the family of polar lines $\pi(\mathcal{C}_M)$ will be denoted by $\mathcal{E}_M$.

Now it naturally arises a list of problems. We shall illustrate some of them over the rational Bézier curves with normal parameterization (7) (see

Figure 1). Some basic geometric properties of these curves are shown in [1, pp.186-187]. Its sphere ($\mathbb{S}^2$-) images and its $\mathbb{S}^3$-Hopf fibres are described in [1, p.189, Theorem 1.4.].

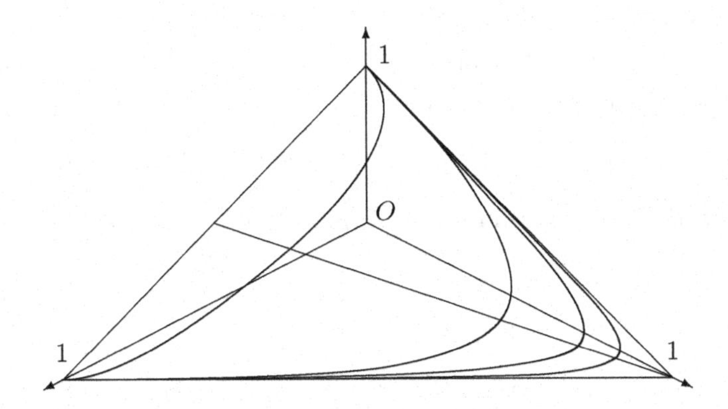

Figure 1. A few examples of rational Bézier curves (7).

In order to calculate the curvature we prepare the following

Theorem 3.1 (curvature in barycentrics). *Let*

$$c : \vec{r}(t) = (p_0(t), p_1(t), p_2(t))$$

be a curve defined on an open interval $\mathcal{J}$ *whose baricentric coordinates* $p_s(t)$, $s = 0, 1, 2$ *with respect to the basic triangle* $A_0A_1A_2$ *are smooth functions satisfying* $\sum_{s=0}^{2} p_s(t) = 1$. *Then its curvature* $\varkappa(t)$ *is given as*

$$\varkappa(t) = \frac{S_{A_0A_1A_2}}{\sqrt{2}} |\det \begin{pmatrix} \dot{p_0} & \dot{p_1} \\ \ddot{p_0} & \ddot{p_1} \end{pmatrix}| (\dot{p_0}^2 + \dot{p_1}^2 + \dot{p_0}\dot{p_1})^{-\frac{3}{2}} ,$$

where $S_{A_0A_1A_2}$ *is the area of the basic triangle.*

Proof. From the representation

$$c : \vec{r}(t) = (p_0(t), p_1(t), 1 - p_0(t) - p_1(t)),$$

we get its derivatives are of the following forms

$$\dot{\vec{r}}(t) = \left(\dot{p_0}(t), \dot{p_1}(t), -\dot{p_0}(t) - \dot{p_1}(t)\right),$$
$$\ddot{\vec{r}}(t) = (\ddot{p_0}(t), \ddot{p_1}(t), -\ddot{p_0}(t) - \ddot{p_1}(t)),$$

hence the cross product is of the form

$$\dot{\vec{r}}(t) \times \ddot{\vec{r}}(t) = \det \begin{pmatrix} \dot{p_0} & \dot{p_1} \\ \ddot{p_0} & \ddot{p_1} \end{pmatrix} (1,1,1).$$

Substituting them in the well known formula $\varkappa(t) = \frac{|\dot{\vec{r}}(t) \times \ddot{\vec{r}}(t)|}{|\dot{\vec{r}}(t)|^3}$ and taking into account that $\triangle A_0A_1A_2$ is equilateral with side length $\sqrt{2}$, we get the result. Obviously, from the representations

$$c : \vec{r}(t) = \big(p_0(t), 1-p_0(t)-p_2(t), p_2(t)\big) = \big(1-p_1(t)-p_2(t), p_1(t), p_2(t)\big),$$

its curvature can be expressed in terms of (p_2, p_0) and of (p_1, p_2), respectively. □

As direct consequences we get the following.

Corollary 3.1. *Any cross-product Bézier curve* (8) *is a segment of a line, and is parallel to the edge* A_0A_2 *of the basic triangle.* (*The cross product* $W_u \times P_t$ *rectifies the standard Bézier parabola.*)

Corollary 3.2. *The curvature of the rational Bézier curve* (7) *is given by*

$$\overset{(u)}{\varkappa}(t) = \frac{uS_{A_0A_1A_2}}{\Big\{t^2[t(u-1)+1]^2 + (t-1)^2[t(u-1)-u]^2 + u^2(1-2t)^2\Big\}^{\frac{3}{2}}}.$$

Further, calculating the affine ratious λ and μ for the rational Bézier curve (7) and applying Theorem 2.1, we get the following

Theorem 3.2. *We take a rational Bézier curve with normal parameterization* (7).

(i) *Its n-th b-trajectoy* ($n \in \mathbb{Z}$) *is the curve*

$$b\overset{(u)}{C_n} = \Big\{\mathcal{P}_n\Big(\frac{(1-\lambda)^n}{(\det A)^n}, \frac{\mu[(1-\lambda)^n-(1-\mu)^n]}{(\lambda-\mu)(\det A)^n}, 1+\frac{\mu(1-\mu)^n-\lambda(1-\lambda)^n}{(\lambda-\mu)(\det A)^n}\Big) : 0<t<1\Big\}.$$

(ii) *Its n-th b-p-trajectory* ($n \in \mathbb{Z}$) *is the curve*

$$p\overset{(u)}{C_n} = \Big\{\mathcal{P}^n\Big(\frac{(1-\lambda)^n}{k_n}, \frac{(-1)^n\mu^n}{k_n}, \frac{\lambda^n\mu^n}{k_n}\Big) : 0<t<1\Big\},$$

where $k_n = (1-\lambda)^n + \mu^n(\lambda^n + (-1)^n)$.

(iii) *Its curve of* $\triangle$*-commutability* $\mathcal{C}_{\mathcal{P}}$ *through its point* $\mathcal{P}$ *is*

$$\mathcal{C}_{\mathcal{P}} = \left\{\left(\frac{1}{1-k\mu}, \frac{-k\mu}{(1-k\lambda)(1-k\mu)}, \frac{k^2\lambda\mu}{(1-k\lambda)(1-k\mu)}\right) : k \in \mathbb{R} \setminus \left\{\frac{1}{\lambda}, \frac{1}{\mu}\right\}\right\}$$

for each fixed t.

In these expressions

$$\lambda = \frac{t}{2u(t-1)}, \quad \mu = -\frac{t^2 + 2ut(1-t)}{(1-t)^2}, \quad \det A = (1-\lambda)(1-\mu).$$

Bibliography

1. M. Hristov, *Bézier type almost complex structures on quaternionic Hermitian vector spaces*, in *Recent progress in differential geometry and its related fields*, T. Adachi, H. Hashimoto & M. Hristov eds., World Sci. Publ., 2011, 177–194.
2. M. Hristov, *Some geometric properties and objects related to Bézier curves*, in *Topics in differential geometry complex analysis and mathematical physics*, K. Sekigawa, V.S. Gerdjikov & S. Dimiev eds. World Sci. Publ., 2009, 109–119.
3. P. Yiu, *The uses of homogeneous barycentric coordinates in plane Euclidean geometry*, Int. J. Math. Educ. Sci. Technol. 31(2000), 569–578.
4. pldx : Translation of the Kimberling's Glossary into barycentrics, December 16, 2011, published under the GNU Free Documentation License.

Received December 17, 2012
Revised March 25, 2013

Proceedings of the 3rd International
Colloquium on Differential Geometry
and its Related Fields
Veliko Tarnovo, September 3–7, 2012

ON GEOMETRICAL STRUCTURES ON $S^3 \times S^3$ IN THE OCTONIONS

Hideya HASHIMOTO *

Department of Mathematics, Meijo University,
Nagoya 468-8502, Japan
E-mail: hhashi@meijo-u.ac.jp

Misa OHASHI †

Department of Mathematics, Meijo University,
Nagoya 468-8502, Japan
Osaka City University, Advanced Mathematical Institute,
Osaka 558-8585, Japan
E-mail: m0851501@ccalumni.meijo-u.ac.jp

We give the geometrical structures related to the orthogonal homogeneous almost complex structure on $S^3 \times S^3$ which is obtained by the multiplication of the octonions $\mathfrak{C}$.

Keywords: Octonions; Orthogonal almost complex structure; $U(3) \times Sp(1)$-structure.

1. Introduction

In the previous paper [7], we calculate the automorphism group of an orthogonal almost complex structure (or an almost Hermitian structure) on several oriented 6-dimensional submanifolds of the octonions $\mathfrak{C}$.

In [1], R.L. Bryant showed that any oriented 6-dimensional submanifold $\varphi : M^6 \to \mathfrak{C}$ of the octonions admits the orthogonal almost complex (Hermitian) structure J defined by

$$\varphi_*(JX) = \varphi_*(X)(\eta \times \xi),$$

where $\{\xi, \eta\}$ is a local oriented orthonormal frame field of the normal bundle of φ over a neighborhood at each point of M^6. Note that this almost

*The first author is partially supported by JSPS KAKENHI Grant Number 24540101.
†The second author is partially supported by the JSPS Institutional Program for Young Researcher Overseas Visits "Promoting international young researchers in mathematics and mathematical sciences led by OCAMI".

complex structure depends only on the orientation of the normal bundle, which is independent of the choice of the orthonormal frame field of the normal bundle. Therefore the orthogonal almost complex structure defines a global section of the tensor bundle $T^*M^6 \otimes TM^6$ over M^6 satisfying the conditions $J^2 = -I_{TM^6}$ and J is compatible with the induced metric.

The fundamental properties of this induced almost complex (Hermitian) structure is a $Spin(7)$-invariant in the following sense: Let $\varphi_1 : M^6, N^6 \to \mathfrak{C}$ and $\varphi_2 : N^6 \to \mathfrak{C}$ be two isometric immersions from 6-dimensional manifolds into the octonions. If there exists an element $g \in Spin(7)$ and the orientation preserving diffeomorphism $\psi : M^6 \to N^6$ satisfying $g \circ \varphi_1 = \varphi_2 \circ \psi$ (up to a parallel translation), then the two maps φ_1, φ_2 are said to be $Spin(7)$-congruent to each other. If the immersions φ_1 and φ_2 are $Spin(7)$-congruent to each other, then the induced orthogonal almost complex structures J_{M^6} of M^6 and J_{N^6} of N^6 are equivalent in the sense that $\psi_* \circ J_{M^6} = J_{N^6} \circ \psi_*$.

In [6], we classified 6-dimensional extrinsic homogeneous almost Hermitian submanifolds of the octonions $\mathfrak{C}$ by making use of the classification of homogeneous isoparametric hypersurfaces of a unit sphere ([8], [10]). Also we give a list of 6-dimensional submanifolds of $\mathfrak{C}$ which are Riemannian homogeneous with respect to the induced metric. In general, the automorphism group of the induced orthogonal almost complex (Hermitian) structure does not act transitively on it. In this article, we give the applications of automorphism group of the homogeneous orthogonal almost complex structure of $S^3 \times S^3$.

2. Preliminaries

We recall the fundamental algebraic properties of the quaternions and octonions. We set $\mathbf{H}$ as the skew field of all quaternions with canonical basis $\{1, i, j, k\}$ satisfying

$$i^2 = j^2 = k^2 = -1, \ ij = -ji = k, \ jk = -kj = i, \ ki = -ik = j.$$

We define the octonions $\mathfrak{C}$ as the direct sum of two fields of quaternions $\mathbf{H} \oplus \mathbf{H} = \mathfrak{C}$ which admits the following multiplication

$$(a + b\varepsilon)(c + d\varepsilon) = ac - \bar{d}b + (da + b\bar{c})\varepsilon$$

with $a, b, c, d \in \mathbf{H}$, where $\varepsilon = (0,1) \in \mathbf{H} \oplus \mathbf{H}$ and the symbol " $^-$ " is the conjugation of the quaternion. For arbitrary $x, y \in \mathfrak{C}$, we can show that

$$\langle xy, xy \rangle = \langle x, x \rangle \langle y, y \rangle.$$

The algebra sayisfying this condition is called a "normed algebra" in [3]. It is well known that the octonions is a non-commutative, non-associative alternative division normed algebra. The important object is the group of automorphisms of the octonions with respect to the above multiplication. We set the exceptional simple Lie group G_2 which consists of all automorphisms of the octonions which preserve the product. That is,

$$G_2 = \{g \in SO(8) \mid g(uv) = g(u)g(v) \text{ for any } u, v \in \mathfrak{C}\}.$$

In this paper, we shall concern ourself with the Lie group $Spin(7)$ which is defined by

$$Spin(7) = \{g \in SO(8) \mid g(uv) = g(u)\chi_g(v) \text{ for any } u, v \in \mathfrak{C}\},$$

where $\chi_g(v) = g(g^{-1}(1)v)$. Note that G_2 is a Lie subgroup of $Spin(7)$:

$$G_2 = \{g \in Spin(7) \mid g(1) = 1\}.$$

The map χ defines a double covering map of $Spin(7)$ onto $SO(7)$ which satisfies the following equivariance

$$g(u) \times g(v) = \chi_g(u \times v),$$

for any $u, v \in \mathfrak{C}$. Here, the "exterior product" $u \times v$ of u and v is given by $u \times v = (1/2)(\bar{v}u - \bar{u}v)$, and $\bar{v} = 2\langle v, 1\rangle - v$ is the conjugation of $v \in \mathfrak{C}$. We note that $u \times v$ is pure-imaginary for any $u, v \in \mathfrak{C}$. To construct $U(3)$ (or $Spin(7)$)–frame field on M^6, we recall that a basis $(N, E_1, E_2, E_3, \bar{N}, \bar{E}_1, \bar{E}_2, \bar{E}_3)$ of the complexification of the octonions $\mathbf{C} \otimes_{\mathbf{R}} \mathfrak{C}$ over $\mathbf{C}$ is given as

$$N = (1/2)(1 - \sqrt{-1}\varepsilon),\ \bar{N} = (1/2)(1 + \sqrt{-1}\varepsilon),$$
$$E_1 = iN,\ E_2 = jN,\ E_3 = -kN,\ \bar{E}_1 = i\bar{N},\ \bar{E}_2 = j\bar{N},\ \bar{E}_3 = -k\bar{N}.$$

We can construct the $U(3)$-frame bundle along a map $\varphi : M \to \mathfrak{C}$ as follows: Let (f_1, f_2, f_3) be a local $U(3)$-frame field of M^6. Then each f_i is a local section of the bundle $T^{(1,0)}M^6 = \{X \in TM^6 \otimes \mathbf{C} \mid JX = \sqrt{-1}X\}$, where $TM^6 \otimes \mathbf{C}$ is a complexification of the tangent bundle of M. We set

$$(\mathfrak{f}_1, \mathfrak{f}_2, \mathfrak{f}_3) = (f_1, f_2, f_3)g$$

where $g \in U(3)$. We can then construct a $U(3)$-principal fibre bundle $\mathscr{F}^{U(3)}_{M^6}$ over M as

$$\mathscr{F}^{U(3)}_{M^6} = \{(m, (\mathfrak{f}_1, \mathfrak{f}_2, \mathfrak{f}_3, \bar{\mathfrak{f}}_1, \bar{\mathfrak{f}}_2, \bar{\mathfrak{f}}_3))\}$$

at any $m \in M^6$, where $\bar{\mathfrak{f}}_i$ is the complex conjugation of $\mathbf{C} \otimes_{\mathbf{R}} \mathfrak{C}$.

We note that $n = (1/2)(\xi - \sqrt{-1}\eta)$ is the local section of the complexified normal bundle satisfying $n(\eta \times \xi) = \sqrt{-1}n$.

3. Automorphism group of the homogeneous orthogonal almost complex structure of $S^3 \times S^3$

Since the oriented isometry group of $S^3 \times S^3$ coincides with $SO(4) \times SO(4)$, we see that the automorphism group preserving the homogeneous orthogonal almost complex structure on $S^3 \times S^3$ is $Sp(1) \times Sp(1) \times Sp(1)/\mathbf{Z}_2$. The action of this automorphism group of $S^3 \times S^3$ on $\mathfrak{C}$ is given by

$$\rho(q_1, q_2, q_3)(a + b\varepsilon) = q_2 a \bar{q}_1 + (q_3 b \bar{q}_1)\varepsilon, \tag{3.1}$$

for $(q_1, q_2, q_3) \in Sp(1) \times Sp(1) \times Sp(1)$, and $a + b\varepsilon \in \mathfrak{C}$. We note that $Sp(1) \times Sp(1) \times Sp(1)/\mathbf{Z}_2 \subset Spin(7)$. Then the orbit through the point $1 + \varepsilon \in \mathfrak{C}$ is isometric to $S^3 \times S^3$. In this case, each S^3 is included in a quaternionic subspace of $\mathfrak{C}$, therefore the automorphism group of $S^3 \times S^3$ is

$$\begin{aligned}&Aut(S^3 \times S^3, J, \langle\ ,\ \rangle)\\&= (Sp(1) \times Sp(1) \times Sp(1))/\mathbf{Z}_2 \subset (SO(4) \times SO(4)) \cap Spin(7),\end{aligned}$$

where J is the right multiplication of the element $(q_3\bar{q}_2)\varepsilon$. We remark that $(q_3\bar{q}_2)\varepsilon \in S^6 \subset \operatorname{Im}\mathfrak{C}$, where $\operatorname{Im}\mathfrak{C} = \{u \in \mathfrak{C} | \langle u, 1\rangle = 0\}$. By (3.1), we have

$$q_2\bar{q}_1 + (q_3\bar{q}_1)\varepsilon = \rho(q_1, q_2, q_3)(1 + \varepsilon),$$

and find

$$\begin{aligned}f_1(q_1, q_2, q_3) &= \frac{1}{2}\Big(q_2 i \bar{q}_1 - \sqrt{-1}(q_3 i \bar{q}_1)\varepsilon\Big) = \rho(q_1, q_2, q_3)(E_1),\\f_2(q_1, q_2, q_3) &= \frac{1}{2}\Big(q_2 j \bar{q}_1 - \sqrt{-1}(q_3 j \bar{q}_1)\varepsilon\Big) = \rho(q_1, q_2, q_3)(E_2),\\f_3(q_1, q_2, q_3) &= -\frac{1}{2}\Big(q_2 k \bar{q}_1 - \sqrt{-1}(q_3 k \bar{q}_1)\varepsilon\Big) = \rho(q_1, q_2, q_3)(E_3),\end{aligned}$$

form a $Sp(1) \times Sp(1) \times Sp(1)/\mathbf{Z}_2(\subset Spin(7))$-frame field. We remark that this frame is a $U(3)$-valued function on $S^3 \times S^3 \times S^3$. Since

$$\begin{aligned}f_1(q_1, q_2, q_3) &= \frac{1}{2}\Big((q_2\bar{q}_1)q_1 i \bar{q}_1 - \sqrt{-1}((q_3\bar{q}_1)q_1 i \bar{q}_1)\varepsilon\Big)\\&= \frac{1}{2}\Big(x q_1 i \bar{q}_1 - \sqrt{-1}(y q_1 i \bar{q}_1)\varepsilon\Big),\end{aligned}$$

the function $(f_1(q_1, q_2, q_3), f_2(q_1, q_2, q_3), f_3(q_1, q_2, q_3))$ is *not* a field on $S^3 \times S^3$. If we fix an element $q_1 \in Sp(1)$, we obtain the complexified frame field on $S^3 \times S^3$.

Proposition 3.1. *Let* $\big(\mathscr{F}^{U(3)}_{S^3 \times S^3}, S^3 \times S^3, \pi, U(3)\big)$ *be the principal* $U(3)$*-bundle over* $S^3 \times S^3$ *which comes from the octonions. Then the total space*

$\mathscr{F}^{U(3)}_{S^3\times S^3}$ can be considered as the fibre bundle structure over $S^3 \times S^3 \times S^3$ with a fibre $U(3)/SO(3)$. The homogeneous space $U(3)/SO(3)$ is the Grassmannian manifold of all Lagrangian subspaces of $\boldsymbol{C}^3$

4. Fibre bundle structures on $S^3 \times S^3$

From the previous section, we have the principal right $Sp(1)$-bundle structure

$$(Sp(1) \times Sp(1) \times Sp(1),\ S^3 \times S^3,\ \pi,\ Sp(1))$$

on $S^3 \times S^3$, where $\pi : Sp(1) \times Sp(1) \times Sp(1) \to S^3 \times S^3$ is the map $\pi(q_1, q_2, q_3) = q_2\bar{q}_1 + (q_3\bar{q}_1)\varepsilon$. The right $Sp(1)$-action on the total space is given by

$$R_q(q_1, q_2, q_3) = (q_1q, q_2q, q_3q),$$

for any $q \in Sp(1)$. We see that $\pi \circ R_q = \pi$. Therefore the map π define the principal right $Sp(1)$-bundle over $S^3 \times S^3$. The vertical unit vector fields are given by

$$\begin{aligned}
V_1(q_1, q_2, q_3) &= \frac{1}{\sqrt{3}}(q_1i,\ q_2i,\ q_3i),\\
V_2(q_1, q_2, q_3) &= \frac{1}{\sqrt{3}}(q_1j,\ q_2j,\ q_3j),\\
V_3(q_1, q_2, q_3) &= \frac{1}{\sqrt{3}}(q_1k,\ q_2k,\ q_3k),
\end{aligned}$$

for any $(q_1, q_2, q_3) \in Sp(1) \times Sp(1) \times Sp(1)$. Then $\{V_\alpha\}$ is an orthonormal frame with respect to the canonical product metric on $Sp(1) \times Sp(1) \times Sp(1) \simeq S^3 \times S^3 \times S^3$. Also we can easily see that

$$\pi_*(V_\alpha) = 0.$$

The metric $\hat{g}_{S^3\times S^3\times S^3}$ compatible with the above fibration is

$$\hat{g}_{S^3\times S^3\times S^3} = \pi^*\left(g^{can}_{S^3\times S^3}\right) + \sum_{\alpha=1}^{3} \eta_\alpha \otimes \eta_\alpha,$$

where η_α is the dual 1-form of V_α such that $\eta_\alpha(V_\beta) = \delta_{\alpha\beta}$. These 1-forms $\{\eta_\alpha\}$ are defined by

$$\eta_\alpha(U) = \langle U,\ V_\alpha\rangle_{S^3\times S^3\times S^3}$$

for any $U \in T_{(q_1,q_2,q_3)}S^3 \times S^3 \times S^3$. Here, $\langle\ ,\ \rangle_{S^3\times S^3\times S^3}$ is the canonical product metric of $S^3 \times S^3 \times S^3$.

The horizontal lift X^H (which is a vector field of $S^3 \times S^3 \times S^3$) of the vector field X on $S^3 \times S^3$ is defined (in the above situation) by

$$\pi_*(X^H) = X \quad \text{and} \quad \eta_\alpha(X^H) = 0,$$

for any $\alpha \in \{1, 2, 3\}$.

Proposition 4.1. *Let $x+y\varepsilon$ be the position vector of $S^3 \times S^3 \subset \boldsymbol{H} \oplus \boldsymbol{H}\varepsilon \simeq \mathfrak{C}$, and $\big(xi,\ xj,\ xk, (yi)\varepsilon,\ (yj)\varepsilon,\ (yk)\varepsilon\big)$ be the global orthonormal frame field of $S^3 \times S^3$. Then the horizontal lift of these vector fields belonging to this orthonormal frame field are given by*

$$\begin{aligned}
(xi)^H_{(q_1,q_2,q_3)} &= \frac{-1}{3}\Big(iq_1, -2q_2(\bar{q}_1 i q_1),\ q_3(\bar{q}_1 i q_1)\Big) \\
&= \frac{-1}{3} R_{\bar{q}_1 i q_1}\big(q_1, -2q_2,\ q_3\big), \\
(xj)^H_{(q_1,q_2,q_3)} &= \frac{-1}{3}\Big(jq_1, -2q_2(\bar{q}_1 j q_1),\ q_3(\bar{q}_1 j q_1)\Big) \\
&= \frac{-1}{3} R_{\bar{q}_1 j q_1}\big(q_1, -2q_2,\ q_3\big), \\
(xk)^H_{(q_1,q_2,q_3)} &= \frac{-1}{3}\Big(kq_1, -2q_2(\bar{q}_1 k q_1),\ q_3(\bar{q}_1 k q_1)\Big) \\
&= \frac{-1}{3} R_{\bar{q}_1 k q_1}\big(q_1, -2q_2,\ q_3\big), \\
((yi)\varepsilon)^H_{(q_1,q_2,q_3)} &= \frac{-1}{3}\Big(iq_1,\ q_2(\bar{q}_1 i q_1), -2q_3(\bar{q}_1 i q_1)\Big) \\
&= \frac{-1}{3} R_{\bar{q}_1 i q_1}\big(q_1,\ q_2, -2q_3\big), \\
((yj)\varepsilon)^H_{(q_1,q_2,q_3)} &= \frac{-1}{3}\Big(jq_1,\ q_2(\bar{q}_1 j q_1), -2q_3(\bar{q}_1 j q_1)\Big) \\
&= \frac{-1}{3} R_{\bar{q}_1 j q_1}\big(q_1,\ q_2, -2q_3\big), \\
((yk)\varepsilon)^H_{(q_1,q_2,q_3)} &= \frac{-1}{3}\Big(kq_1,\ q_2(\bar{q}_1 k q_1), -2q_3(\bar{q}_1 k q_1)\Big) \\
&= \frac{-1}{3} R_{\bar{q}_1 k q_1}\big(q_1,\ q_2, -2q_3\big),
\end{aligned}$$

at $(q_1, q_2, q_3) \in S^3 \times S^3 \times S^3$, where $x + y\varepsilon = \pi(q_1, q_2, q_3) = q_2\bar{q}_1 + (q_3\bar{q}_1)\varepsilon$.

Proof. (1) We can easily see that $\eta_\alpha((xi)^H_{(q_1,q_2,q_3)}) = 0$.
(2) Next we show that $\pi_*\big((xi)^H_{(q_1,q_2,q_3)}\big) = xi$. Note that $\pi_*\big((0, q_2 i, 0)\big) = q_2 i \bar{q}_1$ and $xi = q_2\bar{q}_1 i$, and that they are different from each other. By direct calculation, we can show that

$$\pi_*\big((0,\ q_2(\bar{q}_1 i q_1),\ 0)\big) = q_2\bar{q}_1 i.$$

Therefore, the horizontal lift of the vector field xi coincides with

$$\begin{aligned}(xi)^H_{(q_1,q_2,q_3)} &= (0,\ q_2(\bar{q}_1 i q_1),\ 0) - \sum_{\alpha=1}^{3} \eta_\alpha\big((0,\ q_2(\bar{q}_1 i q_1),\ 0)\big)V_\alpha \\ &= (0,\ q_2(\bar{q}_1 i q_1),\ 0) - \frac{1}{3}\Big\{\langle \bar{q}_1 i q_1,\ i\rangle(q_1 i,\ q_2 i,\ q_3 i) \\ &\qquad + \langle \bar{q}_1 i q_1,\ j\rangle(q_1 j,\ q_2 j,\ q_3 j) + \langle \bar{q}_1 i q_1,\ k\rangle(q_1 k,\ q_2 k,\ q_3 k)\Big\} \\ &= (0,\ q_2(\bar{q}_1 i q_1),\ 0) - \frac{1}{3}\big(q_1(\bar{q}_1 i q_1),\ q_2(\bar{q}_1 i q_1),\ q_3(\bar{q}_1 i q_1)\big) \\ &= \frac{-1}{3}\Big(i q_1,\ -2q_2(\bar{q}_1 i q_1),\ q_3(\bar{q}_1 i q_1)\Big).\end{aligned}$$

Also we have

$$\pi_*\big((0,\ 0,\ q_3(\bar{q}_1 i q_1))\big) = (q_3\bar{q}_1 i)\varepsilon = (yi)\varepsilon.$$

In the same way, since

$$\pi_*\big((0,\ 0,\ q_3(\bar{q}_1 i q_1))\big) = (q_3\bar{q}_1 i)\varepsilon,$$

we get

$$((yi)\varepsilon)^H_{(q_1,q_2,q_3)} = \frac{-1}{3}\Big(i q_1,\ q_2(\bar{q}_1 i q_1),\ -2q_3(\bar{q}_1 i q_1)\Big).$$

We can prove other equalities by the same arguments. □

5. On some geometric structure on $S^3 \times S^3 \times S^3$

By Proposition 4.1, we have the following splitting

$$T_{(q_1,q_2,q_3)}S^3 \times S^3 \times S^3 = H_{(q_1,q_2,q_3)} \oplus V_{(q_1,q_2,q_3)}.$$

Here $H_{(q_1,q_2,q_3)}$ is the horizontal distribution whose basis (over $\mathbf{R}$) are given by

$$\begin{aligned}&(xi)^H_{(q_1,q_2,q_3)}, \quad (xj)^H_{(q_1,q_2,q_3)}, \quad (xk)^H_{(q_1,q_2,q_3)}, \\ &\big((yi)\varepsilon\big)^H_{(q_1,q_2,q_3)}, \quad \big((yj)\varepsilon\big)^H_{(q_1,q_2,q_3)}, \quad \big((yk)\varepsilon\big)^H_{(q_1,q_2,q_3)},\end{aligned}$$

and $V_{(q_1,q_2,q_3)}$ is the vertical distribution which is given by

$$V_{(q_1,q_2,q_3)} = \mathrm{span}_R\Big\{V_1(q_1,q_2,q_3), V_2(q_1,q_2,q_3), V_3(q_1,q_2,q_3)\Big\}.$$

From this splitting, we can define a triplet of $(1,1)$ tensor fields $(\psi_1,\ \psi_2,\ \psi_3)$ as

$$\psi_1 = (J\pi_*)^H + \mathscr{I},\ \psi_2 = (J\pi_*)^H + \mathscr{J},\ \psi_3 = (J\pi_*)^H + \mathscr{K}, \tag{5.1}$$

globally on $S^3 \times S^3 \times S^3$ as

$$\begin{aligned}
\psi_1(X^H + V) &= (J\pi_*(X^H))^H + \mathscr{I}(V) = (JX)^H + (R_i(V))^\top, \\
\psi_2(X^H + V) &= (J\pi_*(X^H))^H + \mathscr{J}(V) = (JX)^H + (R_j(V))^\top, \\
\psi_3(X^H + V) &= (J\pi_*(X^H))^H + \mathscr{K}(V) = (JX)^H + (R_k(V))^\top,
\end{aligned}$$

where X is the vector field of $S^3 \times S^3$ and V is a vertical vector field of $S^3 \times S^3 \times S^3$. Here

$$(R_i(V))^\top = (Vi)^\top = \sum_{\alpha=1}^{3} \langle Vi, V_\alpha \rangle V_\alpha,$$

and $(R_j(V))^\top, (R_k(V))^\top$ are defined by the same way. Then we have

$$\begin{aligned}
{\psi_1}^2 &= \psi_1 \circ \psi_1 = -id_{T(S^3 \times S^3 \times S^3)} + \eta_1 \otimes V_1, \\
{\psi_2}^2 &= \psi_2 \circ \psi_2 = -id_{T(S^3 \times S^3 \times S^3)} + \eta_2 \otimes V_2, \\
{\psi_3}^2 &= \psi_3 \circ \psi_3 = -id_{T(S^3 \times S^3 \times S^3)} + \eta_3 \otimes V_3, \\
-\psi_1 \circ \psi_2 + \eta_2 \otimes V_1 &= id_H + \mathscr{K} \\
\psi_2 \circ \psi_1 - \eta_1 \otimes V_2 &= -id_H + \mathscr{K}, \\
-\psi_2 \circ \psi_3 + \eta_3 \otimes V_2 &= id_H + \mathscr{I}, \\
\psi_3 \circ \psi_2 - \eta_2 \otimes V_3 &= -id_H + \mathscr{I}, \\
-\psi_3 \circ \psi_1 + \eta_1 \otimes V_3 &= id_H + \mathscr{J}, \\
\psi_1 \circ \psi_3 - \eta_3 \otimes V_1 &= -id_H + \mathscr{J},
\end{aligned}$$

where $id_{T(S^3 \times S^3 \times S^3)}$ and id_H are the identity map at each tangent space $T(S^3 \times S^3 \times S^3)$ and the horizontal subspace, respectively. Summing up,

Theorem 5.1. *Let $(S^3 \times S^3 \times S^3,\ S^3 \times S^3,\ \pi,\ Sp(1))$ be the principal $Sp(1)$-bundle over $S^3 \times S^3$. Then there exist a $U(3) \times Sp(1)$-structure $(\psi_1,\ \psi_2,\ \psi_3)$ which is defined as* (5.1).

We note that if we put

$$\begin{aligned}
{f_1}^H &= \frac{1}{2}\Big((xi)^H_{(q_1,q_2,q_3)} - \sqrt{-1}((yi)\varepsilon)^H_{(q_1,q_2,q_3)}\Big) \\
&= \frac{-1}{6}\Big(R_{\bar{q}_1 i q_1}\Big\{\Big(q_1, -2q_2,\ q_3\Big) - \sqrt{-1}\Big(q_1,\ q_2,\ -2q_3\Big)\Big\}\Big), \\
{f_2}^H &= \frac{1}{2}\Big((xj)^H_{(q_1,q_2,q_3)} - \sqrt{-1}((yj)\varepsilon)^H_{(q_1,q_2,q_3)}\Big) \\
&= \frac{-1}{6}\Big(R_{\bar{q}_1 j q_1}\Big\{\Big(q_1, -2q_2,\ q_3\Big) - \sqrt{-1}\Big(q_1,\ q_2,\ -2q_3\Big)\Big\}\Big),
\end{aligned}$$

$$\begin{aligned} f_3{}^H &= -\frac{1}{2}\Big((xk)^H_{(q_1,q_2,q_3)} - \sqrt{-1}((yk)\varepsilon)^H_{(q_1,q_2,q_3)}\Big) \\ &= \frac{1}{6}\Big(R_{\bar{q}_1 k q_1}\Big\{\Big(q_1, -2q_2,\ q_3\Big) - \sqrt{-1}\Big(q_1,\ q_2,\ -2q_3\Big)\Big\}\Big). \end{aligned}$$

Then we see that $\big(f_1{}^H + V_1,\ \ f_2{}^H + V_2,\ \ f_3{}^H + V_3\big)$ is a $U(3) \times Sp(1)$-frame field on $S^3 \times S^3 \times S^3$.

6. Remarks

Note that the homogeneous space

$$\{\rho(q_1, q_2, q_3)(N, E, \bar{N}, \bar{E}) \mid (q_1, q_2, q_3) \in Sp(1) \times Sp(1) \times Sp(1)\}$$

is a $Sp(1) \times Sp(1) \times Sp(1)/Z_2$ orbit which can be considered as a homogeneous Lagrangian frame bundle over $S^3 \times S^3$ with respect to the induced homogeneous orthogonal almost complex structure. However, the principal right $Sp(1)$-fibre bundle $\big(S^3 \times S^3 \times S^3,\ S^3 \times S^3,\ \pi,\ Sp(1)\big)$ does not coincide with the above homogeneous space.

Bibliography

1. R.L. Bryant, *Submanifolds and special structures on the octonions*, J. Diff. Geom. 17(1982), 185–232.
2. T. Fukami and S. Ishihara, *Almost Hermitian structure on S^6*, Tohoku Math. J. 7(1955), 151–156.
3. R. Harvey and H.B. Lawson, *Calibrated geometries*, Acta Math. 148(1982), 47–157.
4. H. Hashimoto, *Characteristic classes of oriented 6-dimensional submanifolds in the octonians*, Kodai Math. J. 16(1993), 65–73.
5. H. Hashimoto, *Oriented 6-dimensional submanifolds in the octonions* ***III***, Internat. J. Math. Math. Sci. 18(1995), 111–120.
6. H. Hashimoto, T. Koda, K. Mashimo and K. Sekigawa, *Extrinsic homogeneous almost Hermitian 6-dimensional submanifolds in the octonions*, Kodai Math. J. 30(2007), 297–321.
7. H. Hashimoto and M. Ohashi, *Orthogonal almost complex structures of hypersurfaces of purely imaginary octonions*, Hokkaido Math. J. 39(2010), 351–387.
8. W.Y. Hsiang and H.B. Lawson, *Minimal submanifolds of low cohomogenity*, J. Diff. Geom. 5 (1971), 1–38.
9. T. Takahashi, *Homogeneous hypersurfaces in space of constant curvature*, J. Math. Soc. Japan (1970) 395–410.
10. R. Takagi and T. Takahashi, *On the principal curvatures of homogeneous hypersurfaces in a sphere*, in *Differential Geometry, in honor of K. Yano*, Kinokuniya, Tokyo, (1972) 469–481.

Received March 28, 2012
Revised April 18, 2013

Proceedings of the 3rd International
Colloquium on Differential Geometry
and its Related Fields
Veliko Tarnovo, September 3–7, 2012

ON A MODULI THEORY OF MINIMAL SURFACES

Norio EJIRI *

Department of Mathematics, Meijo University,
Tempaku, Nagoya, 468-8502 Japan
E-mail: ejiri@meijo-u.ac.jp

Toshihiro SHODA †

Faculty of Culture and Education, Saga University,
1 Honzyo, Saga 840-8502 Japan
E-mail: tshoda@cc.saga-u.ac.jp

In this paper, we report on a moduli theory of compact oriented minimal surfaces in flat tori. A key space of the moduli space can be classified into connected components via the Morse index and nullity, and some connected components have good geometric structures. Moreover, the theory includes a procedure to compute the Morse index of a minimal surface with only trivial Jacobi fields. As its applications, we can compute the Morse index of some examples of minimal surfaces. This is a survey of the previous works [3–5].

Keywords: Minimal surfaces; Flat tori; Moduli; Morse index; Nullity.

1. Introduction

Our object is a properly n-periodic minimal immersion of an oriented surface into an n-dimensional Euclidean space $\mathbb{R}^n$. In particular, many physicists and chemists have studied such a minimal surface for $n = 3$ ([6, 11]). It can be considered as a minimal immersion of a compact oriented surface into an n-dimensional flat torus $\mathbb{R}^n/\Lambda$, and the conformal structure induced by the immersion makes the surface a Riemann surface. It is usually called ***a conformal minimal immersion***. We now refer to backgrounds from two points of view, namely, the Morse theory and moduli theory of minimal surfaces.

The Morse index of a compact oriented minimal surface in a flat torus is defined as the sum of the dimensions of the eigenspaces corresponding

*Partially supported by JSPS Grant-in-Aid for Scientific Research (C) 22540103.
†Partially supported by JSPS Grant-in-Aid for Young Scientists (B) 24740047.

to negative eigenvalues of the second variational operator of area. The nullity is the dimension of the 0-eigenspace. In 1968, J. Simons [16] obtained the second variational formula for an oriented minimal submanifold and computed the Morse index and nullity of a totally geodesic sphere in the sphere. Also, using Simons' technique, we can compute the Morse index and nullity of a totally geodesic subtorus in a flat torus. But there were few other results for a compact oriented minimal surface in a flat torus until the 1990s. One of impressive developments was given by S. Montiel and A. Ros [9] in 1991. By their results, we can show that both Morse index and nullity of Schwarz' CLP surface are 3. After that, in 1992, M. Ross obtained that each Morse index of Schwarz' P surface, D surface, and Schoen's Gyroid is 1. Moreover, Ross' argument implies that each nullity of the three surfaces is 3. No other results are known in the past two decades, however.

Next we mention the moduli theory. The moduli theory of compact oriented minimal surfaces in flat tori has been established by C. Arezzo [1] and G. P. Pirola [12]. They showed a geometric structure of the moduli space and obtained the existence of minimal surfaces. Their main tool is given by the deformation theory and algebraic geometry. On the other hand, N. Ejiri [3, 4], the first author, considered a moduli theory of compact oriented minimal surfaces in flat tori in terms of differential geometry, and studied it from the viewpoint of the Morse index and nullity. He obtained a geometric structure of the moduli space. Arezzo and Pirola considered only nullity in their theory, and so, Ejiri's theory may be one of refined versions. Furthermore, he gave an algorithm to compute the Morse index under some assumptions.

In the previous paper [5], we carried out Ejiri's algorithm and computed the Morse index of the examples which were studied in physics and chemistry [6, 11] (H family, rPD family, tP family, tD family, tCLP family, and so on). Note that our results include Montiel-Ros and Ross' results. In the present paper, we devote to introduce an outline of Ejiri's theory and do not refer to details of arguments on the Morse index.

The paper is organized as follows: §2 contains the moduli theory of minimal surfaces given by Ejiri. We prepare six subsections for explanation. In §3, we state some results related to the Morse index of a minimal surface to impress with a usefulness of Ejiri's algorithm.

2. A moduli theory of minimal surfaces

2.1. *Translations of the Morse index and nullity of a minimal surface*

In this subsection, we translate the Morse index and nullity of a minimal surface into the Morse index and nullity of a critical point of the energy on the Teichmüller space defined by R. Schoen and S. T. Yau [15], respectively. First, we review Schoen and Yau's arguments.

Let R be a compact surface of genus $\gamma \geq 2$. Suppose that ϕ is a smooth map of R to an n-dimensional flat torus $\mathbb{R}^n/\Lambda$, and set

$$\mathcal{F} = \{f \in L_1^2(R, \mathbb{R}^n/\Lambda) \,|\, f_\sharp = C^{-1}\phi_\sharp C \text{ for some curve } C \text{ from } f(*) \text{ to } \phi(*)\},$$

where $L_1^2(R, \mathbb{R}^n/\Lambda)$ is the Hilbert space of maps having square integrable first derivatives in the distribution sense and $*$ is a fixed point on R. Note that, for the case $\dim_{\mathbb{R}} R = 2$, the set $C^0(R, \mathbb{R}^n/\Lambda)$ is dense in $L_1^2(R, \mathbb{R}^n/\Lambda)$. So every map $f \in L_1^2(R, \mathbb{R}^n/\Lambda)$ can be approximated by a continuous map. Using such approximation, we can define a homomorphism $f_\sharp$ of f.

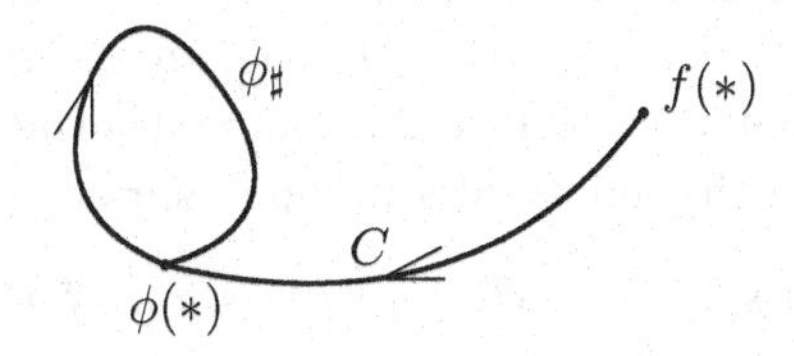

Schoen and Yau showed that there exists an energy minimizing harmonic map s_ϕ in $\mathcal{F}$. We call s_ϕ ***a harmonic map for*** ϕ and s_ϕ is unique up to translations ([10]). We now define the Teichmüller space $\mathcal{T}_\gamma$ and an energy E_ϕ on $\mathcal{T}_\gamma$.

A point in $\mathcal{T}_\gamma$ is a pair (M, h) of a Riemann surface M of genus γ and a diffeomorphism $h : R \to M$. Two pairs (M_1, h_1), (M_2, h_2) represent the same point if and only if $h_2 \circ h_1^{-1}$ is homotopic to a biholomorphism. It is well-known that $\mathcal{T}_\gamma$ is diffeomorphic to $\mathbb{C}^{3\gamma-3}$. By using the isothermal coordinates compatible with a metric, we can consider the Riemann surface M as a pair (R, g) of the fixed compact surface R and a Riemannian metric g on R with constant scalar curvature -1. Given any smooth map $\phi : R \to \mathbb{R}^n/\Lambda$ and a point $p = (M, h)$ on $\mathcal{T}_\gamma$, we apply the result as the above to see the existence of a harmonic map $s_{(\phi \circ h^{-1})} : M \to \mathbb{R}^n/\Lambda$ for $\phi \circ h^{-1}$. Recall that the energy $E(s_{(\phi \circ h^{-1})})$ of $s_{(\phi \circ h^{-1})}$ is given by $\int_R |ds_{(\phi \circ h^{-1})}|^2 dv_g$.

The conformal invariance of $E(\cdot)$ shows that the value of $E(s_{(\phi\circ h^{-1})})$ is independent of the pair (M, h) used to represent p. This construction gives a function $E_\phi : \mathcal{T}_\gamma \to \mathbb{R}$ defined by $E_\phi(p) = E(s_{(\phi\circ h^{-1})})$.

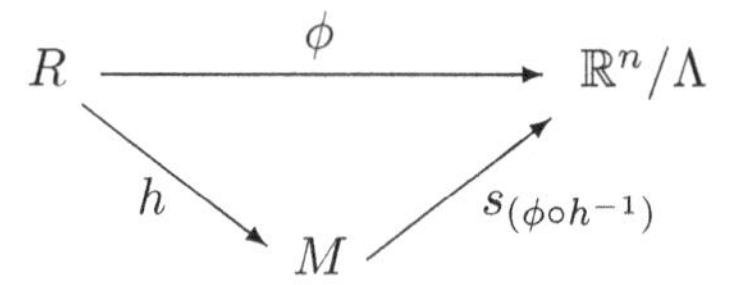

The following theorem suggests that the energy E_ϕ is one of important objects in Differential Geometry.

Theorem 2.1 ([14, 15]). *A point $p = (M, h) \in \mathcal{T}_\gamma$ is critical for E_ϕ if and only if $s_{(\phi\circ h^{-1})}$ is a weakly conformal harmonic map, that is, a conformal (branched) minimal immersion.*

Every minimal surface is a critical point of the area functional and it is corresponding to a critical point of E_ϕ as the above. Next we consider relations between the Morse index (resp. the nullity) of a minimal surface and the Morse index (resp. the nullity) of E_ϕ at the corresponding critical point. Note that E_ϕ is a smooth function on $\mathcal{T}_\gamma \cong \mathbb{R}^{6\gamma-6}$ ([3]), so we can define the Morse index and nullity of E_ϕ in the standard way. Let $index_a$ (resp. $index_E$) denote the Morse index of a minimal surface (resp. E_ϕ) and $nullity_a$ (resp. $nullity_E$) the nullity of a minimal surface (resp. E_ϕ).

Theorem 2.2 ([3]). *If $s_{(\phi\circ h^{-1})} : M \to \mathbb{R}^n/\Lambda$ is a conformal minimal immersion and $p = (M, h) \in \mathcal{T}_\gamma$ is the corresponding critical point of E_ϕ. Then*

$$index_a = index_E, \quad nullity_a - n = nullity_E.$$

Remark 2.1. A normal vector field vanishing the second variational operator of area is called ***Jacobi field***, and the dimension of the space of Jacobi fields is equal to the nullity of a minimal surface. Normal components of the Killing vector fields generated by translation on $\mathbb{R}^n/\Lambda$ induce Jacobi fields. As a consequence, $nullity_a \geq n(= \dim_{\mathbb{R}} \mathbb{R}^n/\Lambda)$. We say that ***it has only trivial Jacobi fields*** if $nullity_a = n$. In this case, the nullity is given by only Killing vector fields generated by translation.

Recall that a compact oriented minimal surface in a flat torus corresponds to a periodic minimal surface in the Euclidean space. Hence the nullity of a compact oriented minimal surface in a flat torus is induced by periodic Jacobi fields in the Euclidean space. Every normal component of translation on the torus can be reconsidered as a periodic Jacobi field but rotation may not. It follows that we treat only translations as Jacobi fields in our case.

Corollary 2.1. *Let $s_{(\phi \circ h^{-1})} : M \to \mathbb{R}^n/\Lambda$ be a conformal minimal immersion and $p = (M, h) \in \mathcal{T}_\gamma$ be the corresponding critical point of E_ϕ. Suppose that it has only trivial Jacobi fields. Then E_ϕ is non-degenerate at p.*

2.2. *A translation of the Weierstrass representation formula via a Riemann matrix and periods*

For a conformal minimal immersion, the following theorem is a basic tool.

Theorem 2.3 (Weierstrass representation formula). *Let $f : M \longrightarrow \mathbb{R}^n/\Lambda$ be a conformal minimal immersion. Then, up to translations, f can be represented by the following path-integrals:*

$$f(p) = \Re \int_{p_0}^{p} (\omega_1, \ldots, \omega_n)^t \mod \Lambda,$$

where p_0 is a fixed point on M and the ω_i's are holomorphic differentials on M satisfying the following three conditions;

$$\omega_1^2 + \cdots + \omega_n^2 = 0, \tag{1}$$

$$\omega_1, \ldots, \omega_n \text{ have no common zeros,} \tag{2}$$

$$\left\{ \Re \int_C (\omega_1, \ldots, \omega_n)^t \,\middle|\, C \in H_1(M, \mathbb{Z}) \right\} \text{ is a sublattice of } \Lambda. \tag{3}$$

Conversely, the real part of path-integrals of holomorphic differentials satisfying the above three conditions defines a conformal minimal immersion.

In this subsection, we introduce a useful way to rewrite the Weierstrass representation formula. First we review the theory of compact Riemann surfaces.

Let $\{(A_j, B_j)\}_{j=1}^{\gamma}$ be a canonical homology basis and $\{\varphi_1, \ldots, \varphi_\gamma\}$ be a basis of the space of holomorphic differentials on M. We set $\Phi = (\varphi_1, \ldots, \varphi_\gamma)^t$. Then there exists a unique basis $\{\varphi_i\}_{i=1}^{\gamma}$ with the following property:

$$\left(\int_{A_1} \Phi \ \cdots \ \int_{A_\gamma} \Phi \ \int_{B_1} \Phi \ \cdots \ \int_{B_\gamma} \Phi \right) = (I_\gamma \ \tau),$$

where I_γ is the identity matrix of degree γ and τ is a complex symmetric matrix of degree γ with $\Im \tau > 0$. This τ is called ***a Riemann matrix of*** M. Let $\mathcal{H}_\gamma$ denote the Siegel upper half space, that is, a set of complex symmetric matrices of degree γ with positive definite imaginary part. Note that $\mathcal{H}_1$ is the upper half plane in $\mathbb{C}$. Then we can define the following map:

$$\begin{aligned} \kappa : \quad \mathcal{T}_\gamma &\longrightarrow \mathcal{H}_\gamma \\ (M, h) &\longmapsto \tau. \end{aligned}$$

Then there exists a unique complex structure on $\mathcal{T}_\gamma$ such that κ is holomorphic.

Recall that a compact Riemann surface is said to be ***hyperelliptic*** if it is a two-sheeted branched covering of the sphere. Compact Riemann surfaces can be classified into hyperelliptic Riemann surfaces and non-hyperelliptic Riemann surfaces. Let

$$H\mathcal{T}_\gamma = \{(M, h) \in \mathcal{T}_\gamma \mid M \text{ is hyperelliptic}\}.$$

The following theorem summarizes classical results.

Theorem 2.4. *The set $H\mathcal{T}_\gamma$ is a $(2\gamma-1)$-dimensional complex submanifold in $\mathcal{T}_\gamma$. Moreover, κ has full rank $3\gamma-3$ at each point on $\mathcal{T}_\gamma - H\mathcal{T}_\gamma$ and $\kappa|_{H\mathcal{T}_\gamma}$ has full rank $2\gamma-1$.*

Setting $RM = \kappa(\mathcal{T}_\gamma)$, $RM_{\mathrm{hyp}} = \kappa(H\mathcal{T}_\gamma)$, and $RM_{\mathrm{non-hyp}} = \kappa(\mathcal{T}_\gamma - H\mathcal{T}_\gamma)$, we have $RM = RM_{\mathrm{hyp}} \cup RM_{\mathrm{non-hyp}}$ and $\dim_{\mathbb{C}} RM_{\mathrm{hyp}} = 2\gamma - 1$. Let $U(\gamma)$ denote the Torelli space. There exist a holomorphic covering $\pi : \mathcal{T}_\gamma \to U(\gamma)$ and a holomorphic map ${}^\sharp\kappa : U(\gamma) \to \mathcal{H}_\gamma$ satisfying $\kappa = {}^\sharp\kappa \circ \pi$. The map ${}^\sharp\kappa$ is $2:1$ on $U(\gamma) - \pi(H\mathcal{T}_\gamma)$ and $1:1$ on $\pi(H\mathcal{T}_\gamma)$. We have an involutive modular transformation Θ on $U(\gamma)$, and $U(\gamma)/\Theta$ can be identified with RM.

Next we construct a conformal minimal immersion via a Riemann matrix and periods. Let $L_{m,n}$ be a set of real (m, n) matrices and $K_{m,n}$ be a set of complex (m, n) matrices. Suppose that $s_{(\phi\circ h^{-1})} : M \to \mathbb{R}^n/\Lambda$ is a conformal minimal immersion and $p = (M, h) \in \mathcal{T}_\gamma$ is the corresponding critical point of E_ϕ. Now we set $\psi = \phi \circ h^{-1}$ for the sake of simplicity. Then ds_ψ is $\mathbb{R}^n$-valued harmonic 1-forms, and hence there exists $Q \in L_{n,2\gamma}$ such that $ds_\psi = Q \begin{pmatrix} \Re\Phi \\ \Im\Phi \end{pmatrix}$ holds. Let

$$L = (\psi_\sharp(A_1), \ldots, \psi_\sharp(A_\gamma), \psi_\sharp(B_1), \ldots, \psi_\sharp(B_\gamma)) \in L_{n,2\gamma}.$$

Since two homomorphisms $(s_\psi)_\sharp : \pi_1(M, *) \to \pi_1(\mathbb{R}^n/\Lambda, s_\psi(*))$ and $\psi_\sharp : \pi_1(M, *) \to \pi_1(\mathbb{R}^n/\Lambda, \psi(*))$ are related by $C^{-1}(s_\psi)_\sharp C = \psi_\sharp$ for some curve

C from $\psi(*)$ to $s_\psi(*)$, we have $(s_\psi)_\sharp = \psi_\sharp$ in $H_1(\mathbb{R}^n/\Lambda, \mathbb{Z})$. So

$$L = \begin{pmatrix} \int_{A_1} ds_\psi & \dots & \int_{A_\gamma} ds_\psi & \int_{B_1} ds_\psi & \dots & \int_{B_\gamma} ds_\psi \end{pmatrix}$$
$$= Q \begin{pmatrix} \Re\int_{A_1}\Phi \dots \Re\int_{A_\gamma}\Phi & \Re\int_{B_1}\Phi \dots \Re\int_{B_\gamma}\Phi \\ \Im\int_{A_1}\Phi \dots \Im\int_{A_\gamma}\Phi & \Im\int_{B_1}\Phi \dots \Im\int_{B_\gamma}\Phi \end{pmatrix} = Q \begin{pmatrix} I_\gamma & \Re\tau \\ O & \Im\tau \end{pmatrix}.$$

Setting $T_\tau = \begin{pmatrix} I_\gamma & \Re\tau \\ O & \Im\tau \end{pmatrix}$, we obtain $T_\tau^{-1} = \begin{pmatrix} I_\gamma & -(\Re\tau)(\Im\tau)^{-1} \\ O & (\Im\tau)^{-1} \end{pmatrix}$ and $Q = LT_\tau^{-1}$. It follows that $ds_\psi = LT_\tau^{-1}\begin{pmatrix} \Re\Phi \\ \Im\Phi \end{pmatrix}$. Note that $(1, 0)$-part of Φ is given by

$$\frac{1}{2}\left\{\begin{pmatrix} \Re\Phi \\ \Im\Phi \end{pmatrix} - i\begin{pmatrix} O & -I_\gamma \\ I_\gamma & O \end{pmatrix}\begin{pmatrix} \Re\Phi \\ \Im\Phi \end{pmatrix}\right\}.$$

Thus

$$ds_\psi^{(1,0)} = \frac{1}{2}LT_\tau^{-1}\left\{\begin{pmatrix} I_\gamma & O \\ O & I_\gamma \end{pmatrix} - i\begin{pmatrix} O & -I_\gamma \\ I_\gamma & O \end{pmatrix}\right\}\begin{pmatrix} \Re\Phi \\ \Im\Phi \end{pmatrix}$$
$$= \frac{1}{2}LT_\tau^{-1}\begin{pmatrix} \Phi \\ (-i)\Phi \end{pmatrix} = K(\tau, L)\Phi,$$

where

$$K(\tau, L) = \frac{1}{2}(L_1 + i[L_1\Re\tau - L_2](\Im\tau)^{-1})$$

for a decomposition $L = (L_1, L_2)$ $(L_i \in L_{n,\gamma})$.

Remark that the above $ds_\psi^{(1,0)}$ corresponds to $(\omega_1, \dots, \omega_n)^t$ in Theorem 2.3, but we do not describe (1) yet. As a result, if we add (1) to the above $ds_\psi^{(1,0)}$, then we obtain a conformal minimal immersion into a flat torus.

2.3. *Translations of the real period matrix and complex period matrix*

The diffeomorphism $L_{n,2\gamma} \cong \mathbb{R}^{2n\gamma}$ yields $T^*L_{n,2\gamma} \cong L_{n,2\gamma} \times L_{n,2\gamma}$, where the latter $L_{n,2\gamma}$ is the base manifold. Moreover, $L_{n,2\gamma} \times L_{n,2\gamma} \cong K_{n,\gamma} \times K_{n,\gamma}$ holds. In fact, we can construct the following diffeomorphisms:

$$\begin{array}{ccccc} T^*L_{n,2\gamma} & \cong & L_{n,2\gamma} \times L_{n,2\gamma} & \cong & K_{n,\gamma} \times K_{n,\gamma} \quad (\cong \mathbb{H}^{n\gamma}) \\ & & ((L_1, L_2), (L_3, L_4)) & \longmapsto & (L_3 - iL_2, L_4 + iL_1). \end{array}$$

For a conformal minimal immersion $s_\psi(p) = \Re\left(\int_{p_0}^p \omega_1, \ldots, \int_{p_0}^p \omega_n\right)^t$, we call

$$\Re\begin{pmatrix} \int_{A_1}\omega_1 \cdots \int_{A_\gamma}\omega_1 \int_{B_1}\omega_1 \cdots \int_{B_\gamma}\omega_1 \\ \cdots \qquad\qquad \cdots \\ \int_{A_1}\omega_n \cdots \int_{A_\gamma}\omega_n \int_{B_1}\omega_n \cdots \int_{B_\gamma}\omega_n \end{pmatrix} \in L_{n,\,2\gamma}$$

a ***real period matrix*** and call

$$\begin{pmatrix} \int_{A_1}\omega_1 \cdots \int_{A_\gamma}\omega_1 \int_{B_1}\omega_1 \cdots \int_{B_\gamma}\omega_1 \\ \cdots \qquad\qquad \cdots \\ \int_{A_1}\omega_n \cdots \int_{A_\gamma}\omega_n \int_{B_1}\omega_n \cdots \int_{B_\gamma}\omega_n \end{pmatrix} \in K_{n,\,\gamma} \times K_{n,\,\gamma} \cong T^* L_{n,\,2\gamma}$$

a ***complex period matrix***.

Now we reconsider the above two period matrices via the Weierstrass representation formula in § 2.2. For the sake of simplicity, we use $\int_{A_1 \ldots A_\gamma B_1 \ldots B_\gamma}$ as path-integrals along $A_1, \ldots, A_\gamma, B_1, \ldots, B_\gamma$. We have

$$\begin{aligned} \Re \int_{A_1 \ldots A_\gamma B_1 \ldots B_\gamma} K(\tau,\, L)\Phi &= \Re\left(K(\tau,\, L)\int_{A_1 \ldots A_\gamma B_1 \ldots B_\gamma} \Phi\right) \\ &= \Re\left(K(\tau,\, L)(I_\gamma,\, \tau)\right) \\ &= \Re(K,\, K\tau) \in L_{n,\,2\gamma}, \\ \int_{A_1 \ldots A_\gamma B_1 \ldots B_\gamma} K(\tau,\, L)\Phi &= (K,\, K\tau) \in K_{n,\,\gamma} \times K_{n,\,\gamma} \cong T^* L_{n,\,2\gamma}. \end{aligned}$$

Using the above diffeomorphism, we obtain

$$\begin{aligned} K_{n,\,\gamma} \times K_{n,\,\gamma} &\cong L_{n,\,2\gamma} \times L_{n,\,2\gamma} \\ (K,\, K\tau) &\longmapsto \Re(-iK\tau,\, iK,\, K,\, K\tau) = \frac{1}{2}(LP(\tau),\, L), \end{aligned}$$

where

$$P(\tau) = \begin{pmatrix} \Im\tau + (\Re\tau)(\Im\tau)^{-1}(\Re\tau) & -(\Re\tau)(\Im\tau)^{-1} \\ -(\Im\tau)^{-1}(\Re\tau) & (\Im\tau)^{-1} \end{pmatrix}.$$

Note that P is an embedding of $\mathcal{H}_\gamma$ into a set of real symmetric matrices of degree 2γ ([3]).

Remark 2.2 (A differential geometric Schottky problem [3]). The original Schottky problem is to characterize the set RM in $\mathcal{H}_\gamma/SP(2\gamma,\, \mathbb{Z})$ (the moduli space of principally polarized Abelian varieties). On the other hand, the first author suggested us to characterize $P(RM)$ instead of RM as ***a differential geometric Schottky problem***. From this point of view, He showed an equivalence between Micallef's result on stable minimal surfaces and Oort-Steenbrink's infinitesimal Torelli Theorem.

2.4. *A new energy on $\mathcal{H}_\gamma$ and a translation of Theorem 2.1*

For each $p = (M, h) \in \mathcal{T}_\gamma$, we defined $E_\phi(p) = E(s_\psi)$. By direct calculations, the energy E_ϕ can be rewritten via the Riemann matrix τ of M and the real period matrix L of s_ψ ([3]). In fact, we have

$$E_\phi = \frac{1}{2}\mathrm{trace}(P(\tau)L^tL).$$

Now, for a fixed $L \in L_{n,2\gamma}$, we consider the following new energy on $\mathcal{H}_\gamma$ derived from E_ϕ:

$$\frac{1}{2}\mathrm{trace}(P(\tau)L^tL).$$

Let $E_L(\tau)$ or $E(\tau, L)$ denote the above energy on $\mathcal{H}_\gamma$.

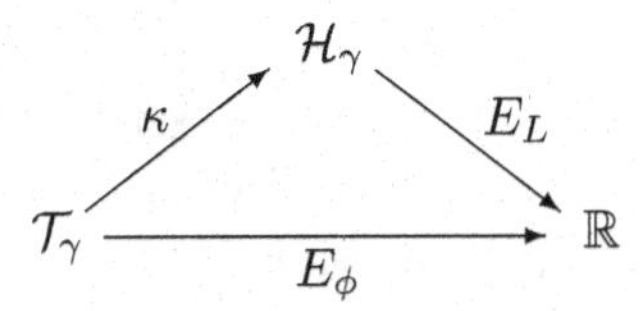

For the original case, τ and L depend on $p = (M, h)$, and so we cannot treat τ and L separately as the above. But we introduce the energy E_L and study its critical points independently of the original case.

Theorem 2.5 ([3]). *Let $L \in L_{n,2\gamma}$ be a fixed matrix.*

(i) *$\tau \in RM_{\mathrm{hyp}}$ is a critical point of $E_L|_{RM_{\mathrm{hyp}}}$ if and only if there exist a unique Riemann surface M and a unique canonical homology basis $\{A_i, B_i\}$ up to automorphisms on M such that $LT_\tau^{-1}(\Re\Phi^t, \Im\Phi^t)^t$ is weakly conformal.*

(ii) *$\tau \in RM_{\mathrm{non-hyp}}$ is a critical point of $E_L|_{RM_{\mathrm{non-hyp}}}$ if and only if there exist a unique Riemann surface M and two canonical homology bases $\{A_i, B_i\}$, $\{-A_i, -B_i\}$ up to automorphisms on M such that $\pm LT_\tau^{-1}(\Re\Phi^t, \Im\Phi^t)^t$ is weakly conformal.*

Corollary 2.2. *Suppose that $L \in L_{n,2\gamma}$ defines a lattice in $\mathbb{R}^n$ and hence determines an n-dimensional flat torus. Then path-integrals of* (i) *$LT_\tau^{-1}(\Re\Phi^t, \Im\Phi^t)^t$ and* (ii) *$\pm LT_\tau^{-1}(\Re\Phi^t, \Im\Phi^t)^t$ in Theorem* 2.5 *give conformal minimal immersions into the torus, respectively.*

Remark 2.3. The set $T_{n,2\gamma} := \{L \in L_{n,2\gamma} \mid L \text{ defines a lattice in } \mathbb{R}^n\}$ is dense in $L_{n,2\gamma}$.

Let $(z_1, \ldots, z_k)$ be a holomorphic coordinate system on RM, where

$$k = \begin{cases} 2\gamma - 1 & (\tau \in RM_{\text{hyp}}) \\ 3\gamma - 3 & (\tau \in RM_{\text{non-hyp}}) \end{cases}. \tag{4}$$

Theorem 2.5 and Corollary 2.2 imply

$$C(E_{RM}) := \left\{ (\tau, L) \in RM \times L_{n,2\gamma} \,\middle|\, \frac{\partial E}{\partial z_1}(\tau, L) = \cdots = \frac{\partial E}{\partial z_k}(\tau, L) = 0 \right\}$$

is a key space for the moduli theory of compact oriented minimal surfaces in flat tori. We call it a ***catastrophe set***. Moreover, Remark 2.3 suggests us that the moduli space of compact oriented minimal surfaces in flat tori may be dense in the catastrophe set (see Theorem 2.6).

2.5. *A geometric structure of the catastrophe set*

The ambient space $RM \times L_{n,2\gamma}$ of the catastrophe set is a complex manifold. The first author [4] proved that the catastrophe set is an analytic subvariety, that is, the common zero locus of a finite collection of holomorphic functions on $RM \times L_{n,2\gamma}$. In fact, $C(E_{RM})$ corresponds to

$$\left\{ (\tau, K) \in RM \times K_{n,\gamma} \,\middle|\, \text{trace}\left(\frac{\partial \tau}{\partial z_1} K^t K \right) = \cdots = \text{trace}\left(\frac{\partial \tau}{\partial z_k} K^t K \right) = 0 \right\}$$

and each $\text{trace}(\frac{\partial \tau}{\partial z_i}(K^t)K)$ is a holomorphic function on $RM \times K_{n,\gamma}$. Thus the geometry of $C(E_{RM})$ is very rich and it is in general made of many different irreducible components containing (possibly empty) singular locus. In this subsection, we introduce a geometric structure of a good irreducible component of the catastrophe set via the period matrix.

Before we state the theorem, we define the following. A point $(\tau, L) \in RM \times L_{n,2\gamma}$ is said to be ***non-degenerate*** if $E|_{RM}$ is non-degenerate at (τ, L). Combining Theorems 2.2 and 2.4, and the implicit function theorem for the natural projection $pr : C(E|_{RM}) \longrightarrow L_{n,2\gamma}$ yields

Theorem 2.6 ([4]). *Suppose that $N \subset C(E|_{RM})$ is an irreducible component with a non-degenerate point. Then, for every connected component $U(\subset N)$ which consists of non-degenerate points, there exists a connected open set $O \subset L_{n,2\gamma}$ such that $U = \{(\tau(L), L) \mid L \in O\}$. It follows that $\dim_{\mathbb{C}} N = n\gamma$. Furthermore, $pr^{-1}(T_{n,2\gamma})$ is dense in N.*

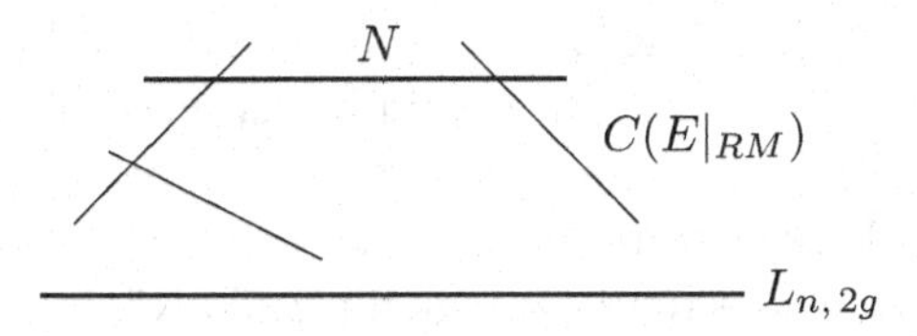

Next we consider an embedding of an irreducible component of the catastrophe set in terms of complex geometry and symplectic geometry. To do this, we first review some fundamental arguments (c.f. [2, 7]).

A ***symplectic manifold*** (Y, ω) is an even dimensional manifold Y endowed with a non-degenerate closed 2-form ω which is called a ***symplectic form***. Non-degeneracy means that the m-fold exterior product $\omega \wedge \cdots \wedge \omega$ never vanishes for $m = \frac{1}{2} \dim Y$. An m-dimensional submanifold X in a $2m$-dimensional symplectic manifold (Y, ω) is said to be ***Lagrangian*** if $\omega|_X = 0$. A typical example of Lagrangian (embedded) submanifolds is the zero section of the cotangent bundle. We can construct many Lagrangian embeddings $X \hookrightarrow T^*X$ by functions $f : X \to \mathbb{R}$ ([8]). In fact, we recall that a special class of Lagrangian submanifolds $X \hookrightarrow T^*X$ are the ones which are graphs of exact 1-forms $X = \text{graph}(df)$.

A ***complex symplectic manifold*** $(Y, \omega_{\mathbb{C}})$ is an even dimensional complex manifold Y endowed with a holomorphic closed 2-form $\omega_{\mathbb{C}}$ which is non-degenerate, i.e. the m-fold exterior product $\omega_{\mathbb{C}} \wedge \cdots \wedge \omega_{\mathbb{C}}$ never vanishes for $m = \frac{1}{2} \dim_{\mathbb{C}} Y$. The real and imaginary parts of $\omega_{\mathbb{C}}$ define real symplectic forms. We call $\omega_{\mathbb{C}}$ a ***complex symplectic form***. A real $2m$-dimensional submanifold X in a $2m$-dimensional complex symplectic manifold $(Y, \omega_{\mathbb{C}})$ is said to be ***complex Lagrangian*** if $\omega_{\mathbb{C}}|_X = 0$.

A ***pseudo Kähler manifold*** (Z, ω, J) is a symplectic manifold (Z, ω) with a complex structure J which is compatible with ω. This ω induces an indefinite Hermitian metric on Z, which is called a ***pseudo Kähler metric***. By a ***pseudo Kähler submanifold***, we mean a complex submanifold of a pseudo Kähler manifold with induced pseudo Kählerian structure. Finally, a (***pseudo***) ***hyper Kähler manifold*** is a $4m$-dimensional (pseudo) Riemannian manifold endowed with three almost complex structures I, J, $K = IJ$ which preserve the metric and satisfy $d\omega_I = d\omega_J = d\omega_K = 0$, where ω_I, ω_J, and ω_K are the fundamental 2-forms of I, J, and K, respectively. Note that each of ω_I, ω_J, ω_K is non-degenerate.

Let $(V, \omega_{\mathbb{C}})$ be a complex symplectic vector space of complex dimension $2m$. Suppose that $\sigma : V \to V$ is a compatible real structure. That is, σ is a $\mathbb{C}$-anti-linear involution such that the restriction of $\omega_{\mathbb{C}}$ to its fixed point set V^σ defines a real symplectic form. Up to isomorphisms, we can assume that $V = T^*\mathbb{C}^m$, $V^\sigma = T^*\mathbb{R}^m$, and $\omega_\mathbb{C}$ is the standard complex symplectic form. For $(V, \omega_\mathbb{C}, \sigma)$, we can obtain a Hermitian form η of signature (m, m) by $\eta(u, v) = -i\omega_\mathbb{C}(u, \sigma v)$. Moreover, $(V, \Im\eta, J)$ is a pseudo Kähler manifold of complex signature (m, m), where J is the complex structure on V. Remark that a connected Lagrangian pseudo Kähler submanifold X in $(V, \Im\eta, J)$ has a well-defined complex signature (p, q) $(p+q = m)$, namely, the signature of the Hermitian form $\eta|_X$. If $q = 0$, then $\eta|_X$ is positive definite and so X is a Kähler submanifold.

We apply the above arguments to our case. Under the hypotheses of Theorem 2.6, we have $\tau = \tau(L)$ on O. Setting $a(L) = E(\tau(L), L)$, we obtain $da(L) = LP(\tau(L))$ ([4]). As we mentioned before, the map

$$\begin{array}{ccc} O(\subset L_{n,2\gamma}) & \longrightarrow & T^*L_{n,2\gamma} \\ L & \longmapsto & (da(L), L)\ (\in \mathrm{graph}(da)) \end{array}$$

is a Lagrangian embedding via a graph of an exact 1-form da. Then, the following diagram follows from the diffeomorphisms in § 2.3.

$$\begin{array}{ccc} N & \xrightarrow{\quad\Pi\quad} & K_{n,\gamma} \times K_{n,\gamma} \cong T^*L_{n,2\gamma} \\ \Cup & & \Cup \\ (\tau(L), L) & \mathcal{I} \nearrow & (K, K\tau(L)) \mapsto (LP(\tau(L)), L) \\ \tau \uparrow & & \\ O & & \\ \Cup & & \\ L & & \end{array}$$

Here Π is defined by the complex period matrix and $\mathcal{I} = \Pi \circ \tau$.

The equality $a(tL) = t^2 a(L)$ was shown by the first author [4]. Thus $\mathcal{I}(O)$ is a Lagrangian cone in $T^*L_{n,2\gamma}$. Furthermore, we now give a procedure to make $\mathcal{I}(O)$ a Lagrangian pseudo Kähler cone in $T^*L_{n,2\gamma}$. Let

$$I\left(L_1\ L_2\ L_3\ L_4\right) = \left(L_1\ L_2\ L_3\ L_4\right)\begin{pmatrix} O & O & O & -I_\gamma \\ O & O & I_\gamma & O \\ O & -I_\gamma & O & O \\ I_\gamma & O & O & O \end{pmatrix},$$

$$J\left(L_1\ L_2\ L_3\ L_4\right) = \left(L_1\ L_2\ L_3\ L_4\right)\begin{pmatrix} O & O & I_\gamma & O \\ O & O & O & I_\gamma \\ -I_\gamma & O & O & O \\ O & -I_\gamma & O & O \end{pmatrix},$$

and $K = IJ$. Suppose that

$$\langle (L_1, L_2, L_3, L_4), (L_1', L_2', L_3', L_4') \rangle = \mathrm{trace} \sum_{i=1}^{4} L_i^t L_i'$$

is the standard Riemannian metric on $T^*L_{n,2\gamma}$ and

$$\omega_{\mathbb{C}}(W_1, W_2) = -\langle KW_1, W_2\rangle + i\langle JW_1, W_2\rangle,$$

where $W_i \in T^*L_{n,2\gamma}$. Then $\omega_{\mathbb{C}}$ defines a complex symplectic form on $T^*L_{n,2\gamma}$. It follows from straightforward calculations that $\omega_{\mathbb{C}}|_{\mathcal{I}(O)} = 0$, and hence $\mathcal{I}(O)$ is a complex Lagrangian cone in $T^*L_{n,2\gamma}$.

Using the diffeomorphism $T^*L_{n,2g} \cong K_{n,\gamma} \times K_{n,\gamma}$ in § 2.3, we arrive at the expression

$$\omega_{\mathbb{C}}((Z_1, Z_2), (Z_1', Z_2')) = \mathrm{trace}(Z_2^t Z_1' - Z_1^t Z_2').$$

Setting

$$\eta((Z_1, Z_2), (Z_1', Z_2')) = -i\omega_{\mathbb{C}}((Z_1, Z_2), \overline{(Z_1', Z_2')}),$$

we can show that $(T^*L_{n,2\gamma}, \Im\eta, J)$ is a pseudo (hyper) Kähler manifold. Therefore, $\mathcal{I} : O \hookrightarrow (T^*L_{n,2\gamma}, \Im\eta, J)$ is a Lagrangian pseudo Kähler cone in $(T^*L_{n,2\gamma}, \Im\eta, J)$. By Cortés' result [2], $\mathcal{I}^*(\Re\eta)$ gives a pseudo Kähler metric on O and its pseudo Kähler potential is a.

Finally, we remark two actions on N, namely, $SO(n)$ and $U(1)$. The ambient space is a flat torus $\mathbb{R}^n/\Lambda$, and $\mathbb{R}^n/\Lambda_1$ is orientation preserving isometric to $\mathbb{R}^n/\Lambda_2$ if and only if $\Lambda_1 = A\Lambda_2$ for some $A \in SO(n)$. Thus $SO(n)$ gives rise to an action on N. On the other hand, considering $e^{i\theta}(\omega_1, \dots, \omega_n)^t$ instead of $(\omega_1, \dots, \omega_n)^t$ in Theorem 2.3, we define an ***associate surface*** of a minimal surface. From this point of view, we obtain an action of $U(1)$ on N. It was shown in [4] that $SO(n) \times U(1)$ acts on U as a holomorphic isometry.

Summing up, we have the following theorem.

Theorem 2.7 ([4]). *Suppose that the hypotheses in Theorem 2.6 hold. Then every connected component $U(\subset N)$ which consists of non-degenerate points induces a Lagrangian pseudo Kähler cone in $T^*L_{n,2\gamma}$ with pseudo Kähler potential a. Furthermore, the signature q of η gives rise to an invariant on U. On the other hand, $SO(n)\times U(1)$ acts on U as a holomorphic isometry.*

2.6. *An algorithm to compute the Morse index of $E(\tau, L)$ and its application to hyperelliptic minimal surfaces*

By Theorem 2.2, it is useful to consider the Hessian matrix of E_ϕ to compute $index_a$ and $nullity_a$. To do this, we studied the Hessian matrix of $E(\tau, L)$ in § 2.4. In this subsection, we introduce an algorithm to compute the Morse index of $E(\tau, L)$ under the hypotheses in Theorem 2.6. Moreover, we give a procedure to compute $index_a$ for a hyperelliptic minimal surface in a flat torus with only trivial Jacobi fields. We use the same notations as in § 2.5.

By straightforward calculations ([4]), we have

$$\begin{aligned}
&\Re\eta\left(\left(\frac{\partial}{\partial L_i}P, \frac{\partial}{\partial L_i}\right), \left(\frac{\partial}{\partial L_j}P, \frac{\partial}{\partial L_j}\right)\right) - \frac{\partial^2 a(L)}{\partial L_i \partial L_j} \\
&= \Re\eta\left(\left(\frac{\partial}{\partial L_i}P, \frac{\partial}{\partial L_i}\right), \left(\frac{\partial}{\partial L_j}P, \frac{\partial}{\partial L_j}\right)\right) - \mathcal{I}^*(\Re\eta)\left(\frac{\partial}{\partial L_i}, \frac{\partial}{\partial L_j}\right) \\
&= \frac{\partial^2 E(\tau, L)}{\partial \tau_a \partial \tau_b}\frac{\partial \tau_a}{\partial L_i}\frac{\partial \tau_b}{\partial L_j},
\end{aligned} \tag{5}$$

where $\{L_i\}$ and $\{\tau_a\}$ are (real) local coordinate systems on $L_{n,\,2\gamma}$ and RM, respectively. To see the behavior of $\dfrac{\partial^2 E(\tau, L)}{\partial \tau_a \partial \tau_b}$, we introduce the following:

Definition 2.1 (surjective condition). Under the hypotheses in Theorem 2.6, we say $(\tau, L) \in N$ satisfies ***surjective condition*** if the differential of the map

$$\begin{aligned}
\tau : L_{n,\,2\gamma} &\longrightarrow RM \\
L &\longmapsto \tau(L)
\end{aligned}$$

is surjective.

From the terms $\left\{\dfrac{\partial \tau_a}{\partial L_i}\right\}, \left\{\dfrac{\partial \tau_b}{\partial L_j}\right\}$ in (5), the 0-eigenspace of

$$\Re\eta\left(\left(\frac{\partial}{\partial L_i}P, \frac{\partial}{\partial L_i}\right), \left(\frac{\partial}{\partial L_j}P, \frac{\partial}{\partial L_j}\right)\right) - \frac{\partial^2 a(L)}{\partial L_i \partial L_j} \tag{6}$$

has the dimension at least $\dim_{\mathbb{R}}(\ker(\tau_*))$. So the dimension is equal to $\dim_{\mathbb{R}}(\ker(\tau_*))$ if and only if $E(\tau, L)$ is non-degenerate. In this case, we conclude that the sum of the dimensions of the eigenspaces which correspond to negative eigenvalues of (6) is the Morse index of $E(\tau, L)$. The fundamental result in the linear algebra yields

$$\begin{aligned}
\dim_{\mathbb{R}}(\ker(\tau_*)) &\geq \begin{cases} \dim_{\mathbb{R}} L_{n,\,2\gamma} - \dim_{\mathbb{R}} RM_{\text{hyp}} & (\tau \in RM_{\text{hyp}}) \\ \dim_{\mathbb{R}} L_{n,\,2\gamma} - \dim_{\mathbb{R}} RM_{\text{non-hyp}} & (\tau \in RM_{\text{non-hyp}}) \end{cases} \\
&= \begin{cases} 2n\gamma - (4\gamma - 2) & (\tau \in RM_{\text{hyp}}) \\ 2n\gamma - (6\gamma - 6) & (\tau \in RM_{\text{non-hyp}}) \end{cases}.
\end{aligned}$$

The equality holds if and only if (τ, L) satisfies surjective condition. Note that the equality

$$\dim_{\mathbb{R}}\{\text{the } 0\text{−eigenspace of } (6)\} = \begin{cases} (2n-4)\gamma + 2 & (\tau \in RM_{\text{hyp}}) \\ (2n-6)\gamma + 6 & (\tau \in RM_{\text{non-hyp}}) \end{cases}$$

implies that (τ, L) satisfies surjective condition and the sum of the dimensions of the eigenspaces corresponding to negative eigenvalues of (6) is the Morse index of $E(\tau, L)$.

Remark 2.4. The zero loci of the determinant of (6) divide N into connected components consist of non-degenerate points. From Theorem 2.7, the signature q, that is, the Morse index of $E(\tau, L)$ gives an invariant on each connected component.

To obtain a basis $\left\{\mathcal{I}_*\left(\frac{\partial}{\partial L_i}\right)\right\}$ of the tangent space of $\mathcal{I}(O)$ at $\mathcal{I}(L)$ in $T^*L_{n,2\gamma}$, we consider two objects, namely, infinitesimal deformations of the complex period matrix $(K, K\tau) \in T^*L_{n,2\gamma}$ and an action of $SO(n) \times U(1)$ on $U(\subset N)$. It follows from the deformation of the Hodge structure that the former can be described by periods of differentials of the second kind, that is, meromorphic differentials with no residues. Let $\delta_1(K, K\tau), \ldots, \delta_k(K, K\tau)$ be infinitesimal deformations of $(K, K\tau)$ for k as in (4). Then these $\delta_i(K, K\tau)$'s define tangent vectors at $(K, K\tau)(L)$ in $K_{n,\gamma} \times K_{n,\gamma}$. For the latter, $SO(n) \times U(1)$ acts on U. Hence

$$\underbrace{\begin{pmatrix} 0 & 1 & \cdots & 0 \\ -1 & 0 & \cdots & 0 \\ & & \cdots\cdots & \\ 0 & & \cdots & 0 \end{pmatrix}(K, K\tau), \ldots, \begin{pmatrix} 0 & \cdots & & 0 \\ 0 & \cdots & & 0 \\ 0 & \cdots\cdots & 0 & 1 \\ 0 & \cdots & -1 & 0 \end{pmatrix}(K, K\tau),}_{\mathcal{SO}(n)-\text{part}} \underbrace{iI_n(K, K\tau)}_{\mathcal{U}(1)-\text{part}}$$

define tangent vectors at $(K, K\tau)(L)$ in $K_{n,\gamma} \times K_{n,\gamma}$, where $\mathcal{SO}(n)$ and $\mathcal{U}(1)$ are the Lie algebras of $SO(n)$ and $U(1)$, respectively. Choosing tangent vectors from the above candidates, we get a basis $\{(\alpha_i, \beta_i)\}$ of the tangent space at $(K, K\tau)(L)$ in $K_{n,\gamma} \times K_{n,\gamma}$. To translate $\{(\alpha_i, \beta_i)\}$ into a basis in $L_{n,2\gamma} \times L_{n,2\gamma}$, we take its real and imaginary parts. Therefore, substituting $\{(\Re\alpha_i, \Re\beta_i)\}$ and $\{(\Im\alpha_i, \Im\beta_i)\}$ into (6), we can compute the Morse index of $E(\tau, L)$ in some cases.

Now we refer to its application to hyperelliptic minimal surfaces, and we use the notations as the aboves. Let $s_{(\phi\circ h^{-1})} : M \to \mathbb{R}^n/\Lambda$ be a conformal minimal immersion of a hyperelliptic Riemann surface M into a flat torus with only trivial Jacobi fields. Let $\{A_i, B_i\}$ be a canonical homology basis on M and $\tau \in RM_{\text{hyp}}$ the Riemann matrix of M. Suppose that $L \in L_{n,2\gamma}$ defines the lattice Λ in $\mathbb{R}^n$. Then the arguments in § 2.2 yield

$$s_{(\phi \circ h^{-1})}(p) = \int_{p_0}^{p} LT_\tau^{-1} \begin{pmatrix} \Re\Phi \\ \Im\Phi \end{pmatrix}.$$

Combining Theorem 2.5 and Corollary 2.2 implies $(\tau, L) \in C(E_{RM})$. Let N be an irreducible component of $C(E_{RM})$ containing (τ, L). From Corollary 2.1, we assert that the hypotheses of Theorem 2.6 are true. Recall that if we decompose $\mathcal{T}_\gamma$ into the tangent and normal component of $H\mathcal{T}_\gamma$, then we arrive at the following expression ([3]):

$$\text{Hess}(E_\phi) = \begin{pmatrix} \left(\dfrac{\partial^2 E(\tau, L)}{\partial\tau_a \partial\tau_b}\right) & & & O & \\ & \lambda_1 & & & \\ & & -\lambda_1 & O & \\ O & & & \cdots & \\ & & O & \lambda_{\gamma-2} & \\ & & & & -\lambda_{\gamma-2} \end{pmatrix}.$$

Therefore,

$$index_a = (\text{the Morse index of } E(\tau, L)) + \gamma - 2.$$

For the case $\dim_{\mathbb{R}}\{\text{the } 0\text{-eigenspace of } (6)\} = (2n-4)\gamma+2$, we can compute the index of $E(\tau, L)$ via (6).

3. List of results related to the Morse index of a hyperelliptic minimal surface in a flat 3-torus

In the previous paper [5], we carried out the procedure in § 2.6. This section contains a summary of our results.

Theorem 3.1 (H family). *For $a \in (0, 1)$, let M be a hyperelliptic Riemann surface of genus 3 defined by $w^2 = z(z^3 - a^3)\left(z^3 - \frac{1}{a^3}\right)$ and f be a conformal minimal immersion given by*

$$f(p) = \Re \int_{p_0}^{p} i(1 - z^2, i(1 + z^2), 2z)^t \frac{dz}{w}.$$

Then there exist $0 < a_1 < a_2 < 1$ satisfying the followings:

(i) *$index_a = 2$, $nullity_a = 3$, and $q = 4$ for $a \in (0, a_1)$,*
(ii) *$index_a = 1$, $nullity_a = 3$, and $q = 5$ for $a \in (a_1, a_2)$,*
(iii) *$index_a = 3$, $nullity_a = 3$, and $q = 3$ for $a \in (a_2, 1)$,*

where $a_1 \sim 0.497010$, $a_2 \sim 0.714792$.

The family of minimal surfaces in the above theorem is said to be a ***H family***. We note that we obtain the similar results for $a > 1$.

Theorem 3.2 (rPD family). *For* $a \in (0, 1]$, *let* M *be a hyperelliptic Riemann surface of genus* 3 *defined by* $w^2 = z(z^3 - a^3)\left(z^3 + \frac{1}{a^3}\right)$ *and* f *be a conformal minimal immersion given by*

$$f(p) = \Re \int_{p_0}^{p} i(1 - z^2,\, i(1 + z^2),\, 2z)^t \frac{dz}{w}.$$

Then there exist $0 < a_1 < 1$ *satisfying the followings:*

(i) $index_a = 2$, $nullity_a = 3$, *and* $q = 4$ *for* $a \in (0, a_1)$,
(ii) $index_a = 1$, $nullity_a = 3$, *and* $q = 5$ *for* $a \in (a_1, 1]$,

where $a_1 \sim 0.494722$.

The family of minimal surfaces in the above theorem is said to be a ***rPD family***. We note that we obtain the similar results for $a \geq 1$.

Theorem 3.3 (tP family, tD family). *For* $a \in (2, \infty)$, *let* M *be a hyperelliptic Riemann surface of genus* 3 *defined by* $w^2 = z^8 + az^4 + 1$ *and* f *be a conformal minimal immersion given by*

$$f(p) = \Re \int_{p_0}^{p} (1 - z^2,\, i(1 + z^2),\, 2z)^t \frac{dz}{w}.$$

Then there exist $2 < a_1 < 14 < a_2 < \infty$ *satisfying the followings:*

(i) $index_a = 2$, $nullity_a = 3$, *and* $q = 4$ *for* $a \in (2, a_1)$,
(ii) $index_a = 1$, $nullity_a = 3$, *and* $q = 5$ *for* $a \in (a_1, a_2)$,
(iii) $index_a = 2$, $nullity_a = 3$, *and* $q = 4$ *for* $a \in (a_2, \infty)$,

where $a_1 \sim 7.40284$, $a_2 \sim 28.7783$.

The family of minimal surfaces in the above theorem is said to be a ***tP family***. Moreover, a ***tD family*** is defined as a family of conjugate surfaces of minimal surfaces in tP family.

Theorem 3.4 (tCLP family). *For* $a \in (-2, 2)$, *let* M *be a hyperelliptic Riemann surface of genus* 3 *defined by* $w^2 = z^8 + az^4 + 1$ *and* f *be a conformal minimal immersion given by*

$$f(p) = \Re \int_{p_0}^{p} (1 - z^2,\, i(1 + z^2),\, 2z)^t \frac{dz}{w}.$$

Then, for an arbitrary $a \in (-2, 2)$, $index_a = 3$, $nullity_a = 3$, *and* $q = 3$.

The family of minimal surfaces in the above theorem is said to be a ***tCLP family***.

Bibliography

1. C. Arezzo and G. P. Pirola, *On the existence of periodic minimal surfaces*, J. Algebraic. Geom. **8** (1999), 765-785.
2. V. Cortés, *On Hyper Kähler Manifolds associated to Lagrangian Kähler Submanifolds of* $T^*\mathbb{C}^n$, Contemporary. Math. **308** (2002), 101-144.
3. N. Ejiri, *A Differential-Geometric Schottky Problem, and Minimal Surfaces in Tori*, Contemporary. Math. **308** (2002), 101-144.
4. N. Ejiri, *A generating function of a complex Lagrangian cone in* $\mathbb{H}^n$, preprint.
5. N. Ejiri and T. Shoda, *in preparation.*
6. A. Fogden, and M. Haeberlein, and S. Lidin, *Generalizations of the gyroid surface*, J. Phys. I. France. **3** (1993), 2371-2385.
7. N. J. Hitchin, *The Moduli space of Complex Lagrangian Submanifolds Asian.* J. Math. **3** (1999), 77-92.
8. D. McDuff and D. Salamon, *Introduction to Symplectic Topology*, second edition, *Oxford Science Publication.*
9. S. Montiel and A. Ros, *Schrödinger operator associated to a holomorphic map*, Global Differential Geometry and Global Analysis, Lecture note in Math. **1481** (1991), 147-174.
10. T. Nagano and B. Smyth, *Minimal Varieties and Harmonic maps in Tori*, Comment. Math. Helvetici. **50** (1975), 249-265.
11. G. E. Schröder-Turk, A. Fogden, and S. T. Hyde, *Bicontinuous geometries and molecular self-assembly: comparison of local curvature and global packing variations in genus-three cubic, tetragonal and rhombohedral surfaces*, Eur. Phys. J. B. **54** (2006), 509-524.
12. G. P. Pirola, *The infinitesimal variation of the spin abelian differentials and periodic minimal surfaces*, Comm. Anal. Geom. **6** (1998), 393-426.
13. M. Ross, *Schwarz' P and D surfaces are stable*, Diff. Geom. App. **2** (1992), 179-195.
14. J. Sacks and K. Uhlenbeck, *The Existence of Minimal Immersions of 2-spheres*, Ann. of Math. **113** (1981), 1-24.
15. R. Schoen and S. T. Yau, *Existence of Incompressible Minimal Surfaces and the Topology of Three Dimensional Manifolds with Non-Negative Scalar Curvature*, Ann. of Math. **110** (1979), 127-142.
16. J. Simons, *Minimal Varieties in Riemannian Manifolds*, Ann. of Math. **88** (1968), 62-105.

Received February 24, 2013
Revised April 15, 2013

Proceedings of the 3rd International
Colloquium on Differential Geometry
and its Related Fields
Veliko Tarnovo, September 3–7, 2012

THE RELATIONSHIPS BETWEEN G_2-INVARIANTS AND $SO(7)$-INVARIANTS OF CURVES IN $\mathrm{Im}\,\mathfrak{C}$

Misa OHASHI *

Department of Mathematics, Nagoya Institute of Technology,
Nagoya 466-8555, Japan
E-mail: ohashi.misa@nitech.ac.jp

For curves in the purely imaginary octonions, we represent their curvatures by using some of their G_2-invariant functions.

Keywords: Exceptional Lie group G_2; Frenet-Serret formula; G_2-congruence theorem; G_2-invariants.

1. Introduction

Let $\gamma : I \to \mathbf{R}^7$ be a regular curve in a 7-dimensional Euclidean space $\mathbf{R}^7$ with canonical metric. The group of orientation preserving isometries of $\mathbf{R}^7$ is the semi-direct product $\mathbf{R}^7 \rtimes SO(7)$. Let $\gamma : I \to \mathbf{R}^7$ and $\tilde{\gamma} : \tilde{I} \to \mathbf{R}^7$ be regular curves with arc-length parameter $s \in I$ and $\tilde{s} \in \tilde{I}$, respectively. These curves γ, $\tilde{\gamma}$ are called (orientation preserving) $SO(7)$-congruent to each other if there exist $(a,\ g) \in \mathbf{R}^7 \rtimes SO(7)$ and some constant $s_0 \in \mathbf{R}$ satisfying $\tilde{\gamma}(s+s_0) = g \circ \gamma(s) + a$ for all $s \in I$ ($\tilde{s} = s + s_0$). The geometry of curves on $\mathbf{R}^7$ under the action of $\mathbf{R}^7 \rtimes SO(7)$ is well known. Two curves are $SO(7)$-congruent to each other if and only if the corresponding 6 curvatures ($SO(7)$-invariant functions) associated two curves coincide each other.

A Euclidean 8-space has a special algebraic structure, which is called the octonions $\mathfrak{C}$ (or the Cayley algebra). The purely imaginary octonions $\mathrm{Im}\,\mathfrak{C}$ of $\mathfrak{C}$ can be identified with $\mathbf{R}^7$ as a vector space. The automorphism group of the octonions is the exceptional simple Lie group G_2 which is a Lie subgroup of $SO(7)$. We consider the geometry of curves on $\mathrm{Im}\,\mathfrak{C}$ under the action of $\mathrm{Im}\,\mathfrak{C} \rtimes G_2$ which is a subgeometry of $\mathbf{R}^7 \rtimes SO(7)$. Two curves $\gamma : I \to \mathrm{Im}\,\mathfrak{C}$, $\tilde{\gamma} : \tilde{I} \to \mathrm{Im}\,\mathfrak{C}$ of unit speed are called G_2-congruent

*This work is partially supported by the JSPS Institutional Program for Young Researcher Overseas Visits "Promoting international young researchers in mathematics and mathematical sciences led by OCAMI".

to each other if there exist $(a,\ g) \in \mathrm{Im}\,\mathfrak{C} \rtimes G_2$ and $s_0 \in \mathbf{R}$ satisfying $\psi(s+s_0) = g \circ \gamma(s) + a$ for all $s \in I$. Besides the geometry of curves on $\mathbf{R}^7 \rtimes SO(7)$ we give the G_2-congruence theorem for curves in $\mathrm{Im}\,\mathfrak{C}$. To do this, we determine G_2-invariant functions. Since G_2 is a closed subgroup of $SO(7)$, it is clear that G_2-invariants are $SO(7)$-invariants, but the converse is not true. The purpose of this paper is to describe the relationships G_2-invariants and $SO(7)$-invariants (Theorem 4.1).

2. Preliminaries

Let $\mathbf{H}$ be the skew field of all quaternions with canonical basis $\{1, i, j, k\}$, satisfying

$$i^2 = j^2 = k^2 = -1,\ ij = -ji = k,\ jk = -kj = i,\ ki = -ik = j.$$

The octonions (or the Cayley) $\mathfrak{C}$ over $\mathbf{R}$ can be considered as a direct sum $\mathbf{H} \oplus \mathbf{H}$ with the following multiplication

$$(a + b\varepsilon)(c + d\varepsilon) = ac - \bar{d}b + (da + b\bar{c})\varepsilon,$$

where $\varepsilon = (0,1) \in \mathbf{H} \oplus \mathbf{H}$ and $a, b, c, d \in \mathbf{H}$. Here any $a \in \mathbf{H}$ we denote by $\bar{a}$ its quaternionic conjugate. For arbitrary $x, y \in \mathfrak{C}$, we have

$$\langle xy, xy \rangle = \langle x, x \rangle \langle y, y \rangle .$$

In [2], an algebra with this condition is said to be a "normed algebra". The octonions is a non-commutative, non-associative alternative division algebra. The group of automorphisms of the octonions is the exceptional simple Lie group

$$G_2 = \{g \in SO(8) \mid g(uv) = g(u)g(v) \text{ for any } u, v \in \mathfrak{C}\}.$$

The "exterior product" of $\mathfrak{C}$ is defined by

$$u \times v = (1/2)(\bar{v}u - \bar{u}v),$$

where $\bar{v} = 2\langle v, 1 \rangle - v$ is the conjugation of $v \in \mathfrak{C}$. We note that $u \times v \in \mathrm{Im}\,\mathfrak{C}$, where

$$\mathrm{Im}\,\mathfrak{C} = \{u \in \mathfrak{C} \mid \langle u, 1 \rangle = 0\}.$$

3. G_2-congruence theorem of curves in $\mathrm{Im}\,\mathfrak{C}$

Let $\gamma : I \to \mathrm{Im}\,\mathfrak{C}$ be a regular curve with an arc-length parameter $s \in I$. We define the G_2-frame field $\{e_1, e_2, e_3, e_4, e_5, e_6, e_7\}$ along γ in the following manner. We put $e_4 = \gamma'(s) = \dfrac{d\gamma}{ds}$. If $k_1 = \|\gamma''(s)\| >$

0, then we set $e_1 = \dfrac{1}{k_1}e_4'$, and put $e_5 = e_1e_4$. If we assume $\kappa_2 = \sqrt{\|e_1'\|^2 - \langle e_1', e_4\rangle^2 - \langle e_1', e_5\rangle^2} > 0$, then we can define $e_2(s)$ as

$$e_2 = \frac{1}{\kappa_2}(e_1' - \langle e_1', e_4\rangle e_4 - \langle e_1', e_5\rangle e_5).$$

We set

$$e_3 = e_1e_2, \qquad e_6 = e_2e_4, \qquad e_7 = e_3e_4.$$

In this paper, we do not deal with curves which satisfy $k_1 = 0$ or $\kappa_2 = 0$. The multiplication table of the above frame field $(e_4,\ e_1, e_2, e_3,\ e_5, e_6, e_7)$ coincides with that of canonical basis $(\varepsilon,\ i, j, k,\ i\varepsilon, j\varepsilon, k\varepsilon)$ of $\operatorname{Im}\mathfrak{C}$. We call $(e_4,\ e_1, e_2, e_3,\ e_5, e_6, e_7)$ G_2-frame field along γ.

We give the Frenet-Serret formulas of type G_2.

Proposition 3.1 ([1, 2]). *Let γ be a curve in $\operatorname{Im}\mathfrak{C}$ with $k_1,\ \kappa_2 > 0$. The G_2-frame field along γ satisfies the following differential equation:*

$$\frac{d}{ds}(e_4,\ e_1, e_2, e_3,\ e_5, e_6, e_7) = (e_4,\ e_1, e_2, e_3,\ e_5, e_6, e_7)\Psi(s),$$

where

$$\Psi(s) = \left(\begin{array}{c|ccc|ccc} 0 & -k_1 & 0 & 0 & 0 & 0 & 0 \\ \hline k_1 & 0 & -\kappa_2 & 0 & -\rho_1 & 0 & 0 \\ 0 & \kappa_2 & 0 & -\alpha & 0 & -\rho_2 & -\beta_1 \\ 0 & 0 & \alpha & 0 & 0 & -\beta_2 & -\rho_3 \\ \hline 0 & \rho_1 & 0 & 0 & 0 & -\kappa_2 & 0 \\ 0 & 0 & \rho_2 & \beta_2 & \kappa_2 & 0 & -\alpha \\ 0 & 0 & \beta_1 & \rho_3 & 0 & \alpha & 0 \end{array}\right)$$

with functions given by

$$\rho_1 = \langle e_1',\ e_5\rangle, \quad \rho_2 = \langle e_2',\ e_6\rangle, \quad \rho_3 = \langle e_3',\ e_7\rangle,$$
$$\alpha = \langle e_2',\ e_3\rangle, \quad \beta_1 = \langle e_2',\ e_7\rangle, \quad \beta_2 = \langle e_3',\ e_6\rangle.$$

These functions satisfy the following relations:

$$\rho_1 + \rho_2 + \rho_3 = 0, \quad \beta_1 - \beta_2 - k_1 = 0. \tag{1}$$

From the construction of G_2-frame field, we find that these 6 functions $(k_1,\ \kappa_2,\ \rho_1,\ \rho_3,\ \alpha,\ \beta_1)$ are invariant under the action of G_2. We call these functions the complete G_2-invariants.

Remark 3.1. For a regular curve $\gamma : I \to \operatorname{Im}\mathfrak{C}$, we define Frenet-frame field $(v_1 \cdots v_7)$ along γ as follows. We put $v_1 = \gamma'(s)$. If $k_1 = \|\gamma''(s)\| > 0$, then we set $v_2 = \frac{1}{k_1}v_1'$. We take

$$v_i = \frac{1}{k_{i-1}}\left(v_{i-1}' - \langle v_{i-1}', v_{i-2}\rangle v_{i-2}\right),$$

by assuming that

$$k_{i-1} = \sqrt{\|v_{i-1}'\|^2 - \langle v_{i-1}', v_{i-2}\rangle^2} > 0, \quad (i = 3, \dots, 7).$$

Then the usual Frenet-Serret formula of $\mathbf{R}^7$ is given by

$$\frac{d}{ds}(v_1 \ \cdots v_7) = (v_1 \ \cdots v_7)\begin{pmatrix} 0 & -k_1 & 0 & \cdots & 0 \\ k_1 & 0 & -k_2 & & \vdots \\ 0 & k_2 & 0 & \ddots & 0 \\ \vdots & & \ddots & \ddots & -k_6 \\ 0 & \cdots & 0 & k_6 & 0 \end{pmatrix}.$$

The 6 functions k_i $(i = 1, \cdots, 6)$ are called curvatures of γ. Note that curvatures are $SO(7)$-invariant functions.

In [2], we proved the following theorem:

Theorem 3.1. *Let $\gamma_1, \gamma_2 : I \to \operatorname{Im}\mathfrak{C}$ be two curves parameterized by their arc-length. If the G_2-invariants $(k_1, \kappa_2, \rho_1, \rho_2, \alpha, \beta_1)$ associated with γ_1 and those $(\tilde{k}_1, \tilde{\kappa}_2, \tilde{\rho}_1, \tilde{\rho}_2, \tilde{\alpha}, \tilde{\beta}_1)$ associated with γ_2 satisfy*

$$k_1 = \tilde{k}_1, \ \kappa_2 = \tilde{\kappa}_2, \ \rho_1 = \tilde{\rho}_1, \ \rho_2 = \tilde{\rho}_2, \ \alpha = \tilde{\alpha}, \ \beta_1 = \tilde{\beta}_1,$$

then these curves are G_2-congruent to each other.

4. The relationship between $SO(7)$-invariants and G_2-invariants

In this section, we shall give the relationships between $SO(7)$-invariants and G_2-invariants. First, we represent our G_2-invariants for a regular curve by using its SO(7)-invariants. From the constructions of its G_2-frame field and Frenet-frame field, the relationship between G_2-frame field $(e_4, \ e_1, e_2, e_3, \ e_5, e_6, e_7)$ and the canonical Frenet frame field $(v_1, \ \cdots, \ v_7)$

is given by

$$\begin{aligned}
&e_4 = v_1, \qquad e_1 = v_2, \qquad e_5 = v_2 \times v_1, \\
&e_2 = \frac{1}{\kappa_2}(k_2 v_3 - \rho_1 v_2 \times v_1), \quad e_3 = \frac{1}{\kappa_2}(k_2 v_2 \times v_3 + \rho_1 v_1), \\
&e_6 = \frac{1}{\kappa_2}(k_2 v_3 \times v_1 + \rho_1 v_2), \quad e_7 = \frac{k_2}{\kappa_2}(v_2 \times v_3) \times v_1.
\end{aligned} \tag{2}$$

In [1], we proved the following:

Proposition 4.1. *Let* $(v_1,\ v_2,\ v_3,\ v_4)$ *be the first 4 vector fields in the Frenet frame field of* γ*. Then* G_2*-invariants* $(k_1,\ \kappa_2,\ \rho_1,\ \rho_3,\ \alpha,\ \beta_1)$ *are give by*

$$\begin{aligned}
k_1 &= \|v_1{}'\|, \\
\rho_1 &= \langle e_1', e_5\rangle = k_2 \langle v_3,\ v_2 \times v_1\rangle, \\
\kappa_2 &= \sqrt{\|e_1'\|^2 - \langle e_1', e_4\rangle^2 - \langle e_1', e_5\rangle^2} = \sqrt{k_2{}^2 - \rho_1{}^2}, \\
\rho_3 &= \langle e_3', e_7\rangle = -\frac{k_2{}^2 k_3}{\kappa_2{}^2}\langle v_4, v_3 \times v_1\rangle, \\
\alpha &= \langle e_2', e_3\rangle = -\frac{k_2{}^2 k_3}{\kappa_2{}^2}\langle v_4, v_3 \times v_2\rangle, \\
\beta_1 &= \langle e_2', e_7\rangle = -\frac{k_2{}^2 k_3}{\kappa_2{}^2}\langle v_4, (v_3 \times v_2) \times v_1\rangle,
\end{aligned}$$

where k_2 *and* k_3 *are 2nd and 3rd curvatures (*$SO(7)$*-invariants) of* γ*, respectively.*

In particular, if we set the associative angle $\sigma(s)$ as

$$\cos\sigma(s) = \langle v_3(s),\ v_2(s) \times v_1(s)\rangle = \frac{1}{\|\gamma'''\|\|\gamma'' \times \gamma'\|}\left\langle \gamma''',\ \gamma'' \times \gamma'\right\rangle,$$

then we have

$$\rho_1 = k_2 \cos\sigma, \quad \kappa_2 = k_2|\sin\sigma|.$$

Remark 4.1. We note that the G_2-invariants are written by the first 4 vector fields in the Frenet frame field of γ and the exterior product of $\mathfrak{C}$. In ortder to calculate the G_2-invariants we need the exterior product. Since the exterior product is not invariant under the action of $SO(7)$ in general, G_2-invariants can not calculate only from $SO(7)$-invariants.

Next, we represent $SO(7)$-invariants for a regular curve by using its G_2-invariants. Let $\gamma : I \to \mathrm{Im}\,\mathfrak{C}$ be a curve in $\mathrm{Im}\,\mathfrak{C}$. Let $(v_1, \cdots, v_7)$ be its Frenet-frame field and $(e_4\ e_1\, e_2\, e_3\ e_5\, e_6\, e_7)$ be its G_2-frame field. We represent the G_2-frame field of a curve γ in $\mathrm{Im}\,\mathfrak{C}$ by its Frenet frame field.

Proposition 4.2. *If we set*

$$a_{i1} = \langle v_{i+3}, v_2 \times v_1 \rangle\,, \quad a_{i2} = \langle v_{i+3}, v_3 \times v_2 \rangle\,,$$
$$a_{i3} = \langle v_{i+3}, v_3 \times v_1 \rangle\,, \quad a_{i4} = \langle v_{i+3}, (v_3 \times v_2) \times v_1 \rangle\,,$$

for any $i \in \{1, \cdots, 4\}$, *then, we obtain*

$$(e_4\ e_1\, e_2\, e_3\ e_5\, e_6\, e_7) = (v_1\ \cdots\ v_7)M, \tag{3}$$

where

$$M = \left(\begin{array}{c|ccc|ccc}
1 & 0 & 0 & 0 & 0 & 0 & 0 \\
\hline
0 & 1 & 0 & 0 & 0 & 0 & 0 \\
0 & 0 & |\sin\sigma| & 0 & \cos\sigma & 0 & 0 \\
0 & 0 & -\dfrac{\cos\sigma}{|\sin\sigma|}a_{11} & -\dfrac{a_{12}}{|\sin\sigma|} & a_{11} & \dfrac{a_{13}}{|\sin\sigma|} & -\dfrac{a_{14}}{|\sin\sigma|} \\
\hline
0 & 0 & -\dfrac{\cos\sigma}{|\sin\sigma|}a_{21} & -\dfrac{a_{22}}{|\sin\sigma|} & a_{21} & \dfrac{a_{23}}{|\sin\sigma|} & -\dfrac{a_{24}}{|\sin\sigma|} \\
0 & 0 & -\dfrac{\cos\sigma}{|\sin\sigma|}a_{31} & -\dfrac{a_{32}}{|\sin\sigma|} & a_{31} & \dfrac{a_{33}}{|\sin\sigma|} & -\dfrac{a_{34}}{|\sin\sigma|} \\
0 & 0 & -\dfrac{\cos\sigma}{|\sin\sigma|}a_{41} & -\dfrac{a_{42}}{|\sin\sigma|} & a_{41} & \dfrac{a_{43}}{|\sin\sigma|} & -\dfrac{a_{44}}{|\sin\sigma|}
\end{array}\right),$$

and $\cos\sigma = \dfrac{\langle e_1{}', e_5 \rangle}{\sqrt{\rho_1{}^2 + \kappa_2{}^2}}$.

Proof. Since $\langle e_5,\ e_4 \rangle = \langle e_5,\ v_1 \rangle = 0$ and $\langle e_5,\ e_1 \rangle = \langle e_5,\ v_2 \rangle = 0$, we can set

$$e_5 = \cos\sigma v_3 + a_{11}v_4 + a_{12}v_5 + a_{13}v_6 + a_{14}v_7. \tag{4}$$

Next we put

$$e_2 = \sum_{i=1}^{7} b_i v_i,$$

where each b_i is a $\mathbf{R}$-valued function on I for any $i \in \{1, \cdots, 7\}$. By (2), we see that

$$b_1 = b_2 = 0,$$

$$b_3 = \langle e_2,\ v_3\rangle = \left\langle \frac{1}{\kappa_2}(k_2 v_3 - \rho_1 v_2 \times v_1),\ v_3 \right\rangle$$
$$= \frac{k_2}{\kappa_2} - \frac{\rho_1}{\kappa_2}\cos\sigma = \frac{1}{|\sin\sigma|} - \frac{\cos^2\sigma}{|\sin\sigma|} = |\sin\sigma|,$$
$$b_4 = \langle e_2,\ v_4\rangle = \left\langle \frac{1}{\kappa_2}(k_2 v_3 - \rho_1 v_2 \times v_1),\ v_4 \right\rangle$$
$$= -\frac{\rho_1}{\kappa_2} a_{11} = -\frac{\cos\sigma}{|\sin\sigma|} a_{11}.$$

In the same way, we have

$$b_5 = -\frac{\cos\sigma}{|\sin\sigma|} a_{12},\quad b_6 = -\frac{\cos\sigma}{|\sin\sigma|} a_{13},\quad b_7 = -\frac{\cos\sigma}{|\sin\sigma|} a_{14}.$$

For other vector fields e_3, e_6 and e_7, applying the same arguments as above, we obtain the desired result. □

Remark 4.2. We note that

$$\mathscr{A}_\sigma(s) = \frac{1}{|\sin\sigma|}\begin{pmatrix} a_{11} & \cdots & a_{14} \\ \vdots & \ddots & \vdots \\ a_{41} & \cdots & a_{44} \end{pmatrix}$$

is an $SO(4)$-valued function.

Theorem 4.1. *Let $(k_1, \cdots, k_6)$ be the curvatures ($SO(7)$-invariants) of γ and $(\sigma\ \mathscr{A}_\sigma)$ be as above. We set $(\vec{a}_1\ \vec{a}_2\ \vec{a}_3\ \vec{a}_4)$ as $\mathscr{A}_\sigma = \dfrac{1}{|\sin\sigma|}(\vec{a}_1\ \vec{a}_2\ \vec{a}_3\ \vec{a}_4)$. Then the following relations hold:*

$$k_2 = \kappa_2|\sin\sigma| + \rho_1\cos\sigma,$$

$$k_3 = -\left(\frac{a_{11}}{\sin\sigma}\sigma' + a_{12}\alpha + a_{13}\rho_3 + a_{14}\beta_1\right),$$

$$k_4 = (\sin\sigma)^{-2}\left\{-k_1\begin{vmatrix} a_{12} & a_{13} \\ a_{22} & a_{23}\end{vmatrix} + k_2\begin{vmatrix} a_{11} & a_{13} \\ a_{21} & a_{23}\end{vmatrix} + k_3 a_{21}\cos\sigma\right\}$$
$$+ \left\langle \left(\frac{\vec{a}_1}{|\sin\sigma|}\right)',\ \frac{\vec{a}_2}{|\sin\sigma|}\right\rangle,$$

$$k_5 = (\sin\sigma)^{-2}\left\{-k_1\begin{vmatrix}a_{22} & a_{23}\\ a_{32} & a_{33}\end{vmatrix} + k_2\begin{vmatrix}a_{21} & a_{23}\\ a_{31} & a_{33}\end{vmatrix} + k_3\begin{vmatrix}a_{12} & a_{13} & a_{14}\\ a_{22} & a_{23} & a_{24}\\ a_{32} & a_{33} & a_{34}\end{vmatrix}\right\}$$

$$+\left\langle\left(\frac{\vec{a}_2}{|\sin\sigma|}\right)', \frac{\vec{a}_3}{|\sin\sigma|}\right\rangle,$$

$$k_6 = (\sin\sigma)^{-2}\left\{-k_1\begin{vmatrix}a_{32} & a_{33}\\ a_{42} & a_{43}\end{vmatrix} + k_2\begin{vmatrix}a_{31} & a_{33}\\ a_{41} & a_{43}\end{vmatrix} + k_3\begin{vmatrix}a_{12} & a_{13} & a_{14}\\ a_{32} & a_{33} & a_{34}\\ a_{42} & a_{43} & a_{44}\end{vmatrix}\right\}$$

$$+\left\langle\left(\frac{\vec{a}_3}{|\sin\sigma|}\right)', \frac{\vec{a}_4}{|\sin\sigma|}\right\rangle.$$

Proof. By (3), Propositions 4.2, 3.1, and by taking the derivative we see that

$$\begin{aligned}\frac{d}{ds}(e_4,\ e_1, e_2, e_3,\ e_5, e_6, e_7) &= (e_4,\ e_1, e_2, e_3,\ e_5, e_6, e_7)\Psi(s)\\ &= (v_1, \cdots, v_7)M(s)\Psi(s).\end{aligned} \tag{5}$$

On the other hand,

$$\begin{aligned}\text{the l. h. s. of (5)} &= \frac{d}{ds}\left((v_1, \cdots, v_7)M(s)\right)\\ &= \left(\frac{d}{ds}(v_1, \cdots, v_7)\right)M(s) + (v_1, \cdots, v_7)\frac{d}{ds}M(s)\\ &= (v_1, \cdots, v_7)\left(\Phi(s)M(s) + \frac{d}{ds}M(s)\right),\end{aligned}$$

where

$$\Phi(s) = \begin{pmatrix} 0 & -k_1 & 0 & \cdots & 0\\ k_1 & 0 & -k_2 & & \vdots\\ 0 & k_2 & 0 & \ddots & 0\\ \vdots & & \ddots & \ddots & -k_6\\ 0 & \cdots & 0 & k_6 & 0\end{pmatrix}.$$

Therefore we obtain

$$\Phi = M\Psi\,{}^tM - M'\,{}^tM. \tag{6}$$

Let n_{ij} be (i,j)-th component of $M\Psi\,{}^tM - M'\,{}^tM$. By (6), we have $n_{ij} = -n_{ji}$. By the definition of k_1, we get $n_{12} = k_1$. We shall show the following:

$$n_{23} = \kappa_2|\sin\sigma| + \rho_1\cos\sigma,$$

$$n_{34} = -\left(\frac{a_{11}}{\sin\sigma}\sigma' + a_{12}\alpha + a_{13}\rho_3 + a_{14}\beta_1\right),$$

$$n_{45} = (\sin\sigma)^{-2}\left\{-k_1\begin{vmatrix}a_{12} & a_{13}\\ a_{22} & a_{23}\end{vmatrix} + k_2\begin{vmatrix}a_{11} & a_{13}\\ a_{21} & a_{23}\end{vmatrix} + k_3 a_{21}\cos\sigma\right\}$$
$$+\left\langle\left(\frac{\vec{a}_1}{|\sin\sigma|}\right)', \frac{\vec{a}_2}{|\sin\sigma|}\right\rangle,$$

$$n_{56} = (\sin\sigma)^{-2}\left\{-k_1\begin{vmatrix}a_{22} & a_{23}\\ a_{32} & a_{33}\end{vmatrix} + k_2\begin{vmatrix}a_{21} & a_{23}\\ a_{31} & a_{33}\end{vmatrix} + k_3\begin{vmatrix}a_{12} & a_{13} & a_{14}\\ a_{22} & a_{23} & a_{24}\\ a_{32} & a_{33} & a_{34}\end{vmatrix}\right\}$$
$$+\left\langle\left(\frac{\vec{a}_2}{|\sin\sigma|}\right)', \frac{\vec{a}_3}{|\sin\sigma|}\right\rangle,$$

$$n_{67} = (\sin\sigma)^{-2}\left\{-k_1\begin{vmatrix}a_{32} & a_{33}\\ a_{42} & a_{43}\end{vmatrix} + k_2\begin{vmatrix}a_{31} & a_{33}\\ a_{41} & a_{43}\end{vmatrix} + k_3\begin{vmatrix}a_{12} & a_{13} & a_{14}\\ a_{32} & a_{33} & a_{34}\\ a_{42} & a_{43} & a_{44}\end{vmatrix}\right\}$$
$$+\left\langle\left(\frac{\vec{a}_3}{|\sin\sigma|}\right)', \frac{\vec{a}_4}{|\sin\sigma|}\right\rangle.$$

To do this, we decompose Ψ and M into blocks

$$\Psi = \left(\begin{array}{c|ccc|ccc} 0 & -k_1 & 0 & 0 & 0 & 0 & 0\\ \hline k_1 & & & & & & \\ 0 & & \Psi_1 & & & -{}^t\Psi_2 & \\ 0 & & & & & & \\ \hline 0 & & & & & & \\ 0 & & \Psi_2 & & & \Psi_1 & \\ 0 & & & & & & \end{array}\right), \quad M = \left(\begin{array}{c|ccc|ccc} 1 & 0 & 0 & 0 & 0 & 0 & 0\\ \hline 0 & & & & & & \\ 0 & & M_{11} & & & M_{12} & \\ 0 & & & & & & \\ \hline 0 & & & & & & \\ 0 & & M_{21} & & & M_{22} & \\ 0 & & & & & & \end{array}\right), \tag{7}$$

where

$$\Psi_1 = \begin{pmatrix} 0 & -\kappa_2 & 0\\ \kappa_2 & 0 & -\alpha\\ 0 & \alpha & 0\end{pmatrix}, \quad \Psi_2 = \begin{pmatrix} \rho_1 & 0 & 0\\ 0 & \rho_2 & \beta_2\\ 0 & \beta_1 & \rho_3\end{pmatrix},$$

$$M_{11} = \begin{pmatrix} 1 & 0 & 0 \\ 0 & |\sin\sigma| & 0 \\ 0 & -\dfrac{\cos\sigma}{|\sin\sigma|}a_{11} & -\dfrac{a_{12}}{|\sin\sigma|} \end{pmatrix}, \quad M_{12} = \begin{pmatrix} 0 & 0 & 0 \\ \cos\sigma & 0 & 0 \\ a_{11} & \dfrac{a_{13}}{|\sin\sigma|} & -\dfrac{a_{14}}{|\sin\sigma|} \end{pmatrix},$$

$$M_{21} = \begin{pmatrix} 0 & -\dfrac{\cos\sigma}{|\sin\sigma|}a_{21} & -\dfrac{a_{22}}{|\sin\sigma|} \\ 0 & -\dfrac{\cos\sigma}{|\sin\sigma|}a_{31} & -\dfrac{a_{32}}{|\sin\sigma|} \\ 0 & -\dfrac{\cos\sigma}{|\sin\sigma|}a_{41} & -\dfrac{a_{42}}{|\sin\sigma|} \end{pmatrix}, \quad M_{22} = \begin{pmatrix} a_{21} & \dfrac{a_{23}}{|\sin\sigma|} & -\dfrac{a_{24}}{|\sin\sigma|} \\ a_{31} & \dfrac{a_{33}}{|\sin\sigma|} & -\dfrac{a_{34}}{|\sin\sigma|} \\ a_{41} & \dfrac{a_{43}}{|\sin\sigma|} & -\dfrac{a_{44}}{|\sin\sigma|} \end{pmatrix}.$$

Also we set the block decomposition

$$M\Psi\,{}^tM - M'\,{}^tM = \left(\begin{array}{c|ccc|ccc} 0 & -k_1 & 0 & 0 & 0 & 0 & 0 \\ \hline k_1 & & & & & & \\ 0 & & N_{11} & & & N_{12} & \\ 0 & & & & & & \\ \hline 0 & & & & & & \\ 0 & & N_{21} & & & N_{22} & \\ 0 & & & & & & \end{array}\right). \tag{8}$$

By (7) and (8), we have

$$\begin{aligned} N_{11} = {} & \{M_{11}\Psi_1\,{}^tM_{11} + M_{12}\Psi_1\,{}^tM_{12} + M_{12}\Psi_2\,{}^tM_{11} - {}^t(M_{12}\Psi_2\,{}^tM_{11})\} \\ & - (M_{11}'\,{}^tM_{11} + M_{12}'\,{}^tM_{12}), \end{aligned} \tag{9}$$

$$\begin{aligned} N_{12} = {} & \{M_{11}\Psi_1\,{}^tM_{21} + M_{12}\Psi_1\,{}^tM_{22} + M_{12}\Psi_2\,{}^tM_{21} - M_{11}\,{}^t\Psi_2\,{}^tM_{22}\} \\ & - (M_{11}'\,{}^tM_{21} + M_{12}'\,{}^tM_{22}), \end{aligned} \tag{10}$$

$$\begin{aligned} N_{21} = {} & \{M_{21}\Psi_1\,{}^tM_{11} + M_{22}\Psi_1\,{}^tM_{12} + M_{22}\Psi_2\,{}^tM_{11} - M_{21}\,{}^t\Psi_2\,{}^tM_{12}\} \\ & - (M_{21}'\,{}^tM_{11} + M_{22}'\,{}^tM_{12}), \end{aligned} \tag{11}$$

$$\begin{aligned} N_{22} = {} & \{M_{21}\Psi_1\,{}^tM_{21} + M_{22}\Psi_1\,{}^tM_{22} + M_{22}\Psi_2\,{}^tM_{21} - {}^t(M_{22}\Psi_2\,{}^tM_{21})\} \\ & - (M_{21}'\,{}^tM_{21} + M_{22}'\,{}^tM_{22}). \end{aligned} \tag{12}$$

We calculate the right hand side of (9) by taking each block individually. We have

$$M_{11}\Psi_1\,{}^tM_{11} = \begin{pmatrix} 0 & -\kappa_2|\sin\sigma| & a_{11}\kappa_2\dfrac{\cos\sigma}{|\sin\sigma|} \\ \kappa_2|\sin\sigma| & 0 & a_{12}\alpha \\ -a_{11}\kappa_2\dfrac{\cos\sigma}{|\sin\sigma|} & a_{12}\alpha & 0 \end{pmatrix},$$

$$M_{12}\Psi_1\,{}^tM_{12} = \begin{pmatrix} 0 & 0 & 0 \\ 0 & 0 & -a_{13}\kappa_2\dfrac{\cos\sigma}{|\sin\sigma|} \\ 0 & a_{13}\kappa_2\dfrac{\cos\sigma}{|\sin\sigma|} & 0 \end{pmatrix},$$

hence obtain

$$\begin{aligned} &M_{11}\Psi_1{}^tM_{11} + M_{12}\Psi_1{}^tM_{12} \\ &= \begin{pmatrix} 0 & -\kappa_2|\sin\sigma| & a_{11}\kappa_2\dfrac{\cos\sigma}{|\sin\sigma|} \\ \kappa_2|\sin\sigma| & 0 & a_{12}\alpha - a_{13}\kappa_2\dfrac{\cos\sigma}{|\sin\sigma|} \\ -a_{11}\kappa_2\dfrac{\cos\sigma}{|\sin\sigma|} & -a_{12}\alpha + a_{13}\kappa_2\dfrac{\cos\sigma}{|\sin\sigma|} & 0 \end{pmatrix}; \end{aligned}$$

we have

$$M_{12}\Psi_2{}^tM_{11} - {}^t\big(M_{12}\Psi_2{}^tM_{11}\big) = \begin{pmatrix} 0 & -\rho_1\cos\sigma & -a_{11}\rho_1 \\ \rho_1\cos\sigma & 0 & -(a_{13}\rho_2 - a_{14}\beta_1) \\ a_{11}\rho_1 & a_{13}\rho_2 - a_{14}\beta_1 & 0 \end{pmatrix};$$

also we have

$$\begin{aligned} &M_{11}{}'\,{}^tM_{11} \\ &= \begin{pmatrix} 0 & 0 & 0 \\ 0 & |\sin\sigma|'|\sin\sigma| & a_{11}|\sin\sigma|'\dfrac{\cos\sigma}{|\sin\sigma|} \\ 0 & -|\sin\sigma|\left(\dfrac{a_{11}\cos\sigma}{|\sin\sigma|}\right)' & \dfrac{a_{11}\cos\sigma}{|\sin\sigma|}\left(\dfrac{a_{11}\cos\sigma}{|\sin\sigma|}\right)' + \dfrac{a_{12}}{|\sin\sigma|}\left(\dfrac{a_{12}}{|\sin\sigma|}\right)' \end{pmatrix}, \end{aligned}$$

$$\begin{aligned} &M_{12}{}'\,{}^tM_{12} \\ &= \begin{pmatrix} 0 & 0 & 0 \\ 0 & (\cos\sigma)'\cos\sigma & a_{11}(\cos\sigma)' \\ 0 & a_{11}\cos\sigma & a_{11}a_{11}' + \dfrac{a_{13}}{|\sin\sigma|}\left(\dfrac{a_{13}}{|\sin\sigma|}\right)' + \dfrac{a_{14}}{|\sin\sigma|}\left(\dfrac{a_{14}}{|\sin\sigma|}\right)' \end{pmatrix}. \end{aligned}$$

Therefore we get

$$\begin{aligned} n_{23} &= -n_{32} = -\left(\kappa_2|\sin\sigma| + \rho_1\cos\sigma\right), \\ n_{34} &= -n_{43} \\ &= a_{12}\alpha - a_{13}\kappa_2\frac{\cos\sigma}{|\sin\sigma|} - (a_{13}\rho_2 - a_{14}\beta_1) \\ &\quad + a_{11}|\sin\sigma|'\frac{\cos\sigma}{|\sin\sigma|} + a_{11}(\cos\sigma)' \end{aligned}$$

$$\begin{aligned}&= a_{12}\alpha - a_{13}(\rho_1+\rho_2) + a_{14}\beta_1 + \frac{a_{11}\sigma'}{\sin\sigma}\\&= a_{12}\alpha + a_{13}\rho_3 + a_{14}\beta_1 + \frac{a_{11}\sigma'}{\sin\sigma}.\end{aligned}$$

Also we get

$$\begin{aligned}k_2 &= \kappa_2|\sin\sigma| + \rho_1\cos\rho,\\ k_3 &= -\left(\frac{a_{11}}{\sin\sigma}\sigma' + a_{12}\alpha + a_{13}\rho_3 + a_{14}\beta_1\right).\end{aligned}$$

Since

$$\begin{aligned}&\left(\frac{a_{21}\cos\sigma}{|\sin\sigma|}\right)'\left(\frac{a_{11}\cos\sigma}{|\sin\sigma|}\right) + a'_{21}a_{11}\\&= \frac{\cos\sigma}{\sin^2\sigma}\left(a'_{21}a_{11}\cos\sigma - \frac{a_{21}a_{11}\sigma'}{\sin\sigma}\right) + a'_{21}a_{11}\\&= \left(\frac{a_{21}}{|\sin\sigma|}\right)'\left(\frac{a_{11}}{|\sin\sigma|}\right),\end{aligned}$$

by (11), we obtain

$$\begin{aligned}n_{45} = &-\frac{\cos\sigma}{\sin^2\sigma}(a_{23}a_{12} - a_{22}a_{31})\,\alpha\\&+\frac{1}{|\sin\sigma|}\left\{-a_{21}a_{13}\kappa_2 + a_{23}\left(a_{11}\kappa_2 + \frac{a_{14}\alpha}{|\sin\sigma|}\right) - \frac{a_{24}a_{13}\alpha}{|\sin\sigma|}\right\}\\&+\frac{1}{\sin^2\sigma}\{-a_{23}(a_{11}\rho_2\cos\sigma + a_{12}\beta_2) + a_{24}(a_{11}\beta_1\cos\sigma + a_{12}\rho_3)\}\\&+\frac{1}{\sin^2\sigma}\{a_{21}(a_{13}\rho_2 - a_{14}\beta_1)\cos\sigma + a_{22}(a_{13}\beta_2 - a_{14}\rho_3)\}\\&-\left\{\left(\frac{a_{21}\cos\sigma}{|\sin\sigma|}\right)'\left(\frac{a_{11}\cos\sigma}{|\sin\sigma|}\right) + \left(\frac{a_{22}}{|\sin\sigma|}\right)'\left(\frac{a_{12}}{|\sin\sigma|}\right)\right.\\&\quad\left.+a'_{21}a_{11} + \left(\frac{a_{23}}{|\sin\sigma|}\right)'\left(\frac{a_{13}}{|\sin\sigma|}\right) + \left(\frac{a_{24}}{|\sin\sigma|}\right)'\left(\frac{a_{14}}{|\sin\sigma|}\right)\right\}\\= &-\frac{\kappa_2}{|\sin\sigma|}(a_{13}a_{21} - a_{11}a_{23})\\&-\frac{1}{\sin^2\sigma}\Big\{\alpha\left(\cos\sigma(a_{12}a_{21} - a_{11}a_{22}) + (a_{13}a_{24} - a_{13}a_{23})\right)\\&-\rho_2\cos\sigma(a_{13}a_{21} - a_{11}a_{23}) - \beta_1\cos\sigma(a_{11}a_{24} - a_{14}a_{21})\\&-\rho_3(a_{12}a_{24} - a_{14}a_{22}) - \beta_2(a_{22}a_{13} - a_{23}a_{12})\Big\}\\&-\left\langle\left(\frac{\vec{a}_2}{|\sin\sigma|}\right)', \frac{\vec{a}_1}{|\sin\sigma|}\right\rangle.\end{aligned}$$

By (1) and Proposition 4.1,

$$
\begin{aligned}
n_{45} = &-\frac{\kappa_2}{|\sin\sigma|}(a_{13}a_{21} - a_{11}a_{23}) \\
&+ \frac{k_3}{\sin^4\sigma}\{a_{12}\big(\cos\sigma(a_{12}a_{21}-a_{11}a_{22}) + (a_{13}a_{24}-a_{13}a_{23})\big) \\
&\qquad + a_{13}\cos\sigma\,(a_{13}a_{21}-a_{11}a_{23}) - a_{14}\cos\sigma\,(a_{11}a_{24}-a_{14}a_{21}) \\
&\qquad - a_{13}\,(a_{12}a_{24}-a_{14}a_{22}) - a_{14}\,(a_{22}a_{13}-a_{23}a_{12})\,\} \\
&+ \frac{\rho_1\cos\sigma}{\sin^2\sigma}(a_{13}a_{21} - a_{11}a_{23}) - \frac{k_1}{\sin^2\sigma}(a_{22}a_{13} - a_{23}a_{12}) \\
&- \left\langle\left(\frac{\vec{a}_2}{|\sin\sigma|}\right)', \frac{\vec{a}_1}{|\sin\sigma|}\right\rangle \\
= &-\frac{\kappa_2}{|\sin\sigma|}(a_{13}a_{21} - a_{11}a_{23}) \\
&+ \frac{k_3\cos\sigma}{\sin^4\sigma}\{a_{21}\sum_{i=2}^{4} a_{1i}{}^2 - a_{11}\sum_{i=2}^{4} a_{1i}a_{2i}\} \\
&+ \frac{\rho_1\cos\sigma}{\sin^2\sigma}(a_{13}a_{21} - a_{11}a_{23}) - \frac{k_1}{\sin^2\sigma}(a_{22}a_{13} - a_{23}a_{12}) \\
&- \left\langle\left(\frac{\vec{a}_2}{|\sin\sigma|}\right)', \frac{\vec{a}_1}{|\sin\sigma|}\right\rangle.
\end{aligned}
$$

Since $\mathscr{A}_\sigma$ is an $SO(4)$-valued function,

$$
\begin{aligned}
n_{45} = &-\frac{\kappa_2}{|\sin\sigma|}(a_{13}a_{21} - a_{11}a_{23}) \\
&+ \frac{k_3\cos\sigma}{\sin^4\sigma}\left\{a_{21}(\sin^2\sigma - a_{11}{}^2) - a_{11}(-a_{11}a_{21})\right\} \\
&+ \frac{k_2}{\sin^2\sigma}\begin{vmatrix} a_{11} & a_{13} \\ a_{21} & a_{23}\end{vmatrix} - \frac{k_1}{\sin^2\sigma}\begin{vmatrix} a_{12} & a_{13} \\ a_{22} & a_{23}\end{vmatrix} + \left\langle\left(\frac{\vec{a}_1}{|\sin\sigma|}\right)', \frac{\vec{a}_2}{|\sin\sigma|}\right\rangle \\
= &(\sin\sigma)^{-2}\left\{-k_1\begin{vmatrix} a_{12} & a_{13} \\ a_{22} & a_{23}\end{vmatrix} + k_2\begin{vmatrix} a_{11} & a_{13} \\ a_{21} & a_{23}\end{vmatrix} + k_3 a_{21}\cos\sigma\right\} \\
&+ \left\langle\left(\frac{\vec{a}_1}{|\sin\sigma|}\right)', \frac{\vec{a}_2}{|\sin\sigma|}\right\rangle.
\end{aligned}
$$

By the similar arguments, we can prove the other equalities. Hence we obtain the desired result. □

From Theorem 4.1, we can calculate $SO(7)$-invariants from the complete G_2-invariants and $(\sigma,\ \mathscr{A}_\sigma)$. The function $(\sigma,\ \mathscr{A}_\sigma)$ would be represented by the complete G_2-invariants. However we still do not obtain the relationship.

Bibliography

1. M. Ohashi, *On G_2-invariants of curves of purely imaginary octonions*, Recent progress in Differential Geometry and its Related Fields, T. Adachi, H. Hashimoto & M. Hristov eds., World Sci. Publ., New Jersey, 2011, 25–40.
2. M. Ohashi, *G_2-congruence theorem for curves in purely imaginary octonions and its application*, Geometriae Dedicata. 163(2013), 1–17.

Received April 19, 2013
Revised April 30, 2013

Proceedings of the 3rd International
Colloquium on Differential Geometry
and its Related Fields
Veliko Tarnovo, September 3–7, 2012

STATISTICAL MANIFOLDS AND GEOMETRY OF ESTIMATING FUNCTIONS

Hiroshi MATSUZOE

Department of Computer Science and Engineering,
Graduate School of Engineering, Nagoya Institute of Technology,
Nagoya, Aichi 466-8555 Japan
E-mail: matsuzoe@nitech.ac.jp

We give a survey on geometry of statistical manifolds in terms of estimating functions. A statistical model naturally has a statistical manifold structure. In particular, a q-exponential family which is a generalization of an exponential family admits several statistical manifold structures. An estimating function can be regarded as a tangent vector of a statistical model, and it gives rise to dualistic structures on a statistical manifold. In this paper, we construct statistical manifold structures on statistical models and divergence functions from the viewpoint of estimating functions. We also study geometry of non-integrable estimating functions.

Keywords: Statistical manifold; Estimating function; Divergence; q-exponential family; Information geometry; Tsallis statistics.

1. Introduction

In geometric theory of statistical inferences, it is known that a statistical model naturally has a Riemannian manifold structure with dualistic affine connections ([1]) Such a geometric theory is called information geometry, and is applied various fields of statistical sciences. The notion of statistical manifold was introduced by Lauritzen [10] to develop such dualistic structures from the viewpoint of differential geometry. Geometry of statistical manifold is also applied to statistical sciences. For example, generalized conformal structures on statistical manifolds are studied in anomalous statistical physics and asymptotic theory of sequential estimations ([3, 6, 12]).

An estimating function is an important object in statistical inference. For example, a score function is a typical estimating function which characterizes the maximum likelihood method. Geometrically speaking, an estimating function is regarded as a tangent vector of statistical model.

In this paper, we discuss geometry of statistical manifolds from the viewpoint of estimating functions. After giving a brief summary of geometry of q-exponential family, we discuss geometry of estimating functions and construct divergence functions on statistical models. Estimating functions are not integrable in general ([19]). We also discuss geometry of non-integrable estimating functions and statistical manifolds admitting torsion ([5, 9])

2. Preliminaries

In this paper, we assume that all objects are smooth, and a manifold M is an open domain in $\boldsymbol{R}^n$. Let us review geometry of dual connections and statistical manifolds.

Let (M, h) be a semi-Riemannian manifold, and ∇ be an affine connection on M. We can define another affine connection ∇^* by

$$Xh(Y, Z) = h(\nabla_X Y, Z) + h(Y, \nabla^*_X Z),$$

where X, Y and Z are arbitrary vector fields on M. We call ∇^* the *dual connection* of ∇ with respect to h. It is easy to check that $(\nabla^*)^* = \nabla$. Suppose that R and R^* are curvature tensors of ∇ and ∇^*, respectively. Then we have

$$h(R(X, Y)Z, V) = -h(Z, R^*(X, Y)V).$$

In this section, we assume that ∇ is torsion-free.

Definition 2.1. Let (M, h) be a semi-Riemannian manifold, and let ∇ be a torsion-free affine connection on M. We say that the triplet (M, ∇, h) is a *statistical manifold* [7] if ∇h is totally symmetric, that is,

$$(\nabla_X h)(Y, Z) = (\nabla_Y h)(X, Z).$$

For a statistical manifold (M, ∇, h), the dual connection ∇^* of ∇ with respect to h is torsion-free, and $\nabla^* h$ is totally symmetric. Hence (M, ∇^*, h) is also a statistical manifold. We call (M, ∇^*, h) the *dual statistical manifold* of (M, ∇, h).

For a statistical manifold, a totally symmetric $(0, 3)$-tensor field C can be defined by $C(X, Y, Z) = (\nabla_X h)(Y, Z)$. This tensor field C is called the *cubic form* for (M, ∇, h). Conversely, for a semi-Riemannian manifold (M, h) with a totally symmetric $(0, 3)$-tensor field C, we can define mutually dual torsion-free affine connections ∇ and ∇^* by

$$h(\nabla_X Y, Z) = h(\nabla^{(0)}_X Y, Z) - \frac{1}{2}C(X, Y, Z),$$
$$h(\nabla^*_X Y, Z) = h(\nabla^{(0)}_X Y, Z) + \frac{1}{2}C(X, Y, Z),$$

where $\nabla^{(0)}$ is the Levi-Civita connection with respect to h. In this case, ∇h and $\nabla^* h$ are totally symmetric. Hence (M, ∇, h) and (M, ∇^*, h) are statistical manifolds.

The notion of statistical manifold was originally given by Lauritzen ([10]). He called the triplet (M, g, C) a statistical manifold, where (M, g) is a Riemannian manifold and C is a totally symmetric $(0, 3)$-tensor field. From the above arguments, our definition is essentially equivalent to Lauritzen's definition.

We say that a statistical manifold (M, ∇, h) is *flat* if ∇ is flat. In this case, the pair (∇, h) is a Hessian structure on M ([16]) and the quadruplet (M, h, ∇, ∇^*) is a dually flat space ([1]). For a flat statistical manifold (M, ∇, h), we take a ∇-affine coordinate system $\{\theta^i\}$ on M. Then there exists a ∇^*-affine coordinate system such that

$$h\left(\frac{\partial}{\partial \theta^i}, \frac{\partial}{\partial \eta_j}\right) = \delta^i_j.$$

We call $\{\eta_i\}$ the *dual coordinate system* of $\{\theta^i\}$ with respect to h.

Proposition 2.1. *Let (M, ∇, h) be a flat statistical manifold. Suppose that $\{\theta^i\}$ is a ∇-affine coordinate system, and $\{\eta_i\}$ is its dual coordinate system. Then there exist functions ψ and ϕ on M such that*

$$\frac{\partial \psi}{\partial \theta^i} = \eta_i, \quad \frac{\partial \phi}{\partial \eta_i} = \theta^i, \quad \psi(p) + \phi(p) - \sum_{i=1}^{n} \theta^i(p)\eta_i(p) = 0,$$

where p is an arbitrary point in M. In addition, the following formulas hold:

$$h_{ij} = \frac{\partial^2 \psi}{\partial \theta^i \partial \theta^j}, \quad h^{ij} = \frac{\partial^2 \phi}{\partial \eta_i \partial \eta_j},$$

where (h_{ij}) is the component matrix of a semi-Riemannian metric h with respect to $\{\theta^i\}$, and (h^{ij}) is its inverse matrix.

The cubic form C for (M, ∇, h) is given by

$$C_{ijk} = \frac{\partial^3 \psi}{\partial \theta^i \partial \theta^j \partial \theta^j}.$$

The functions ψ and ϕ are called the *θ-potential* and the *η-potential*, respectively. Under the same assumptions of Proposition 2.1, we define a function D on $M \times M$ by

$$D(p, r) = \psi(p) + \phi(r) - \sum_{i=1}^{n} \theta^i(p)\eta_i(r), \quad (p, r \in M).$$

We call D the *canonical divergence* of (M, ∇, h).

Let us review generalized conformal equivalence of statistical manifolds. For a fixed $\alpha \in \boldsymbol{R}$, we say that two statistical manifolds (M, ∇, h) and $(M, \bar{\nabla}, \bar{h})$ are *α-conformally equivalent* to each other if there exists a function λ on M such that

$$\bar{h}(X,Y) = e^{\lambda} h(X,Y),$$
$$\bar{\nabla}_X Y = \nabla_X Y - \frac{1+\alpha}{2} h(X,Y)\mathrm{grad}_h \lambda + \frac{1-\alpha}{2}\left\{d\lambda(Y)\,X + d\lambda(X)\,Y\right\}.$$

A statistical manifold (M, ∇, h) is said to be *α-conformally flat* if it is locally α-conformally equivalent to some flat statistical manifold [7, 8].

3. Statistical models and estimating functions

Let $(\Omega, \mathcal{F}, \mathcal{P})$ be a probability space and let Ξ be a open subset of $\boldsymbol{R}^n$. We say that S is a *statistical model* or a *parametric model* if S is a set of probability density functions on Ω with parameter $\xi \in \Xi$ such that

$$S = \left\{ p(x;\xi) \,\middle|\, \int_\Omega p(x;\xi)dx = 1,\ p(x;\xi) > 0,\ \xi = (\xi^1, \ldots, \xi^n) \in \Xi \subset \boldsymbol{R}^n \right\}.$$

In this paper, we regard S as a manifold with local coordinate system $\{\xi^i\}$ ([1]).

For a statistical model S, we define a function g^F_{ij} on Ξ by

$$\begin{aligned} g^F_{ij}(\xi) &= \int_\Omega \left(\frac{\partial}{\partial \xi^i}\log p(x;\xi)\right)\left(\frac{\partial}{\partial \xi^j}\log p(x;\xi)\right) p(x;\xi)\,dx \\ &= E_p[\partial_i l_\xi \partial_j l_\xi] \ = \ -E_p[\partial_i\partial_j l_\xi], \end{aligned}$$

where $\partial_i = \partial/\partial\xi^i$, $l_\xi = l(x;\xi) = \log p(x;\xi)$ is the log-likelihood of $p(x;\xi)$, and

$$E_p[f] = \int_\Omega f(x)p(x;\xi)dx$$

is the expectation of $f(x)$ with respect to $p(x;\xi)$. Under suitable conditions, $g^F = (g^F_{ij})$ is a Riemannian metric on S. We call g^F the *Fisher metric* on S. This metric g^F has the following representations:

$$g^F_{ij}(\xi) = \int_\Omega \left(\frac{\partial}{\partial \xi^i}\log p(x;\xi)\right)\left(\frac{\partial}{\partial \xi^j}\log p(x;\xi)\right) p(x;\xi)\,dx$$

$$= \int_\Omega \left(\frac{\partial}{\partial \xi^i} p(x;\xi)\right)\left(\frac{\partial}{\partial \xi^j}\log p(x;\xi)\right) dx \tag{1}$$

$$= \int_\Omega \frac{1}{p(x;\xi)}\left(\frac{\partial}{\partial \xi^i} p(x;\xi)\right)\left(\frac{\partial}{\partial \xi^j} p(x;\xi)\right) dx \tag{2}$$

We define an $\boldsymbol{R}^n$ valued function $s(x;\xi) = (s^1(x;\xi), \ldots, s^n(x;\xi))^T$ by

$$s^i(x;\xi) = \frac{\partial}{\partial \xi^i} \log p(x;\xi).$$

We call $s(x;\xi)$ the *score function* of $p(x;\xi)$ with respect to ξ. In information geometry, the score function $s(x;\xi)$ is called the *e-(exponential) representation* of $p(x;\xi)$, and $(\partial_1 p(x;\xi), \ldots, \partial_n p(x;\xi))^T$ is called the *m-(mixture) representation*. From Equation (1), the Fisher metric g^F is regarded as an L_2-inner product of e- and m-representations of $p(x;\xi)$.

For a fixed $\alpha \in \boldsymbol{R}$, we define the *$\alpha$-connection* $\nabla^{(\alpha)}$ on S by

$$\Gamma^{(\alpha)}_{ij,k}(\xi) = E_p\left[\left(\partial_i\partial_j l_\xi + \frac{1-\alpha}{2}\partial_i l_\xi \partial_j l_\xi\right)(\partial_k l_\xi)\right].$$

It is known that $\nabla^{(0)}$ is the Levi-Civita connection with respect to the Fisher metric g^F. On the other hand, we call $\nabla^{(e)} = \nabla^{(1)}$ the *exponential connection* and $\nabla^{(m)} = \nabla^{(-1)}$ the *mixture connection*.

We can check that the α-connection $\nabla^{(\alpha)}$ is torsion-free and that $\nabla^{(\alpha)} g^F$ is totally symmetric. These imply that $(S, \nabla^{(\alpha)}, g^F)$ forms a statistical manifold. We call $(S, \nabla^{(\alpha)}, g^F)$ an *invariant statistical manifold*. The cubic form C^F of the invariant statistical manifold $(S, \nabla^{(e)}, g^F)$ is given by

$$C^F_{ijk} = \Gamma^{(m)}_{ij,k} - \Gamma^{(e)}_{ij,k}.$$

An *exponential family* S_e is a typical statistical model defined by

$$S_e = \left\{ p(x;\theta) \;\middle|\; p(x;\theta) = \exp\left[\sum_{i=1}^n \theta^i F_i(x) - \psi(\theta)\right],\ \theta \in \Theta \subset \boldsymbol{R}^n \right\},$$

where $F_1(x), \ldots, F_n(x)$ are random variables, $\theta = \{\theta^1, \ldots, \theta^n\}$ is a parameter, and $\psi(\theta)$ is the normalization term with respect to the parameter θ. We assume that the parameter space Θ is an open domain in $\boldsymbol{R}^n$.

Proposition 3.1. *For an exponential family S_e, the following hold:*

(1) *$(S_e, g^F, \nabla^{(e)}, \nabla^{(m)})$ is a dually flat space;*
(2) *$\{\theta^i\}$ is a $\nabla^{(e)}$-affine coordinate system on S_e;*
(3) *$\psi(\theta)$ is the potential of g^F with respect to $\{\theta^i\}$, that is,*

$$g^F_{ij}(\theta) = \partial_i\partial_j\psi(\theta), \quad (\partial_i = \partial/\partial\theta^i);$$

(4) *If we set the expectation of $F_i(x)$ by $\eta_i = E_p[F_i(x)]$, then $\{\eta_i\}$ is the dual affine coordinate system of $\{\theta^i\}$ with respect to g^F;*
(5) *If we set $\phi(\eta) = E_p[\log p(x;\theta)]$, then it is the potential of g^F with respect to $\{\eta_i\}$.*

Since $(S_e, \nabla^{(e)}, g^F)$ is a Hessian manifold, the formulas in Proposition 2.1 hold. In particular, the cubic form C^F of $(S_e, \nabla^{(e)}, g^F)$ is given by

$$C^F_{ijk}(\theta) = \partial_i \partial_j \partial_k \psi(\theta).$$

For a statistical model S, we define the *Kullback-Leibler divergence* (or the *relative entropy*) by

$$\begin{aligned} D_{KL}(p,r) &= \int_\Omega p(x) \log \frac{p(x)}{r(x)} dx \\ &= E_p[\log p(x) - \log r(x)], \quad (p(x), r(x) \in S). \end{aligned}$$

The Kullback-Leibler divergence is an asymmetric distance like function on S, which measures a dissimilarity between $p(x)$ and $r(x)$. In the case of an exponential family S_e, the Kullback-Leibler divergence D_{KL} coincides with the canonical divergence D on $(S_e, \nabla^{(m)}, g^F)$.

The way of construction of a Kullback-Leibler divergence from a score function is as follows. By integrating the score function at $r(x;\xi) \in S$ with respect to ξ, and by taking the expectation with respect to $p(x)$, we define a *cross entropy* $d_{KL}(p,q)$ by

$$d_{KL}(p,r) = -E_p[\log r(x;\xi')].$$

Then we have a Kullback-Leibler divergence by

$$D_{KL}(p,r) = -d_{KL}(p,p) + d_{KL}(p,r).$$

Let us review statistical inference. The maximum likelihood method is a typical one for parameter estimation. We assume that the true probability distribution for observation belongs to a statistical model S. We then infer the true parameter which maximizes the log-likelihood $l(x,\xi) = \log p(x;\xi)$. The estimator can be obtained by solving the logarithmic likelihood equation $s(x;\xi) = 0$, where $s(x;\xi) = (s^1(x;\xi), \dots, s^n(x;\xi))^T$ is the score function of $p(x;\xi)$.

However, the maximum likelihood method is inappropriate or difficult sometimes. An alternative method is to consider an $\boldsymbol{R}^n$ valued function

$$u(x;\xi) = (u^1(x;\xi), \dots, u^n(x;\xi))^T.$$

Then we obtain an estimator by solving $u(x;\xi) = 0$.

Here, we say that $u(x;\xi)$ is an *estimating function* if it satisfies the following conditions:

$$E_p[u(x;\xi)] = 0, \quad E_p\left[||u(x;\xi)||^2\right] < \infty,$$

$$\det\left(E_p\left[\frac{\partial u}{\partial \xi}(x,\xi)\right]\right) \neq 0 \quad ({}^\forall \xi \in \Xi).$$

In particular, an estimating function $u(x;\xi)$ is *integrable* if

$$\frac{\partial u^i}{\partial \xi^j}(x;\xi) = \frac{\partial u^j}{\partial \xi^i}(x;\xi),$$

that is, $u(x;\xi)$ is locally given by the differential of some function on S.

The score function $s(x;\xi)$ is a typical example of integrable estimating functions. We can check the fact as follows. The score function is unbiased:

$$E_p[s^i(x;\xi)] = \int_\Omega (\partial_i l_\xi)\, p_\xi dx = \int_\Omega (\partial_i p_\xi)\, dx = \partial_i \int_\Omega p_\xi dx = 0.$$

Since we assumed that the statistical model has its Fisher metric, and $g^F_{ij} = E_p[\partial_i l_\xi \partial_j l_\xi] = -E_p[\partial_i \partial_j l_\xi] = -E_p[\partial_i s^j(x;\xi)]$, we have

$$E_p\left[||s(x;\xi)||^2\right] < \infty, \quad \det(g_{ij}) = \det\left(-E_p\left[\partial_i s^j(x;\xi)\right]\right) \neq 0.$$

The score function $s(x;\xi)$ is integrable since $\partial_j s^i(x;\xi) = \partial_j\partial_i l_\xi = \partial_i \partial_j l_\xi = \partial_i s^j(x;\xi)$.

For further details of estimating functions, see [17] and [20].

4. The q-exponential family

In this section, we review the q-exponential family and divergences on statistical models. See [2, 12, 14, 18] for more details for q-exponential families.

For a fixed positive number q, the *q-exponential function* is defined by the following formula (cf. [14])

$$\exp_q x = \begin{cases} (1+(1-q)x)^{\frac{1}{1-q}}, & (q \neq 1, \text{ and } 1+(1-q)x > 0), \\ \exp x, & (q = 1), \end{cases}$$

and the *q-logarithm function* by

$$\log_q x = \begin{cases} \dfrac{x^{1-q}-1}{1-q}, & (q \neq 1, \text{ and } x > 0), \\ \log x, & (q = 1, \text{ and } x > 0). \end{cases}$$

By taking the limit $q \to 1$, the standard exponential and the standard logarithm are recovered, respectively.

Definition 4.1. A statistical model S_q is called a *q-exponential family* if

$$S_q = \left\{ p(x,\theta) \,\middle|\, p(x;\theta) = \exp_q\left[\sum_{i=1}^n \theta^i F_i(x) - \psi(\theta)\right],\ \theta \in \Theta \subset \boldsymbol{R}^n \right\},$$

where $F_1(x), \ldots, F_n(x)$ are random variables on the sample space Ω, $\theta = \{\theta^1, \ldots, \theta^n\}$ is a parameter, and $\psi(\theta)$ is the normalization with respect to the parameter θ.

For a q-exponential probability density $p(x;\theta) \in S_q$, we define the *escort distribution* $P(x;\theta)$ by

$$P(x;\theta) = \frac{1}{Z_q(\theta)} p(x;\theta)^q, \quad Z_q(\theta) = \int_\Omega p(x;\theta)^q dx,$$

and the *q-expectation* $E_{q,p}[f]$ of $f(x)$ with respect to $P(x;\theta)$ by

$$E_{q,p}[f] = \int_\Omega f(x)P(x;\theta)\, dx.$$

Here, we summerize divergences on a statistical model for the later use. See [4, 11, 13] for further details and induced geometry derived from divergences.

Fix real numbers α and $q \in \boldsymbol{R}$. For a statistical model S, we define the *α-divergence* $D^{(\alpha)}$ and the *Tsallis relative entropy of Csiszár type* D_q^C by the following formulas:

$$D^{(\alpha)}(p,r) = \frac{4}{1-\alpha^2}\left\{1 - \int_\Omega p(x)^{\frac{1-\alpha}{2}} r(x)^{\frac{1+\alpha}{2}} dx\right\},$$

$$D_q^C(p,r) = \frac{1}{1-q}\left\{1 - \int_\Omega p(x)^q r(x)^{1-q} dx\right\},$$

where $p(x), r(x) \in S$. These divergences coincide with each other when $\alpha = 1 - 2q$, except a constant multiplication. That is, the following relation holds:

$$D^{(\alpha)}(p,r) = \frac{1}{q} D_q^C(p,r).$$

By taking the limit $q \to 1$ (or $\alpha \to -1$), each of these divergences recovers the KL-divergence. In this paper, we define the *Tsallis relative entropy* D_q^T by

$$\begin{aligned} D_q^T(p,r) &= \int_\Omega \frac{p(x)^q}{Z_q(p)} \left(\log_q p(x) - \log_q r(x)\right) dx \\ &= E_{q,p}[\log_q p(x) - \log_q r(x)]. \end{aligned}$$

On a q-exponential family S_q, the Tsallis relative entropy D_q^T is written as

$$\begin{aligned} D_q^T(p,r) &= \frac{1}{Z_q(p)} \int_\Omega \left(\frac{p(x) - p(x)^q}{1-q} - \frac{p(x)^q r(x)^{1-q} - p(x)^q}{1-q} \right) dx \\ &= \frac{1}{(1-q)Z_q(p)} \left\{1 - \int_\Omega p(x)^q r(x)^{1-q} dx\right\} \\ &= \frac{1}{Z_q(p)} D_q^C(p,r). \end{aligned}$$

Fix a real number $\beta \in \boldsymbol{R}$. We define the *β-divergence* D_β and the *Tsallis relative entropy of Bregman type* D_q^B by

$$D_\beta(p,r) = \int_\Omega \left\{ p(x)\frac{p(x)^\beta - r(x)^\beta}{\beta} - \frac{p(x)^{\beta+1} - r(x)^{\beta+1}}{\beta+1} \right\} dx$$

$$= \frac{1}{\beta(\beta+1)} \int_\Omega p(x)^{\beta+1} dx - \frac{1}{\beta} \int_\Omega p(x) r(x)^\beta dx + \frac{1}{\beta+1} \int_\Omega r(x)^{\beta+1} dx,$$

$$D_q^B(p,r) = \int_\Omega \left\{ p(x)\frac{p(x)^{q-1} - r(x)^{q-1}}{q-1} - (p(x) - r(x)) r(x)^{q-1} \right\} dx.$$

These divergences coincide with each other when $\beta = q-1$, except a constant multiplication. Taking the limit $\beta \to 0$ (or $q \to 1$), each of these divergence recovers the KL-divergence. We remark that D_q^T is called simply the "Tsallis relative entropy" in this paper

For a fixed $\gamma \in \boldsymbol{R}$, we define *$\gamma$-divergence* $\tilde{D}_\gamma$ by the following formula, which belongs neither to the Csiszár type nor to the Bregman type:

$$\tilde{D}_\gamma(p,r) = \frac{1}{\gamma(\gamma+1)} \log \int_\Omega p(x)^{\gamma+1} dx - \frac{1}{\gamma} \log \int_\Omega p(x) r(x)^\gamma dx + \frac{1}{\gamma+1} \log \int_\Omega r(x)^{\gamma+1} dx.$$

5. Geometry of q-exponential family derived from the β-divergence

In geometric theory of statistical inference, the dualistic structure of affine connections originates from the duality between the e-representation and the m-representation of probability distributions. In this section, we generalize the e-representation by the q-logarithm function, then consider geometry of q-exponential family in terms of estimating functions.

Let S_q be a q-exponential family. We say that an $\boldsymbol{R}^n$ valued function $s^q(x;\theta) = \left((s^q)^1(x;\theta), \ldots, (s^q)^n(x;\theta)\right)^T$ is the *q-score function* of $p(x;\theta)$ if

$$(s^q)^i(x;\theta) = \frac{\partial}{\partial \theta^i} \log_q p(x;\theta), \quad (i = 1, \ldots, n).$$

We define a Riemannian metric g^M on S_q by

$$g_{ij}^M = \int_\Omega \left(\frac{\partial}{\partial \theta^i} p(x;\theta)\right) \left(\frac{\partial}{\partial \theta^j} \log_q p(x;\theta)\right) dx.$$

On the other hand, we define another Riemann metric g^N in terms of the escort distribution by

$$g_{ij}^N = \int_\Omega \frac{1}{P(x;\theta)} \left(\frac{\partial}{\partial\theta^i} p(x;\theta)\right) \left(\frac{\partial}{\partial\theta^j} p(x;\theta)\right) dx.$$

Riemannian metrics g^M and g^N are generalizations in terms of the representation (1) and (2) of the Fisher metric, respectively. The Riemannian metric g^M is well-known in information geometry ([13, 15]). On the other hand, g^N is known in statistical physics ([14]), which is called the *generalized information metric.*

Proposition 5.1. *For a q-exponential family S_q, Riemannian metrics g^M and g^N are conformally equivalent to each other, that is, the following holds:*

$$Z_q(p) g^M(p) = g^N(p), \quad (p \in S_q),$$

where Z_q is the normalization of the escort distribution $P(x;\theta)$ with respect to $p(x;\theta)$ in S_q.

Proof. From the definition of the q-logarithm function, we have

$$\begin{aligned} g_{ij}^M &= \int_\Omega \left(\frac{\partial}{\partial\theta^i} p(x;\theta)\right) \left(\frac{\partial}{\partial\theta^j} \log_q p(x;\theta)\right) dx \\ &= \int_\Omega \left(\frac{\partial}{\partial\theta^i} p(x;\theta)\right) \left(\frac{\partial}{\partial\theta^j} p(x;\theta)\right) \frac{1}{p(x;\theta)^q} dx \\ &= \frac{1}{Z_q} g^N. \end{aligned}$$

From now on, we consider information geometric structures with respect to g^M. We define affine connections $\nabla^{M(e)}$ and $\nabla^{M(m)}$ by

$$\Gamma_{ij,k}^{M(e)}(\theta) = \int_\Omega \partial_k p(x;\theta) \partial_i \partial_j \log_q p(x;\theta) dx,$$

$$\Gamma_{ij,k}^{M(m)}(\theta) = \int_\Omega \partial_i \partial_j p(x;\theta) \partial_k \log_q p(x;\theta) dx.$$

These affine connections $\nabla^{M(e)}$ and $\nabla^{M(m)}$ are mutually dual with respect to g^M. From the definitions of the q-exponential family and the q-logarithmic function, $\Gamma_{ij,k}^{M(e)}$ vanishes identically. Hence the connection $\nabla^{M(e)}$ is flat, and $(\nabla^{M(e)}, g^M)$ is a Hessian structure on S_q.

Here, we define a *generalized Massieu potential* by

$$\Psi(\theta) = \phi(\theta) + \psi(\theta),$$

where ψ is the normalization with respect to the q-exponential family and ϕ is defined by

$$\phi(\theta) = \frac{1}{2-q}\int_{\Omega}(p(x;\theta)^{2-q} - p(x;\theta))dx.$$

By taking the limit $q \to 1$, we have $\phi \to 0$.

Proposition 5.2. *For a q-exponential family S_q, the following hold:*

(1) $(S_q, g^M, \nabla^{M(e)}, \nabla^{M(m)})$ *is a dually flat space;*
(2) $\{\theta^i\}$ *is a* $\nabla^{M(e)}$*-affine coordinate system on* S_q*;*
(3) $\Psi(\theta)$ *is the potential of* g^M *with respect to* $\{\theta^i\}$*, that is,*

$$g^M_{ij}(\theta) = \partial_i\partial_j\Psi(\theta); \tag{3}$$

(4) *If we set the expectation of* $F_i(x)$ *by* $\eta_i = E_p[F_i(x)]$*, then* $\{\eta_i\}$ *is a* $\nabla^{M(m)}$*-affine coordinate system on* S_q *and the dual coordinate system of* $\{\theta^i\}$ *with respect to* g^M*;*
(5) *If we set* $\Phi(\eta) = E_p[\frac{1}{2-q}\log_q p(x;\theta)]$*, then it is the potential of* g^M *with respect to* $\{\eta_i\}$*.*

It is known that the canonical divergence D on the flat statistical manifold $(S_q, \nabla^{M(m)}, g^M)$ coincides with the β-divergence D_β with $\beta = 1-q$ ([15]).

Let us construct the β-divergence from an estimating function. We define a weighted score function $u_q(x;\theta) = (u_q^1(x;\theta), \ldots, u_q^n(x;\theta))^T$ by

$$u_q^i(x;\theta) = p(x;\theta)^{1-q}s^i(x;\theta) - E_\theta[p(x;\theta)^{1-q}s^i(x;\theta)].$$

From the definition of q-logarithm function, $u_q(x;\theta)$ is written as

$$\begin{aligned} u_q^i(x;\theta) &= \frac{\partial}{\partial\theta^i}\left\{\frac{1}{1-q}p(x;\theta)^{1-q} - \frac{1}{2-q}\int_{\Omega}p(x;\theta)^{2-q}dx\right\} \\ &= \frac{\partial}{\partial\theta^i}\log_q p(x;\theta) - E_\theta\left[\frac{\partial}{\partial\theta^i}\log_q p(x;\theta)\right]. \end{aligned}$$

Hence the estimating function $u_q(x;\theta)$ is nothing else but the bias corrected q-score function of $p(x;\theta)$.

By integrating the estimating function at $r(x;\theta) \in S_q$ with respect to θ, and by taking the standard expectation with respect to $p(x)$, a *cross entropy* $d_{1-q}(p,r)$ is defined by

$$d_{1-q}(p,r) = -\frac{1}{1-q}\int_{\Omega}p(x)r(x)^{1-q}dx + \frac{1}{2-q}\int_{\Omega}r(x)^{2-q}dx.$$

Finally, we obtain the β-divergence ($\beta = 1 - q$) by

$$\begin{aligned} D_{1-q}(p,r) &= -d_{1-q}(p,p) + d_{1-q}(p,r) \\ &= \frac{1}{(1-q)(2-q)} \int_\Omega p(x)^{2-q} dx \\ &\quad - \frac{1}{1-q} \int_\Omega p(x) r(x)^{1-q} dx + \frac{1}{2-q} \int_\Omega r(x)^{2-q} dx. \end{aligned}$$

6. Geometry of q-exponential family derived from the Tsallis relative entropy

In this section, we assume that a q-exponential family has its Fisher metric g^F and its α-connection $\nabla^{(\alpha)}$.

Since S_q is linearizable by q-logarithm function, we can define geometric structures from the potential function ψ as in the case of the standard exponential family.

We define the *q-Fisher metric* g^q and the *q-cubic form* C^q by

$$g^q_{ij}(\theta) = \partial_i \partial_j \psi(\theta), \quad C^q_{ijk}(\theta) = \partial_i \partial_j \partial_k \psi(\theta), \quad (\partial_i = \partial / \partial \theta^i),$$

respectively ([12]). In this case, its *q-exponential connection* $\nabla^{q(e)}$ and its *q-mixture connection* $\nabla^{q(e)}$ is defined by

$$g^q(\nabla^{q(e)}_X Y, Z) = g^q(\nabla^{q(0)}_X Y, Z) - \frac{1}{2} C^q(X,Y,Z),$$

$$g^q(\nabla^{q(m)}_X Y, Z) = g^q(\nabla^{q(0)}_X Y, Z) + \frac{1}{2} C^q(X,Y,Z),$$

respectively, where $\nabla^{q(0)}$ is the Levi-Civita connection with respect to the q-Fisher metric g^q.

Proposition 6.1. *For a q-exponential family S_q, the following hold:*

(1) *$(S_q, g^g, \nabla^{q(e)}, \nabla^{q(m)})$ is a dually flat space;*
(2) *$\{\theta^i\}$ is a $\nabla^{q(e)}$-affine coordinate system on S_q;*
(3) *ψ is the potential of g^q with respect to $\{\theta^i\}$;*
(4) *If we set the q-expectation of $F_i(x)$ by $\eta_i = E_{q,p}[F_i(x)]$, then $\{\eta_i\}$ is a $\nabla^{q(m)}$-affine coordinate system on S_q and the dual coordinate system of $\{\theta^i\}$ with respect to g^g;*
(5) *If we set $\phi(\eta) = E_{q,p}[\log_q p(x;\theta)]$, then it is the potential of g^q with respect to $\{\eta_i\}$.*

Proof. Statements (1), (2) and (3) are obtained from well-known facts in Hessian geometry ([10]) and the definition of q-Fisher metric. Statements (4) and (5) are obtained from the following:

$$E_{q,p}[\log_q p(x;\theta)] = E_{q,p}\left[\sum_{i=1}^{n} \theta^i F_i(x) - \psi(\theta)\right] = \sum_{i=1}^{n} \theta^i \eta_i(x) - \psi(\theta).$$

Let us elucidate the relation between the invariant statistical manifold $(S_q, \nabla^{(\alpha)}, g^F)$ and the flat statistical manifold $(S_q, \nabla^{q(e)}, g^q)$.

Proposition 6.2. *Let S_q be a q-exponential family, and $Z_q(p)$ be the normalization of the escort distribution with respect to $p(x;\theta) \in S_q$. Denote by C^F the cubic form with respect to the invariant statistical manifold $(S_q, \nabla^{(1)}, g^F)$. Then the following hold:*

$$g^q_{ij}(\theta) = \frac{q}{Z_q(p)} g^F_{ij}(\theta),$$

$$C^q_{ijk}(\theta) = \frac{q(2q-1)}{Z_q(p)} C^F_{ijk}(\theta) - \frac{q}{(Z_q(p))^2}\left\{g^F_{ij}\partial_k Z_q(p) + g^F_{jk}\partial_i Z_q(p) + g^F_{ki}\partial_j Z_q(p)\right\}.$$

From the above proposition, we have the following theorem.

Theorem 6.1 ([12]). *For a q-exponential family S_q, the invariant statistical manifold $(S_q, \nabla^{(2q-1)}, g^F)$ and the flat statistical manifold $(S_q, \nabla^{q(e)}, g^q)$ are 1-conformally equivalent to each other. In addition, the invariant statistical manifold $(S_q, \nabla^{(2q-1)}, g^F)$ is 1-conformally flat.*

We remark that the invariant statistical manifold $(S_q, \nabla^{(2q-1)}, g^F)$ is induced from the α-divergence with $\alpha = 2q-1$, and that the flat statistical manifold $(S_q, \nabla^{q(e)}, g^q)$ is induced from the dual of the Tsallis relative entropy.

For a q-exponential family S_q, divergences have the following relations:

$$D(p,r) = D^T_q(r,p) = \frac{q}{Z_q(r)} D^{(2q-1)}(p,r),$$

where D is the canonical divergence of the flat statistical manifold $(S_q, \nabla^{q(e)}, g^q)$, D^T_q is the Tsallis relative entropy, and $D^{(2q-1)}$ is the α-divergence with $\alpha = 2q-1$.

Let us consider construction of Tsallis relative entropy from an estimating function. Suppose that $s^q(x;\theta) = ((s^q)^1(x;\theta), \ldots, (s^q)^n(x;\theta))^T$ is the

q-score function:

$$(s^q)^i(x;\theta) = \frac{\partial}{\partial\theta^i}\log_q p(x;\theta).$$

The q-score function $s^q(x;\theta)$ is not an estimating function. However, $s^q(x;\theta)$ is unbiased with respect to the q-expectation, that is $E_{q,p}[(s^q)^i(x;\theta)] = 0$. Hence we regard $s^q(x;\theta)$ as a generalization of estimating function.

By integrating the q-score function at $r(x;\theta) \in S_q$ with respect to θ, and by taking the q-expectation, we define the *q-cross entropy* by

$$d^q(p,r) = -\int_\Omega P(x)\log_q r(x)dx.$$

Finally, we obtain the Tsallis relative entropy by

$$\begin{aligned} D_q^T(p,r) &= -d^q(p,p) + d^q(p,r) \\ &= E_{q,p}[\log_q p(x) - \log_q r(x)]. \end{aligned}$$

7. Statistical manifolds admitting torsion and non-integrable estimating functions

We saw constructions of divergence functions by integrating estimating functions. However, estimating functions are not integrable with respect to parameters in general. Hence divergence functions may not be obtained. In this section, we summerize geometry of non-integrable estimating functions, briefly.

Definition 7.1 ([9]). Let (M,h) be a semi-Riemannian manifold, and ∇ be an affine connection on M. Denote by T^∇ the torsion tensor field of ∇. We say that the triplet (M,∇,h) is a *statistical manifold admitting torsion* if

$$(\nabla_X h)(Y,Z) - (\nabla_Y h)(X,Z) = -h(T^\nabla(X,Y),Z).$$

We remark that the dual connection ∇^* of ∇ with respect to h is torsion free. However, (M,∇^*,h) is not a statistical manifold in general since $\nabla^* h$ may not be symmetric.

Let us consider geometry of estimating functions. In general, some modification for an estimating function is required, since each component of an estimating function does not necessarily satisfy the coordinate transformation rule for a basis of a tangent space of S. Hence we standardize the estimating function to satisfy the coordinate transformation rule.

Definition 7.2. For a statistical model S, suppose that $u(x;\theta)$ is an estimating function for the parameter θ, and $s(x;\theta)$ is the score function. Then the estimating function $\tilde{u}(x;\theta)$ defined by

$$\tilde{u}(x;\theta) = E_\theta[s(x;\theta)u(x;\theta)^T]\, E_\theta[u(x;\theta)u(x;\theta)^T]^{-1}u(x;\theta)$$

is said to be the *standardization* of $u(x;\theta)$.

The standardized estimating function $\tilde{u}(x;\theta)$ is regarded as the orthogonal projection of the score function $s(x;\theta)$ onto the linear space spanned by the estimating function $u(x;\theta)$ in the Hilbert space

$$\mathcal{H}_\theta = \{a(x)|E_\theta[a(x)] = 0,\ E_\theta[||a(x)||^2] < \infty\}.$$

For a standardized estimating function $\tilde{u}(x,\theta)$, we define a function ρ_u on $TS \times S$ by

$$\rho_u((\partial_i)_\theta, p(x;\theta')) = -\int_\Omega \tilde{u}^i(x,\theta)p(x;\theta')dx.$$

It is known that ρ_u is a pre-contrast function on S. A pre-contrast function is regarded as a differential version of a contrast function or a divergence function, and it induces a statistical manifold admitting torsion ([5]) In this case, the torsion tensor is regarded as non-integrability of an estimating function.

Conversely, if the torsion tensor on the statistical manifold admitting torsion vanishes, that is, a statistical manifold is induced from a pre-contrast function, then the estimating function is integrable and the pre-contrast function is given by the differential of a contrast function.

Acknowledgments

The author would like to express his sincere gratitude to the referee for giving insightful comments to improve this paper. This work was partially supported by JSPS KAKENHI No. 23740047.

Bibliography

1. S. Amari and H. Nagaoka, *Method of Information Geometry*, Amer. Math. Soc., Providence, Oxford University Press, Oxford, 2000.
2. S. Amari and A. Ohara *Geometry of q-exponential family of probability distributions*, Entropy, **13**(2011), 1170-1185.
3. S. Amari, A. Ohara and H. Matsuzoe, *Geometry of deformed exponential families: invariant, dually-flat and conformal geometry*, Physica A., **391**(2012), 4308-4319.

4. S. Eguchi, *Geometry of minimum contrast*, Hiroshima Math. J., **22**(1992), 631–647.
5. M. Henmi and H. Matsuzoe, *Geometry of pre-contrast functions and non-conservative estimating functions*, AIP Conference Proceedings Volume 1340: International Workshop on Complex Structures, Integrability and Vector Fields, Amer. Inst. of Physics, **1340**(2011), 32-41.
6. M. Kumon, A. Takemura, and K. Takeuchi, *Conformal geometry of statistical manifold with application to sequential estimation*, Sequential Anal., **30**(2011), 308-337.
7. T. Kurose, *On the divergences of 1-conformally flat statistical manifolds*, Tôhoku Math. J., **46**(1994), 427–433.
8. T. Kurose, *Conformal-projective geometry of statistical manifolds*, Interdiscip. Inform. Sci., **8**(2002), 89–100.
9. T Kurose, *Statistical manifolds admitting torsion*, Geometry and something, Fukuoka University, 2007 (in Japanese).
10. S. L. Lauritzen, *Statistical manifolds*, Differential geometry in statistical inferences, IMS Lecture Notes Monograph Series 10, Institute of Mathematical Statistics, Hayward California, (1987), 96–163.
11. H. Matsuzoe, *Geometry of contrast functions and conformal geometry*, Hiroshima Math. J., **29**(1999), 175 – 191.
12. H. Matsuzoe and A. Ohara, *Geometry for q-exponential families*, in *Recent progress in Differential Geometry and its Related Fields*, T. Adachi, H. Hashimoto and M. Hristov eds, World Sci. Publ., New Jersey, (2011), 55-71.
13. N. Murata, T. Takenouchi, T. Kanamori and S. Eguchi, *Information geometry of U-boost and Bregman divergence*, Neural Comput., **16**(2004), 1437-1481.
14. J. Naudts, *Generalised Thermostatistics*, Springer-Verlag, 2011.
15. A. Ohara and T. Wada, *Information geometry of q-Gaussian densities and behaviors of solutions to related diffusion equations*, J. Phys. A: Math. Theor., **43**(2010) No.035002.
16. H. Shima, *The Geometry of Hessian Structures*, World Scientific, 2007.
17. M. Sorensen, *On asymptotics of estimating functions*, Brazilian J. Probab. Statist. **13**(1999), 111–136.
18. C. Tsallis, *Introduction to Nonextensive Statistical Mechanics: Approaching a Complex World*, Springer, New York, 2009.
19. J. Wang, *Nonconservative estimating functions and approximate quasi-likelihoods*, Ann. Inst. Stat. Math., **51**(1999), 603–619.
20. A.W. van der Vaart, *Asymptotic Statistics*, Cambridge University Press, 1998.

Received February 12, 2013
Revised May 1, 2013

Proceedings of the 3rd International
Colloquium on Differential Geometry
and its Related Fields
Veliko Tarnovo, September 3–7, 2012

α-CONNECTIONS ON LEVEL SURFACES IN A HESSIAN DOMAIN

Keiko UOHASHI

Department of Mechanical Engineering and Intelligent Systems,
Faculty of Engineering, Tohoku Gakuin University,
Tagajo, Miyagi 985-8537, Japan
E-mail: uohashi@tjcc.tohoku-gakuin.ac.jp

A Hessian domain is a flat statistical manifold, and its level surfaces are 1-conformally flat statistical submanifolds. In addition, a level surface with an α-connection is α-conformally equivalent to a statistical manifold with an α-transitively flat connection for each real number α. In this study, we investigate these facts and present details related to α-connections.

Keywords: Hessian domain; Level surface; Foliation; Statistical manifold; α-connection.

1. Introduction

In [11], we proved that n-dimensional level surfaces of an $(n+1)$-dimensional Hessian domain are 1-conformally flat statistical submanifolds, considering the Hessian domain to be a flat statistical manifold. In [11], we also showed that a 1-conformally flat statistical manifold can be locally realized as a submanifold of a flat statistical manifold. However, our statement held only for one 1-conformally flat statistical manifold. Then, in [10], we presented conditions required for the realization of many 1-conformally flat statistical manifolds as level surfaces of their common Hessian domain.

In this study, we review the above facts and show similar results for statistical manifolds with α-connections in the case of $\alpha \neq 1$. In Sec. 2, we investigate statistical manifolds and level surfaces of Hessian domains. In Sec. 3, we show a relation between a Hessian domain and a foliation constructed by 1-conformally flat statistical manifolds; we also provide remarks regarding affine immersions, which were not contained in [10]. In Sec. 4, we discuss a relation between a Hessian domain and a foliation constructed by statistical manifolds which are α-conformally equivalent to statistical manifolds with α-transitively flat connections.

2. Preliminaries

The following theorems hold on level surfaces of Hessian domains and on the realization of submanifolds.

Theorem 2.1 **([11]).** *Let M be a simply connected n-dimensional level surface of φ on an $(n+1)$-dimensional Hessian domain $(\Omega, D, g = Dd\varphi)$ with a Riemannian metric g, and suppose that $n \geq 2$. If we consider (Ω, D, g) to be a flat statistical manifold, (M, D^M, g^M) is a 1-conformally flat statistical submanifold of (Ω, D, g), where D^M and g^M are the connection and the Riemannian metric on M induced by D and g, respectively.*

Theorem 2.2 **([11]).** *An arbitrary 1-conformally flat statistical manifold of* $\dim n \geq 2$ *with a Riemannian metric can be locally realized as a submanifold of a flat statistical manifold of* $\dim\,(n+1)$.

Theorem 2.3 **([9]).** *Let $(N, \nabla^{((\alpha))}, h)$ be a statistical manifold of* $\dim n \geq 2$ *which is α-conformally equivalent to a statistical manifold with an α-transitively flat connection for non-zero $\alpha \in \mathbf{R}$, where h is the Riemannian metric. Then $(N, \nabla^{((\alpha))}, h)$ can be locally realized as a submanifold of a statistical manifold of* $\dim\,(n+1)$ *with an α-transitively flat connection.*

Theorem 2.2 describes Theorem 2.3 for $\alpha = 1$. We study the definitions and properties to understand the above theorems.

2.1. *α-connections of statistical manifolds*

For a torsion-free affine connection ∇ and a pseudo-Riemannian metric h on a manifold N, the triple (N, ∇, h) is said to be a *statistical manifold* if ∇h is symmetric. For a statistical manifold (N, ∇, h), let ∇' be an affine connection on N such that

$$Xh(Y, Z) = h(\nabla_X Y, Z) + h(Y, \nabla'_X Z), \quad \text{for } X, Y \text{ and } Z \in TN,$$

where TN is the set of all tangent vector fields on N. The affine connection ∇' is torsion free and $\nabla' h$ is symmetric. Then, ∇' is called the *dual connection* of ∇ and the triple (N, ∇', h) is called the *dual statistical manifold* of (N, ∇, h). For a real number α, an affine connection defined by

$$\nabla^{(\alpha)} := \frac{1+\alpha}{2}\nabla + \frac{1-\alpha}{2}\nabla'$$

is called an *α-connection* of (N, ∇, h). The triple $(N, \nabla^{(\alpha)}, h)$ is also a statistical manifold, and $\nabla^{(-\alpha)}$ is the dual connection of $\nabla^{(\alpha)}$. The 1-connection, the (-1)-connection, and the 0-connection coincide with ∇, ∇', and the Levi-Civita connection of (N, h), respectively. When the curvature tensor R of $\nabla^{(\alpha)}$ vanishes, the statical manifold $(N, \nabla^{(\alpha)}, h)$ is said to be *flat.* If (N, ∇, h) is flat, (N, ∇', h) is flat and $(N, \nabla^{(\alpha)}, h)$ is not always flat [1]. If (N, ∇, h) is a flat statistical manifold, $\nabla^{(\alpha)}$ said to be an *α-transitively flat connection* of (N, ∇, h) ([9]).

In his paper [4], Kurose defined the α-conformal equivalence and α-conformal flatness of statistical manifolds. For $\alpha \in \mathbf{R}$, statistical manifolds (N, ∇, h) and $(N, \bar{\nabla}, \bar{h})$ are said to be *α-conformally equivalent* to each other if there exists a function ϕ on N such that

$$\begin{aligned} \bar{h}(X, Y) &= e^{\phi} h(X, Y), \\ h(\bar{\nabla}_X Y, Z) &= h(\nabla_X Y, Z) - \frac{1+\alpha}{2} d\phi(Z) h(X, Y) \\ &\quad + \frac{1-\alpha}{2} \{d\phi(X) h(Y, Z) + d\phi(Y) h(X, Z)\} \end{aligned}$$

for X, Y and $Z \in TN$. A statistical manifold (N, ∇, h) is said to be *α-conformally flat* if (N, ∇, h) is locally α-conformally equivalent to a flat statistical manifold. Statistical manifolds (N, ∇, h) and $(N, \bar{\nabla}, \bar{h})$ are *α-conformally equivalent* to each other if and only if the dual statistical manifolds (N, ∇', h) and $(N, \bar{\nabla}', \bar{h})$ are $(-\alpha)$-conformally equivalent to each other.

Theorem 2.3 relates α-conformal equivalence and the realization into a statistical manifold with a α-transitively flat connection.

2.2. *Dualistic structure of Hessian domains*

Let D and $\{x^1, \ldots, x^{n+1}\}$ be the canonical flat affine connection and the canonical affine coordinate system for $\mathbf{A}^{n+1}$, respectively, i.e., $Ddx^i = 0$. If the Hessian $Dd\varphi = \sum_{i,j} (\partial^2 \varphi / \partial x^i \partial x^j) dx^i dx^j$ is non-degenerate for a function φ on a domain Ω in $\mathbf{A}^{n+1}$, the triple $(\Omega, D, g = Dd\varphi)$ is said to be a *Hessian domain,* which is a flat statistical manifold. Conversely, a flat statistical manifold is locally a Hessian domain. Let $\mathbf{A}^*_{n+1}$ and $\{x^*_1, \ldots, x^*_{n+1}\}$ be the dual affine space of $\mathbf{A}^{n+1}$ and the dual affine coordinate system of $\{x^1, \ldots, x^{n+1}\}$, respectively. We define the *gradient mapping* ι from Ω to $\mathbf{A}^*_{n+1}$ by

$$x^*_i \circ \iota = -\frac{\partial \varphi}{\partial x^i},$$

and a flat affine connection D' on Ω by

$$\iota_*(D'_X Y) = D^*_X \iota_*(Y) \quad \text{for} \ \ X, Y \in T\Omega,$$

where $D^*_X \iota_*(Y)$ is the covariant derivative along ι induced by the canonical flat affine connection D^* on $\mathbf{A}^*_{n+1}$. Then, (Ω, D', g) is the dual statistical manifold of (Ω, D, g) if (Ω, D, g) is a statistical manifold [1, 7, 8].

2.3. *Statistical manifolds and level surfaces*

Henceforth, we suppose that g is positive definite.

Now, we give properties of the level surface of a Hessian domain as a statistical manifold [11].

Let $\tilde{E}$ be the gradient vector field of φ on Ω defined by

$$g(X, \tilde{E}) = d\varphi(X) \quad \text{for } X \in T\Omega,$$

where $T\Omega$ is the set of all tangent vector fields on Ω. We set

$$E = -d\varphi(\tilde{E})^{-1}\tilde{E} \quad \text{on } \Omega_o = \{p \in \Omega \mid d\varphi_p \neq 0\}.$$

A key point for the proofs of Theorems 2.1 and 2.2 is to make use of the normalized gradient vector field E.

For $p \in \Omega_o$, the vector E_p is perpendicular to T_pM with respect to g, where $M \subset \Omega_o$ is a level surface of φ containing p, and T_pM is the set of all tangent vectors at p on M. Let x be a canonical immersion of an n-dimensional level surface M into Ω. For D and an affine immersion (x, E), we can define the induced affine connection D^E and the affine fundamental form g^E on M by

$$D_X Y = D^E_X Y + g^E(X, Y)E \quad \text{for } X, Y \in TM.$$

A submanifold that itself is a statistical manifold is called a *statistical submanifold.* Let D^M and g^M be the connection and the Riemannian metric obtained by the restriction of D and g to the submanifold M, respectively. Then, $D^M = D^E$ and $g^M = g^E$ hold. Hence, the triple (M, D^M, g^M) is the statistical submanifold realized in (Ω, D, g) which coincides with the manifold (M, D^E, g^E) induced by an affine immersion (x, E). In addition, an immersion (x, E) is equiaffine, i.e.,

$$D_X E = S^E(X) \in TM \quad \text{for } X \in TM,$$

where S^E is the shape operator. These facts lead to Theorems 2.1 and 2.2.

3. Foliations constructed by 1-conformally flat statistical manifolds

Theorems 2.2 and 2.3 describe the realization of one statistical manifold. Then, the next problem arises.

Problem 3.1. *What are the conditions for a foliation constructed by 1-conformally flat statistical manifolds to coincide with a flat statistical manifold by a Hessian domain, and for a foliation constructed by statistical manifolds that are α-conformally equivalent to statistical manifolds with α-transitively flat connections?*

In this section, we solve Problem 3.1 for the case of foliations constructed by 1-conformally flat statistical manifolds.

Let $\mathcal{F}$ be a foliation on a differentiable manifold N of dimension $n \geq 2$ and co-dimension 1, and the triple (M, ∇^M, h^M) be a 1-conformally flat statistical manifold for each leaf $M \in \mathcal{F}$. We suppose that a non-degenerate affine immersion (x^M, E^M) realizes (M, ∇^M, h^M) in $\mathbf{A}^{n+1}$, and that a mapping $x : N \to \Omega$ defined by $x(p) = x^M(p)$ for $p \in M$ is a diffeomorphism, where $\Omega = \bigcup_{M \in \mathcal{F}} x^M(M) \subset \mathbf{A}^{n+1}$ is a domain diffeomorphic to N.

We set ι^M as the conormal immersion for x^M, i.e., denoting by $\langle a, b\rangle$ a pairing of $a \in \mathbf{A}^*_{n+1}$ and $b \in \mathbf{A}^{n+1}$,

$$\langle \iota^M(p), Y_p\rangle = 0 \;\text{ for } Y_p \in T_pM, \;\; \langle \iota^M(p), E^M_p\rangle = 1 \tag{1}$$

for $p \in M$, considering $T_p\mathbf{A}^{n+1}$ with $\mathbf{A}^{n+1}$ [5, 6]. The immersion ι^M satisfies

$$\langle \iota^M_*(Y), E^M\rangle = 0, \;\; \langle \iota^M_*(Y), X\rangle = -h^M(Y, X) \;\text{ for } X, Y \in TM.$$

Let S^{E^M} be the shape operator of (x^M, E^M). Then, we can describe

$$D_XY = \nabla^M_XY + h^M(X, Y)E^M \quad \text{for } X, Y \in TM,$$
$$D_XE^M = S^{E^M}(X) \in TM.$$

In general, for statistical manifolds $\{(\bar{M}, \nabla^{\bar{M}}, h^{\bar{M}}) | \bar{M} \in \bar{\mathcal{F}}\}$, where $\bar{\mathcal{F}}$ is a foliation on a differentiable manifold, we call $(\bar{M}, \nabla^{\bar{M}}, h^{\bar{M}})$ a statistical leaf of $\bar{\mathcal{F}}$. The following theorem holds.

Theorem 3.1 ([10]). *Each 1-conformally flat statistical leaf (M, ∇^M, h^M) of $\mathcal{F}$ is locally realized as a level surface of the common Hessian domain, and $N = \bigcup_{M \in \mathcal{F}} M$ is locally diffeomorphic to the Hessian domain if and only if a foliation $\mathcal{F}$ satisfies the following conditions:*

(i) *A mapping $E : N \to \mathbf{A}^{n+1}$ defined by $E(p) = E^M(p)$ for $p \in M$ is differentiable;*
(ii) *A mapping $\iota : N \to \Omega^*$ defined by $\iota(p) = \iota^M(p)$ for $p \in M$ is a diffeomorphism, where $\Omega^* = \cup_{M \in \mathcal{F}} \iota^M(M) \subset \mathbf{A}^*_{n+1}$;*
(iii) *There exists a function $\rho : M \to \mathbf{R}^+$ such that*

$$X\rho = -h^M((D_EE)|_{TM}, X),$$
$$S^{E^M}(X) = -\rho X \quad \textit{for } X \in TM,$$

where we set $D_E E = (D_E E)|_{TM} + \mu E^M$, $(D_E E)|_{TM} \in TM$, $\mu \in \mathbf{R}$;

(iv) $D_E X = 0$ *for* $X \in T\Omega$, *where* $X_p = X_{\hat{p}} \in T_p M = T_{\hat{p}} \hat{M}$ *for* $e^{\lambda(\hat{p})}\iota(p) = \iota(\hat{p})$, $p \in U \cap M$, $\hat{p} \in U \cap \hat{M}$, $\hat{M} \in \mathcal{F}$ $(U \subset \Omega$: *a small open set) and* $\lambda(\hat{p}) \in \mathbf{R}$.

Outline of proof. For more details, see Theorem 4.1 in [10].

We consider a manifold N as a domain $\Omega \subset \mathbf{A}^{n+1}$, and define a metric g on Ω by

$$g(Y, X) = h^M(Y, X), \ g(E, E) = \rho, \ g(Y, E) = 0 \quad \text{for } X, Y \in TM \subset TN = T\Omega.$$

Then, it holds that (D, g) satisfies the Codazzi equation

$$(D_X g)(Y, Z) = (D_Y g)(X, Z) \quad \text{for all } X, Y, \text{and } Z \in T\Omega. \tag{2}$$

Thus, g is a Hessian metric based on Proposition 2.1 in [8]. From the definition of g, we can consider that each leaf (M, ∇^M, h^M) of $\mathcal{F}$ is a level surface of the Hessian domain (Ω, D, g).

For the proof of Equation (2), we make use of the Codazzi equation and the Ricci equation for an equiaffine immersion (x^M, E^M), i.e.,

$$(\nabla^M_X h^M)(Y, Z) = (\nabla^M_Y h^M)(X, Z),$$
$$h^M(S^{E^M}(X), Y) = h^M(X, S^{E^M}(Y)) \quad \text{for } X, Y \in TM$$

[6], and compair the left-hand side and the right-hand side of Equation (2) in the case of X, Y, $Z \in TM$ or $= E$, respectively.

For example, in the case of $X, Z \in TM$ and Y= E, the left-hand side of Equation (2) is deformed by the condition (iii) as the following;

$$(D_X g)(E, Z) = -\rho h^M(X, Z) - h^M(S^{E^M}(X), Z) = 0.$$

For the condition (ii), any conormal immersion ι^M is a centro-affine hypersurface that possesses a projective transformation to $\iota^{\hat{M}}$ for $\hat{M} \in \mathcal{F}$, $p \in M$, $\hat{p} \in \hat{M}$ locally. By the first equation of (1), we identify $T_{\hat{p}}\hat{M}$ with $T_p M$ for $e^{\lambda(\hat{p})}\iota(p) = \iota(\hat{p})$, $\lambda(\hat{p}) \in \mathbf{R}$. Then, it holds that $h^{\hat{M}}|_{\hat{p}} = h^M|_p$ (cf. pp.16–17 in [6]), and that $g_{\hat{p}} = h^M|_p$ for $e^{\lambda(\hat{p})}\iota(p) = \iota(\hat{p})$, $\lambda(\hat{p}) \in \mathbf{R}$ locally. We here suppose that the parallel displacement along E with respect to D maps $T_p M$ to $T_{\bar{p}}\bar{M}$ for some $p \in M$ and $\bar{p} \in \bar{M}$ so that $e^{\lambda}\iota(p) \neq \iota(\bar{p})$ for any $\lambda \in \mathbf{R}$. Since ι coincides with ι^M on M and with $\iota^{\bar{M}}$ on $\bar{M}$, and since we have (1) for ι^M and for $\iota^{\bar{M}}$, we find that $(D_E X)|_p$ contains a vertical component to $T_p M$ for a certain $X \in T\Omega$ satisfying $X_p = X_{\hat{p}} \in T_p M = T_{\hat{p}}\hat{M}$ for $e^{\lambda(\hat{p})}\iota(p) = \iota(\hat{p})$. This statement contradicts to the condition (iv). Thus the parallel displacement along E with respect to D maps T_pM to $T_{\hat{p}}\hat{M}$ only for $e^{\lambda(\hat{p})}\iota(p) = \iota(\hat{p})$, $\lambda(\hat{p}) \in \mathbf{R}$. Hence, $E(g(X,Z))|_p = 0$ holds for $p \in M$. Thus it holds that

$$(D_E g)(X, Z) = E(g(X, Z)) - g(D_E X, Z) - g(X, D_E Z) = 0.$$

Thus we have the Codazzi equation

$$(D_X g)(E, Z) = (D_E g)(X, Z).$$

In other cases, we are also able to show the Codazzi equation for (D, g).

The need for the conditions (i)–(iv) follows from Lemma 3.1 described below. □

It is known that the conormal immersion $\iota^M : M \to \mathbf{A}^*_{n+1}$ coincides with the gradient mapping $\iota : \Omega \to \Omega^*$ on a level surface M of a Hessian domain (Ω, D, g) [12]. Thus, a mapping $\iota : N \to \Omega^*$ defined by Theorem 3.1 (ii) coincides with the gradient mapping ι on Ω identified by N.

The next lemma holds based on the need for Theorem 3.1, where we denote by $\mathcal{F}$ a foliation on $\Omega_o = \{p \in \Omega \mid d\varphi_p \neq 0\}$ by level surfaces of a given Hessian domain $(\Omega, D, g = Dd\varphi)$.

Lemma 3.1 ([10]). *An $(n+1)$-dimensional Hessian domain $(\Omega, D, g = Dd\varphi)$ and n-dimensional level surfaces $\{(M, D^M, g^M) | M \in \mathcal{F}\}$ $(n \geq 2)$ satisfy the following conditions:*

(i) *A mapping $E : \Omega_o \to \mathbf{A}^{n+1}$ defined by $E(p) = E^M(p)$ for $p \in M$ is differentiable;*
(ii) *The gradient mapping $\iota : \Omega \to \Omega^* = \iota(\Omega) \subset \mathbf{A}^*_{n+1}$ is a diffeomorphism;*
(iii) *$X(g(E, E)) = -g(D_E E, X)$ for $X \in TM$;*
(iv) *$S^{E^M}(X) = -g(E, E)X$ for $X \in TM$;*
(v) *$D_E X = 0$ for $X \in T\Omega_o$, where $X_p = X_{\hat{p}} \in T_pM = T_{\hat{p}}\hat{M}$ for $e^{\lambda(\hat{p})}\iota(p) = \iota(\hat{p})$, $p \in U \cap M$, $\hat{p} \in U \cap \hat{M}$, $\hat{M} \in \mathcal{F}$ ($U \subset \Omega_o$: a small open set) and $\lambda(\hat{p}) \in \mathbf{R}$.*

Outline of proof. For more details, see Lemma 3.1 in [10].

The conditions (i) and (ii) are clear.

In their papers [2, 8], Hao and Shima calculated $(D_X g)(\tilde{E}, \tilde{E})$ and $(D_{\tilde{E}} g)(\tilde{E}, X)$ for $(x, \tilde{E})$, not for (x, E), and showed that the transversal connection form $\tau^{\tilde{E}}$ vanishes if and only if $D_{\tilde{E}}\tilde{E} = \mu\tilde{E}$, $\mu \in \mathbf{R}$. For the proof of condition (iii), we calculate $(D_X g)(E, E)$ and $(D_E g)(E, X)$ for

(x, E) and $X \in TM$ by using their technique. From the Codazzi equation for (D, g) on a Hessian domain, i.e.,

$$(D_X g)(E, E) = (D_E g)(E, X)$$

[8], the condition (iii) holds.

To obtain the proofs of (iv) and (v), we calculate $(D_X g)(E, Z)$ and $(D_E g)(X, Z)$ for $X, Z \in TM$. Because M is a level surface of a Hessian domain, any conormal immersion ι^M is a centro-affine hypersurface. Then, it holds that $g^{\hat{M}} = g^M$ for $e^{\lambda(\hat{p})}\iota(p) = \iota(\hat{p})$, $p \in M$, $\hat{p} \in \hat{M} \in \mathcal{F}$ and $\lambda(\hat{p}) \in \mathbf{R}$ locally (cf. pp.16-17 in [6]). This fact and several calculations lead to condition (v). From (v) and the Codazzi equation for (D, g), i.e.,

$$(D_X g)(E, Z) = (D_E g)(X, Z),$$

we obtain the condition (iv). □

Finally, in this section, we discuss about a projectively flat connection and a dual-projectively flat connection. In their papers Kurose [4] and Ivanov [3] proved the following theorems, respectively.

Theorem 3.2 ([4]). *A statistical manifold (N, ∇, h) is 1-conformally flat if and only if the dual connection ∇' is a projectively flat connection with a symmetric Ricci tensor.*

Theorem 3.3 ([3]). *A statistical manifold (N, ∇, h) is 1-conformally flat if and only if ∇ is a dual-projectively flat connection with a symmetric Ricci tensor.*

Thus, we can describe Theorem 3.1 as follows.

Corollary 3.1. *Let ∇^M be a dual-projectively flat connection with a symmetric Ricci tensor for each $M \in \mathcal{F}$. Then, each statistical leaf (M, ∇^M, h^M) of a foliation $\mathcal{F}$ is locally realized as a level surface of the common Hessian domain and $N = \cup_{M \in \mathcal{F}} M$ is locally diffeomorphic to the Hessian domain if and only if $\mathcal{F}$ satisfies the conditions (i)–(iv) of Theorem 3.1.*

4. α-conformal equivalence and Hessian domains

In this section, we solve Problem 3.1 for a foliation constructed by statistical manifolds which are α-conformally equivalent to statistical manifolds with α-transitively flat connections.

The following theorem holds.

Theorem 4.1. *Let $\mathcal{F}$ be a foliation on a differentiable manifold N of dimension $n \geq 2$ and co-dimension 1. For a fixed non-zero parameter $\alpha \in \mathbf{R}$ and for each leaf $M \in \mathcal{F}$, let the triple $(M, \nabla^{((\alpha))M}, h^M)$ be a statistical manifold that is α-conformally equivalent to a statistical manifold with an α-transitively flat connection. Then, each statistical leaf $(M, \nabla^{((\alpha))M}, h^M)$ of $\mathcal{F}$ is locally realized as a statistical submanifold by a level surface of the common Hessian domain with an α-connection, and $N = \bigcup_{M\in\mathcal{F}} M$ is locally diffeomorphic to the Hessian domain if and only if a foliation $\mathcal{F}$ satisfies the conditions (i)-(iv) of Theorem 3.1 with respect to leaves $\{(M, \nabla^M, h^M) | M \in \mathcal{F}\}$, where*

$$\nabla^M := \nabla^{(LC)M} + \frac{1}{\alpha}(\nabla^{((\alpha))M} - \nabla^{(LC)M}) \tag{3}$$

and $\nabla^{(LC)M}$ is the Levi-civita connection for h^M.

Proof. Based on the definition of α-transitive flatness, there exist statistical manifolds $(M, \bar{\nabla}^{M(\alpha)}, \bar{h}^M)$ and $(M, \bar{\nabla}^M, \bar{h}^M)$ such that $(M, \nabla^{((\alpha))M}, h^M)$ is α-conformally equivalent to $(M, \bar{\nabla}^{M(\alpha)}, \bar{h}^M)$, where $\bar{\nabla}^{M(\alpha)}$ is an α-connection of a flat statistical manifold $(M, \bar{\nabla}^M, \bar{h}^M)$. From Lemma 4.3 of [9], (M, ∇^M, h^M) is 1-conformally equivalent to a flat statistical manifold $(M, \bar{\nabla}^M, \bar{h}^M)$. Thus, leaves $\{(M, \nabla^M, h^M) | M \in \mathcal{F}\}$ are locally realized as a level surface of the common Hessian domain, and $N = \bigcup_{M\in\mathcal{F}} M$ is locally diffeomorphic to the Hessian domain if and only if a foliation $\mathcal{F}$ satisfies the conditions (i)–(iv) of Theorem 3.1 with respect to leaves $\{(M, \nabla^M, h^M) | M \in \mathcal{F}\}$.

We suppose that a Hessian domain (Ω, D, g) locally realizes leaves $\{(M, \nabla^M, h^M) | M \in \mathcal{F}\}$. Let $D^{(\alpha)}$ be an α-connection of (Ω, D, g), i.e.,

$$D^{(\alpha)} := \frac{1+\alpha}{2} D + \frac{1-\alpha}{2} D',$$

where D' is the dual connection of (Ω, D, g). We denote by D^M, D'^M, and $D^{(\alpha)M}$ the induced connections on M as a submanifold of Ω from D, D', and $D^{(\alpha)}$, respectively. Then, it holds that

$$D^{(\alpha)M} = \frac{1+\alpha}{2} D^M + \frac{1-\alpha}{2} D'^M \tag{4}$$

Considering the definition of ∇^M by Equation (3), we have

$$\nabla^{((\alpha))M} = (1-\alpha)\nabla^{(LC)M} + \alpha\nabla^M = \frac{1+\alpha}{2}\nabla^M + \frac{1-\alpha}{2}\nabla'^M, \tag{5}$$

where ∇'^M is the dual connection of ∇^M. Because $D^M = \nabla^M$ and $D'^M = \nabla'^M$, the connection $\nabla^{((\alpha))M}$ coincides with $D^{(\alpha)M}$ on M by Equations (4) and (5).

Thus, $(M, \nabla^{((\alpha))M}, h^M)$ is locally realized as a submanifold of $(\Omega, D^{(\alpha)}, g)$ for each leaf $M \in \mathcal{F}$, and Theorem 4.1 holds. □

If $\alpha = 1$, Theorem 4.1 describes Theorem 3.1.

Acknowledgments

The author thanks the referees and the editors for their helpful comments.

Bibliography

1. S. Amari and H. Nagaoka, *Method of information geometry*, American Mathematical Society, Providence, Oxford University Press, Oxford, (2000).
2. J.H. Hao and H. Shima, *Level surfaces of non-degenerate functions in* $\mathbf{R}^{n+1}$, Geom. Dedicata **50** (1994), 193–204.
3. S. Ivanov, *On dual-projectively flat affine connections*, J. Geom. **53** (1995), 89–99.
4. T. Kurose, *On the divergence of 1-conformally flat statistical manifolds*, Tôhoku Math. J. **46** (1994), 427–433.
5. K. Nomizu and U. Pinkal, *On the geometry and affine immersions*, Math. Z. **195** (1987), 165–178.
6. K. Nomizu and T. Sasaki, *Affine Differential Geometry: Geometry of Affine Immersions*, Cambridge University Press, Cambridge, (1994).
7. H. Shima, *Harmonicity of gradient mapping of level surfaces in a real affine space*, Geom. Dedicata **56** (1995), 177–184.
8. H. Shima, *The geometry of Hessian Structures*, World Scientific, Singapore, (2007).
9. K. Uohashi, *On α-conformally equivalence of statistical submanifolds*, J. Geom. **75** (2002), 179–184.
10. K. Uohashi, *A Hessian domain constructed with a foliation by 1-conformally flat statistical manifolds*, International Mathematical Forum **7** (2012), 2363–2371.
11. K. Uohashi, A. Ohara and T. Fujii, *1-conformally flat statistical submanifolds*, Osaka J. Math. **37** (2000), 501–507.
12. K. Uohashi, A. Ohara and T. Fujii, *Foliations and divergences of flat statistical manifolds*, Hiroshima Math. J. **30** (2000), 403–414.

Received December 25, 2012
Revised May 24, 2013

Proceedings of the 3rd International
Colloquium on Differential Geometry
and its Related Fields
Veliko Tarnovo, September 3–7, 2012

RULED COMPLEX LAGRANGIAN SUBMANIFOLDS OF DIMENSION TWO IN $\mathbf{C}^4$

Norio EJIRI *

Department of Mathematics, Meijo University,
Shiogamaguchi, 1-501, Tempaku-ku,
Nagoya, 468-8502 Japan
E-mail: ejiri@meijo-u.ac.jp

We construct examples of ruled complex Lagrangian submanifolds in $\mathbf{C}^4$ as deformations of the complex Lagrangian cone in $\mathbf{C}^4$.

Keywords: Ruled submanifolds; Complex Lagrangian cone; Complex symplectic form.

1. Introduction

Let $\mathbf{C}^n$ be the set of all $1 \times n$ complex matrices. We denote by $\mathbf{C}^{2n}$ the set $\{(Z_1, Z_2) : Z_1, Z_2 \in \mathbf{C}^n\}$ with the Hermitian inner product $(\ ,\ )$ defined by $((Z_1, Z_2), (Z_1', Z_2')) = Z_1 {}^t\overline{Z_1'} + Z_2 {}^t\overline{Z_2'}$ and the Euclidean inner product $\langle\ ,\ \rangle$ given by $\langle\ ,\ \rangle = \mathrm{Re}(\ ,\ )$.

We define three orthogonal complex structures I, J, K on $\mathbf{C}^{2n}$ by

$$I(Z_1, Z_2) = (iZ_1, iZ_2),\ J(Z_1, Z_2) = (i\overline{Z}_2, -i\overline{Z}_1),\ K(Z_1, Z_2) = (-\overline{Z}_2, \overline{Z}_1).$$

Then these I, J, K satisfy $IJ = K$, $JI = -IJ$ and define a quaternion structure on $\mathbf{C}^{2n}$. Therefore $\mathbf{C}^{2n}$ can be considered as an n-dimensional quaternion vector space $\mathbf{H}^n$. We define a complex symplectic form ω on $\mathbf{C}^{2n}$ as

$$\omega\big((Z_1, Z_2), (Z_1', Z_2')\big) = -(Z_1 {}^tZ_2' - Z_2 {}^tZ_1').$$

We then have $\omega = dZ_2 \wedge {}^t dZ_1$. If we denote by ω_J and ω_K two Kähler 2-forms corresponding to J and K, respectively, then we have $\omega = -\omega_K + i\omega_J$.

Let M be a real q-dimensional manifold and f be an immersion from M into $\mathbf{C}^{2n}$. If $f^*\omega = 0$, then M is called a *complex isotropic submanifold.* The condition $f^*\omega = 0$ is equivalent to the condition that both $f^*\omega_J = 0$

*The author is partially supported by JSPS KAKENHI Grant Number 22540103.

and $f^*\omega_K = 0$ hold. Thus M is isotropic with respect to J, K and $q \leq 2n$. We call M *complex Lagrangian* if $q = 2n$. When M is complex Lagrangian, it is Lagrangian with respect to J and K. So each tangent space of M is invariant under I, and I induces a complex structure on M. Therefore M becomes a complex submanifold in $\mathbf{C}^{2n}$. The condition that both $f^*\omega_J = 0$ and $f^*\omega_K = 0$ hold is called "bilagrangian condition" by Hitchin [11].

If $q < 2n$, then generally the condition $f^*\omega = 0$ does not imply the existence of a complex structure on M. In this paper, we assume that M is complex manifold with respect to I.

We give some examples of complex Lagrangian submanifolds. For a complex $n \times n$ symmetric matrix A, the n-dimensional subspace $\{(Z, ZA)|Z \in \mathbf{C}^n\} \subset \mathbf{C}^{2n}$ is complex Lagrangian subspace. For a holomorphic function f on a domain $\Omega \subset \mathbf{C}^n$, the graph

$$\left(z_1, ..., z_n, \frac{\partial f}{\partial z_1}, ..., \frac{\partial f}{\partial z_n}\right), \quad (z_1, ..., z_n) \in \Omega$$

of the gradient of f is a complex Lagrangian submanifold in $\mathbf{C}^{2n}$, We call this a complex Lagrangian graph. Note that any holomorphic curve in $\mathbf{C}^2$ is a complex Lagrangian submanifold in $\mathbf{C}^2$.

Remark 1.1. A complex Lagrangian submanifold in $\mathbf{C}^{2n}$ is a special Lagrangian submanifold with respect to J. In particular, a special Lagrangian surface in $\mathbf{C}^2$ is a complex Lagrangian surface [13].

Let $Lag^C = U(n)\backslash Sp(n)$ be the Grassmann manifold of complex Lagrangian subspace in $\mathbf{C}^{2n}$. When $n = 1$, it is a 1-dimensional complex projective space $\mathbf{C}P^1$, When $n = 2$, it is a 3-quadric Q^3. Generally, Lag^C is a Hermitian symmetric space of compact type whose rank is n. Let V be the tautological vector bundle on Lag^C and Φ be the corresponding map of V onto $\mathbf{C}^{2n}$.

We construct an $Sp(n, \mathbf{C})$-invariant holomorphic 1-form Ξ which satisfies $d\Xi = -2\Phi^*\omega$. Since Ξ is invariant under the action of $\mathbf{C}^* = \mathbf{C} \setminus \{0\}$, We have the following.

Theorem 1.1. *Let M be a complex submanifold in Lag^C and $V|_M$ be the induced bundle on M. Let β be a closed holomorphic 1-form on $V|_M$.*

(1) *Let the subset S of $V|_M$ be solutions of the equation $\Xi - \beta = 0$. Then the image $\Phi(S)$ which may have singularities is complex isotropic. In the case that $\beta = 0$, the image $\Phi(S)$ is a complex isotropic cone.*
(2) *If we assume that the projection of S into M is a submersion and that the image $\Phi(S)$ is a complex submanifold N in M, then there exists a*

closed holomorphic 1*-form* α *such that* S *is contained in the subset of* $V|_N$ *defined by the equation* $\Xi - \alpha = 0$.

If the analytic set of the solutions of the equation $\Xi = 0$ of $V|_N$ is a complex Lagrangian cone, then the subset $\Xi - t\alpha = 0$ of $V|_N$ for a small t, is a deformation of a complex Lagrangian cone.

Let $\mathbf{C}P_I^{2n-1}$ be a $(2n-1)$-dimensional complex projective space with respect to I. Then it admits a holomorphic contact structure [2]. We can consider a holomorphic horizontal submanifold with respect to the holomorphic contact structure in $\mathbf{C}P_I^{2n-1}$. We show that a holomorphic horizontal submanifold in $\mathbf{C}P_I^{2n-1}$ is given by the image of a complex Lagrangian cone in $\mathbf{C}^{2n}$ through the natural projection of $(\mathbf{C}^{2n}, I)$ onto $\mathbf{C}P_I^{2n-1}$ ([9]).

Next, we define a null submanifold in Lag^C as the Gauss image of a complex Lagrangian cone into Lag^C. We reconstruct a complex Lagrangian cone and a holomorphic horizontal submanifold from a null submanifold. The way of this reconstruction may be a generalization of the Lie transform between a null curve in Q^3 and a horizontal holomorphic curve in $\mathbf{C}P^3$ [1]. By Theorem 1.1, we can construct a ruled complex Lagrangian submanifold of dimension 2 in $\mathbf{C}^4$ as a deformation of a complex Lagrangian cone.

Theorem 1.2. *Let* M *be a null curve of genus* 1 *in* Q^3 *and* $\beta \in H^{(1,0)}(M)$.

(I) *If* M *be an immersed null curve of genus* 1 *in* Q^3*, we have the following.*
   (1) *The complex Lagrangian cone which is defined by* $\Xi = 0$ *in* $\mathbf{C}^4$ *has a singularity which is different from* 0.
   (2) *If the number of irreducible components of the solutions of the equations* $\Xi - \beta = 0$ *is* 2*, then the complex Lagrangian cone is deformed to two ruled and asymptotical conical complex Lagrangian submanifolds with rate* 0 *without a singularity.*
   (3) *If the number of irreducible components of the solutions of the equations* $\Xi - \beta = 0$ *is* 1*, then the complex Lagrangian cone is deformed to an asymptotical conical complex Lagrangian submanifold with rate* 0 *without a singularity.*

(II) *If* M *is a branched null curve of genus* 1 *in* Q^3*, then the obtained ruled submanifold is regular, which is not asymptotical conical complex Lagrangian submanifold with rate* 0.

We here note that we have many examples of algebraic minimal surfaces. In his paper [17], Small gave an algebraic minimal surface of genus 1 derived from a charge 2 Monopole spectral curve. Unfortunately, this surface is branched. On the other hand, Pirola [16] proved that there is a complete algebraic minimal immersion $F : X \setminus Z \longrightarrow \mathbf{R}^3$ with finite total Gaussian curvature for an arbitrary compact Riemann surface X and an arbitrary non-empty finite subset Z of X.

We note also that Harvey and Lawson [12], Bryant [3], Joyce [14] studied ruled special Lagrangian 3-folds in $\mathbf{C}^3$. Theorem 1.2 gives some ruled special Lagrangian 4-folds in $\mathbf{C}^4$ with respect to J.

2. Proof of Theorem 1.1

In this section we give a proof of Theorem 1.1.

Let S^2_C denote the set of all $n \times n$ complex symmetric matrices. For $\tau \in S^2_C$, we set $V_\tau = \{(Z, Z\tau) | Z \in \mathbf{C}^n\}$, and put $Lag^C_o = \{V_\tau \mid \tau \in S^2_C\}$, which is a subset of Lag^C. We can hence identify Lag^C_o with S^2_C. We note that Lag^C_o is a Zariski open subset in Lag^C. A restriction of the tautological vector bundle V onto Lag^C_o is given by $\{(\tau, (Z, Z\tau)) \mid \tau \in S^2_C, Z \in \mathbf{C}^n\} \simeq S^2_C \times \mathbf{C}^n$.

Let $Sp(n, \mathbf{C})$ be a symplectic group which is given as

$$Sp(n,\mathbf{C}) = \left\{ \begin{pmatrix} a & b \\ c & d \end{pmatrix} \in M_{2n\times 2n}(\mathbf{C}) \;\middle|\; \begin{matrix} a\,{}^t d - b\,{}^t c = E_n, \\ {}^t(a\,{}^t b) = a\,{}^t b,\ {}^t(c\,{}^t d) = c\,{}^t d \end{matrix} \right\},$$

where $a, b, c, d \in M_{n\times n}(\mathbf{C})$. Then the right action $Sp(n, \mathbf{C})$ on $\mathbf{C}^{2n}$ is given by

$$(Z_1, Z_2) \begin{pmatrix} a & b \\ c & d \end{pmatrix} = (Z_1 a + Z_2 c, Z_1 b + Z_2 d).$$

It is a holomorphic symplectic transformation with respect to ω. Since $Sp(n, \mathbf{C})$ acts on complex Lagrangian subspaces, it acts also on Lag^C and on its tautological vector bundle. Let Φ be a map from the tautological vector bundle on Lag^C into $\mathbf{C}^{2n}$. The restricion of the map Φ on the induced bundle on Lag^C_o is given by $\Phi(\tau, K) = (K, K\tau)$ for $(\tau, K) \in S^2_C \times \mathbf{C}^n$. We can show that the map Φ is $Sp(n, \mathbf{C})$-equivariant in the following manner. For any $(\tau, K) \in S^2_C \times \mathbf{C}^n$ and $\begin{pmatrix} a & b \\ c & d \end{pmatrix} \in Sp(n, \mathbf{C})$, we set

$$(\tau, K) \begin{pmatrix} a & b \\ c & d \end{pmatrix} = \big((a + \tau c)^{-1}(b + \tau d), K(a + \tau c)\big).$$

On the other hand, we have

$$(Z, Z\tau) \begin{pmatrix} a & b \\ c & d \end{pmatrix} = \big(Z(a + \tau C), Z(b + \tau d)\big).$$

Hence, if we set $W = Z(a + \tau C)$, we see

$$(Z, Z\tau)\begin{pmatrix} a & b \\ c & d \end{pmatrix} = \big(W, W(a+\tau c)^{-1}(b+\tau d)\big).$$

We therefore find that Φ is $Sp(n,\mathbf{C})$-equivariant. Moreover, by the definition of the map Φ, we get

$$\begin{aligned}\Phi^*\omega &= \big((dK)\tau + K d\tau\big) \wedge {}^t dK = K d\tau \wedge {}^t dK \\ &= \operatorname{tr}(d\tau \wedge^t K dK) = -\frac{1}{2}\operatorname{dtr}(d\tau\, {}^t K K).\end{aligned}$$

Therefore we see that $\Phi^*\omega$ is an $Sp(n,\mathbf{C})$-invariant closed 2-form on the tautological vector bundle.

We here give the $Sp(n,\mathbf{C})$-invariant Ξ in §1 explicitly. We define that 1-form Ξ by $\Xi = \operatorname{tr}(d\tau\, {}^t K K)$ on S^2_C. Since $\Phi^*\omega$ is an $Sp(n,\mathbf{C})$-invariant, we can show that Ξ is an $Sp(n,\mathbf{C})$-invariant holomorphic 1-form of V on Lag^C. Thus the subset of the solutions of the equation $\Xi = 0$ is C^*-invariant, hence we get a cone.

Since each point of $\mathbf{C}^{2n} \setminus \{0\}$ is a regular value of Φ, the inverse image W by Φ of a complex isotropic submanifold in $\mathbf{C}^{2n} \setminus \{0\}$ in $S^2_C \times \mathbf{C}^n$ is a complex submanifold in $S^2_C \times \mathbf{C}^n$.

Proof of Theorem 1.1. Let $\hat{\pi}$ be the natural projection $\hat{\pi} : S^2_C \times \mathbf{C}^n \to S^2_C$. We consider a point $p \in W$ which satisfies $\operatorname{rank}(\hat{\pi}|_p) = \max\big(\operatorname{rank}(\hat{\pi}|_W)\big)$. Then there exists some neighborhood at the point which is expressed by a holomorphic map $\phi : \mathcal{U} \times \mathcal{V} \to S^2_C \times \mathbf{C}^n$ defined by

$$\phi(z^1, ..., z^p, k^1, ..., k^q) = \big(\tau(z^1, ..., z^p), K(z^1, ..., z^p, k^1, ..., k^q)\big),$$

where $\mathcal{U}$ and $\mathcal{V}$ are neighborhoods of $0 \in \mathbf{C}^p$ and $0 \in \mathbf{C}^q$, respectively, and $(z^1, ..., z^p)$ and $(k^1, ..., k^q)$ are the canonical coordinate system of $\mathbf{C}^p$ and of $\mathbf{C}^q$, respectively, which satisfy $\phi(0,0) = p$.

If we calculate $\phi^*\Phi^*\omega$, we have

$$\begin{aligned}\phi^*\Phi^*\omega &= \operatorname{tr}\Big(\frac{\partial\tau}{\partial z^k}dz^k\Big) \wedge {}^t K\Big(\frac{\partial K}{\partial z^\ell}dz^\ell + \frac{\partial K}{\partial k^m}dk^m\Big) \\ &= \operatorname{tr}\Big(\frac{\partial\tau}{\partial z^k}\,{}^t K\frac{\partial K}{\partial z^\ell}\Big)dz^k \wedge dz^\ell + \operatorname{tr}\Big(\frac{\partial\tau}{\partial z^k}\,{}^t K\frac{\partial K}{\partial k^m}\Big)dz^k \wedge dk^m.\end{aligned}$$

Hence we find that $\phi^*\Phi^*\omega_1 = 0$ if and only if

$$\frac{\partial}{\partial z^k}\operatorname{tr}\Big(\frac{\partial\tau}{\partial z^\ell}\,{}^t KK\Big) - \frac{\partial}{\partial z^\ell}\operatorname{tr}\Big(\frac{\partial\tau}{\partial z^k}\,{}^t KK\Big) = 0, \quad \frac{\partial}{\partial k^m}\operatorname{tr}\Big(\frac{\partial\tau}{\partial z^k}\,{}^t KK\Big) = 0,$$

and that Ξ is the induced form of a holomorphic closed 1-form on $\mathcal{U}$. □

We are interested in the problem whether any complex isotropic submanifold is constructed in the above way.

Corollary 2.1. *Let M be a complex submanifold in S^2_C.*

(1) *For any $K \in \mathbf{C}^n$, the function $M \ni \tau \mapsto \mathrm{tr}(\tau\, {}^tKK) \in \mathbf{C}$ is holomorphic.*
(2) *For $K \in \mathbf{C}^n$ let $C_K(M)$ be the set of critical points of the function $M \ni \tau \mapsto \mathrm{tr}(\tau\, {}^tKK) \in \mathbf{C}$. The image $\Phi(\bigcup_{K \in \mathbf{C}^n} C_K(M))$ is a complex isotropic cone in $\mathbf{C}^n$.*
(3) *For each holomorphic function f on M, the image of the union $\bigcup_{K \in \mathbf{C}^n} C_K(M; f)$ of the set $C_K(M; f)$ of critical points of the function $M \ni \tau \mapsto \mathrm{tr}(\tau\, {}^tKK) + f(\tau) \in \mathbf{C}$ through Φ is a complex isotropic submanifold.*

Remark 2.1. In [7], the author studied about real generating functions more generally.

3. The Klein correspondence and the Lie transform

Let T be a complex isotropic subspace in $\mathbf{C}^{2n}$. Then complex Lagrangian subspaces containing T gives a locus in $Lag^C = U(n)\backslash Sp(n)$. For each $\tau \in S^2_C$, there exist a unique subspace E_τ in $\mathbf{C}^n$ so that the complex isotropic subspace T is a graph, that is, $\{(X, X\tau) \in \mathbf{C}^n | X \in E_\tau\} = T$, For $A \in S^2_C$ we set $S = \{(X, XA) \in \mathbf{C}^{2n} \mid X \in \mathbf{C}^n\}$, which is a complex Lagrangian subspace. Then S contains T if and only if $X(A - \tau) = 0$ for any $X \in E_\tau$. There exists a basis $A_1, ..., A_p \in S^2_C$ such that $XA_k = 0$ for any $X \in E_\tau$. The locus of S which includes T in S^2_C is given by $t_1A_1 + \cdots + t_pA_p + \tau$.

We say an affine subspace in S^2_C to be a *null subspace* if there exists $\tau \in S^2_C$ such that $X(A - A') = 0$ for an arbitrary $X \in E_\tau$ and for arbitrary A, A' in the affine subspace. For a null subspace, we only consider τ whose subspace E_τ is maximal. The dimension of E_τ is called the *nullity* of the null subspace.

Associated with this null space we get a complex isotropic subspace $\{(X, X\tau) \in \mathbf{C}^{2n} \mid X \in E_\tau\}$. Note that τ is independent of the choice of points of the null subspace. This correspondence between a null subspace and a complex isotropic subspace is called the *Klein correspondence*. An null space of *nullity* $= r$ corresponds to an isotropic subspace of dimension r. For example, a point of S^2_C is a null space of nullity$= n$ which corresponds to a complex Lagrangian subspace, and a null space of nullity$= 1$ corresponds to a complex line passing through 0 of $\mathbf{C}^{2n}$. Note that these notions are $Sp(n, \mathbf{C})$-invariant.

Let N be a complex submanifold in S^2_C. We call N a *null submanifold* of nullity r if each tangent space of N is a null space of nullity r.

For $\tau \in N$ and E_τ, we construct a complex vector bundle E of rank r on N. When we consider a set $\{(X, X\tau), X \in E_\tau\}$ as a fibre, each r-dimensional complex isotropic subspace in $\mathbf{C}^{2n}$ defines a subbundle of the trivial bundle $N \times \mathbf{C}^{2n}$. So we get a map $E \ni \big(\tau, (X, X\tau)\big) \mapsto (X, X\tau) \in \mathbf{C}^{2n}$. Let $(z^1, ..., z^p)$ be a local coordinate system of N and $(\xi_1, ..., \xi_r)$ is a local frame field of cross sections of E. The image of the map is locally given by $\big(k_1\xi_1 + \cdots + k_r\xi_r, (k_1\xi_1 + \cdots + k_r\xi_r)\tau\big)$. Since N is a null submanifold, we find

$$(k_1\xi_1 + \cdots + k_r\xi_r)\frac{\partial \tau}{\partial z_i} = 0$$

holds. Thus the basis of the image of the differential of that map of E into $\mathbf{C}^{2n}$ is given by

$$\left(\frac{\partial(k_1\xi_1 + \cdots + k_r\xi_r)}{\partial z_i}, \left(\frac{\partial(k_1\xi_1 + \cdots + k_r\xi_r)}{\partial z_i}\tau\right)\right), \quad 1 \leq i \leq p,$$

$$(\xi_j, \xi_j\tau), \qquad 1 \leq j \leq r,$$

which span a complex isotropic subspace. This complex isotropic subspace is included in the complex Lagrangian subspace with respect to τ. Thus E gives a complex isotropic cone. In particular, if ξ_i, $1 \leq i \leq p$ and their derivatives span $\mathbf{C}^n$, then E gives a complex Lagrangian cone whose Gauss image is N. By the same argument as in the §2, we find that the converse is also true. We call this correspondence *Lie transform*.

4. A complex Lagrangian cone in $\mathbf{C}^4$

In this section, we concretely construct examples of complex Lagrangian cones in $\mathbf{H}^2 = \mathbf{C}^4$. For two holomorphic maps F_1, $F_2 : \mathbf{C} \to \mathbf{C}^2$, we define a map $\mu : \mathbf{C}^2 \to \mathbf{C}^4$ by $\mu(z, k) = k(F_1(z), F_2(z))$, This map μ gives a complex Lagrangian cone if and only if $F_1\,{}^tF_2' - F_2\,{}^tF_1' = 0$. We set $F_1 = (f_1, f_2), F_2 = (f_3, f_4)$. Then each f_i is a holomorphic function on $\mathbf{C}$ with respect to z. The condition $F_1\,{}^tF_2' - F_2\,{}^tF_1' = 0$ turns to $f_1f_3' + f_2f_4' - f_3f_1' - f_4f_2' = 0$. By changing the parameter k to kf_1, we may set $(f_1, f_2) = (1, g)$. So we obtain $f_3' + gf_4' - f_4g' = 0$. If we put $f = f_3 + gf_4$, we have $f_3 = f - \frac{gf'}{2g'}, f_4 = \frac{f'}{2g'}$. This is the Bryant representation formula given in [2] on horizontal holomorphic curves with respect to the holomorphic contact structure on

$\mathbf{C}P^3$. We consider the local representation $(g,(1,g)H(z))$. Namely

$$k\big((1,g),(1,g)H(z)\big),\quad H(z)=\begin{pmatrix} f & -\dfrac{f'}{2g'} \\ -\dfrac{f'}{2g'} & \dfrac{f'}{gg'}\end{pmatrix}.$$

Let $\hat{H}(z)$ be the (tangential) Gauss map of the complex Lagrangian cone $k\big((1,g),(1,g)H(z)\big)$. Then we see that

$$\big((1,g),(1,g)H(z)\big)=\big((1,g),(1,g)\hat{H}(z)\big)$$

and

$$\big((0,g'),(0,g')H(z)\big)+\big(0,(1,g)H(z)'\big)=\big((0,g'),(0,g')\hat{H}(z)\big)$$

are satisfied. Thus we obtain

$$\begin{pmatrix}1 & g\\ 0 & g'\end{pmatrix}H(z)+\begin{pmatrix}0 & 0\\ 1 & g\end{pmatrix}H(z)'=\begin{pmatrix}1 & g\\ 0 & g'\end{pmatrix}\hat{H}(z),$$

which implies

$$\hat{H}(z)=H(z)+\frac{1}{g'}\begin{pmatrix}-g & -g^2\\ 1 & g\end{pmatrix}H(z)'.$$

Thus we get

$$\hat{H}(z)=\begin{pmatrix} f-\dfrac{gf'}{g'}+\dfrac{g^2(f''g'-f'g'')}{2g'^3} & \dfrac{f'}{2g'}-\dfrac{g(f''g'-f'g'')}{2g'^3} \\ \dfrac{f'}{2g'}-\dfrac{g(f''g'-f'g'')}{2g'^3} & \dfrac{f''g'-f'g''}{2g'^3}\end{pmatrix}.$$

The matrix valued function $\hat{H}(z)$ appears in the Darboux representation formula on a null curve in $\mathbf{C}^3$ (cf. [1]).

Next, we consider a complex Lagrangian graph $\big(z^1,z^2,\frac{\partial\phi}{\partial z^1},\frac{\partial\phi}{\partial z^2}\big)$. This complex Lagrangian graph is a cone if and only if the function ϕ satisfies the condition $t^2\phi(z^1,z^2)=\phi(tz^1,tz^2)$ for any $t\in\mathbf{C}$. If we set $\psi(w)=\phi(1,w)$, $\phi(z^1,z^2)=(z^1)^2\psi(\frac{z^2}{z^1})$, then its Gauss map is given by $\tau(w)=\mathrm{Hessian}\,\phi$. Here

$$\tau(w)=\begin{pmatrix}2\psi(w)-2w\psi'(w)+w^2\psi''(w) & \psi'(w)-w\psi''(w)\\ \psi'(w)-w\psi''(w) & \psi''(w)\end{pmatrix}.$$

We then get a null curve in S^2_C and a null vector $(1,w)$. If we set $g=w,\ f=2\psi$ in $\hat{H}(w)$, we obtain $\tau(w)=\hat{H}(w)$.

Let ξ be a branched minimal immersion of a Riemann surface M into $\mathbf{R}^3$. The Riemann surface M is an algebraic minimal surface if there exists a holomorphic null map $F = (F_1, F_2, F_3)$ of M into $\mathbf{C}^3$ such that $\xi = \mathrm{Re}F$. Here, we say F to be a null map if it satisfies $(F_1')^2 + (F_2')^2 + (F_2')^2 = 0$. It is well known that all simply connected minimal surface are algebraic. If we set $\hat{\tau}$ as

$$\hat{\tau} = \begin{pmatrix} F_1 - iF_2 & iF_3 \\ iF_3 & F_1 + iF_2 \end{pmatrix},$$

then it is a null curve in $S^2_C \subset Q^3$. This corresponds to the above construction (see [7]).

Remark 4.1. A null curve in Q^3 is related to a surface of constant mean curvature 1 in a hyperbolic space $H^3(-1)$ of sectional curvature -1 (see [5, 15]). A horizontal holomorphic curve in $\mathbf{C}P^3$ has a relation to a superminimal surface in the sphere $S^4(1)$ of sectional curvature 1 given in [2] and a flat surface in $H^3(-1)$ given in [15] (see [8]).

We shall consider a fractional linear transformation of a algebraic minimal surface in $\mathbf{C}^3$: For any $\begin{pmatrix} a & b \\ c & d \end{pmatrix} \in Sp(2, \mathbf{C})$, we define a fractional linear transformation of $\hat{\tau}$ by

$$\begin{aligned} &\begin{pmatrix} F_1 - iF_2 & iF_3 \\ iF_3 & F_1 + iF_2 \end{pmatrix} \begin{pmatrix} a & b \\ c & d \end{pmatrix} \\ &= \left(a + \begin{pmatrix} F_1 - iF_2 & iF_3 \\ iF_3 & F_1 + iF_2 \end{pmatrix} c\right)^{-1} \left(b + \begin{pmatrix} F_1 - iF_2 & iF_3 \\ iF_3 & F_1 + iF_2 \end{pmatrix} d\right). \end{aligned}$$

If we put $\psi(w) = w^3$, we obtain $\phi(z^1, z^2) = (z^2)^3/z^1$ and a null curve

$$\hat{\tau} = \begin{pmatrix} 2z^3 & -3z^2 \\ -3z^2 & 6z \end{pmatrix}$$

of genus 0, which is associated to the Enneper surface This is an immersed null curve of genus 0 and of degree 4 in Q^3 whose Lie transform is an immersed horizontal holomorphic curve of degree 3 ([1]). In the same paper [1], Bryant classified immersed null curves of genus 0 and of degree 4 and proved that the moduli space is $SO(5, \mathbf{C})/\rho_5(SL(2, \mathbf{C}))$. Note that $Sp(2, \mathbf{C})$ is the double cover of $SO(5, \mathbf{C})$. The corresponding complex Lagrangian cones are regular except 0. A null curve which is transverse at the infinity $Q^2 = Q^3 \setminus S^2_C$ gives a complete minimal surface in $\mathbf{R}^3$ with embedded flat ends. Its Gauss map is of degree 3 (see [1, 4]). We consider the following fractional linear transformation of the Enneper surface.

$$\begin{pmatrix} 2z^3 & -3z^2 \\ -3z^2 & 6z \end{pmatrix} \begin{pmatrix} -1/2 & 0 & -1 & 0 \\ 0 & 0 & 0 & -1 \\ 1 & 0 & 0 & 0 \\ 0 & 1 & 0 & 0 \end{pmatrix} = -\frac{1}{3z(z^3-1)} \begin{pmatrix} 6z & 3z^2 \\ 3z^2 & 2z^3 - 1/2 \end{pmatrix}.$$

Then we obtain a complete minimal surface with flat ends in $\mathbf{R}^3$ as follows

$$\mathrm{Re}\left(-\frac{2z^3+6z-1/2}{6z(z^3-1)},\ -i\frac{2z^3-6z-1/2}{6z(z^3-1)},\ i\frac{z}{z^3-1} \right).$$

Remark 4.2. The degree of the Gauss map of the Enneper surface is 1. This comes from the non-transversality of the mull curve at infinity [1].

5. Proof Theorem 1.2

In this section we give a proof of Theorem 1.2 by considering the deformation of a complex Lagrangian cone as an application of Theorem 1.1. We take a deformation of $\Xi = 0$ by $\beta \in H^{1,0}(M)$ if $\gamma = 2$, $n = 1$ and if M is a compact null surface in Q^3. By $Sp(2, \mathbf{C})$-action, we may consider that the null curve is locally given by the following form $\tau(z) = \begin{pmatrix} \tau_{11} & \tau_{12} \\ \tau_{12} & \tau_{22} \end{pmatrix} \in S^2_C$. Then it satisfies the differential equation $\tau'_{11}\tau'_{22} - \tau'^2_{12} = 0$, which is the null curve condition. We set $\beta = \alpha dz$ for a complex coordinate system z of M. Then $\Xi = \beta$ implies $k_1^2\tau'_{11} + 2k_1k_2\tau'_{12} + k_2^2\tau'_{22} = \alpha$. For a point such that τ'_{11} or τ'_{22} is different from 0, we obtain

$$k_1 = -\frac{k_2\tau'_{12}}{\tau'_{11}} + \frac{\sqrt{\alpha\tau'_{11}}}{\tau'_{11}}, \quad k_2 = -\frac{k_1\tau'_{12}}{\tau'_{22}} + \frac{\sqrt{\alpha\tau'_{22}}}{\tau'_{22}}$$

by $\tau'_{11}\tau'_{22} - \tau'^2_{12} = 0$. There are two ways of choosing of root. So the corresponding complex Lagrangian submanifold admits two representations.

$$\begin{aligned} C_1(k_1, z) &= k_1 \left(1,\ -\frac{\tau'_{12}}{\tau'_{22}},\ \tau_{11} - \frac{\tau'_{12}}{\tau'_{22}}\tau_{12}, \tau_{12} - \frac{\tau'_{12}}{\tau'_{22}}\tau_{22}\right) \\ &\quad + \left(0,\ \frac{\sqrt{\alpha\tau'_{22}}}{\tau'_{22}},\ \frac{\sqrt{\alpha\tau'_{22}}}{\tau'_{22}}\tau_{12},\ \frac{\sqrt{\alpha\tau'_{22}}}{\tau'_{22}}\tau_{22}\right), \\ C_2(k_2, z) &= k_2 \left(-\frac{\tau'_{12}}{\tau'_{11}},\ 1,\ \tau_{12} - \frac{\tau'_{12}}{\tau'_{11}}\tau_{11},\ \tau_{22} - \frac{\tau'_{12}}{\tau'_{11}}\tau_{12}\right) \\ &\quad + \left(\frac{\sqrt{\alpha\tau'_{11}}}{\tau'_{11}}, 0,\ \frac{\sqrt{\alpha\tau'_{11}}}{\tau'_{11}}\tau_{11},\ \frac{\sqrt{\alpha\tau'_{11}}}{\tau'_{11}}\tau_{12}\right). \end{aligned}$$

In each of these representations, the first part gives the complex Lagrangian cone, and the second part shows a perturbation. The set of points satisfying $\Xi = \beta$ on $V|_M$ consists of sets of two affine lines parallel to the line of the cone on a fibre of M except zero set of β, the line of the cone on zero set of β. This may have singularities. Furthermore, the number of irreducible components are 1 or 2.

Assume that M is immersed. Then τ'_{11} and τ'_{22} are not simultaneously 0 by $\tau'_{11}\tau'_{22} - \tau'^2_{12} = 0$. Then we can use one of two representations $C_1(k_1, z)$ and $C_2(k_2, z)$. The condition $\tau'_{11}\tau'_{22} - \tau'^2_{12} = 0$ implies

$$\begin{aligned}\frac{\partial C_1(k_1,z)}{\partial z} &= k_1\left(0,\ 0,\ \tau'_{11} - \frac{\tau'_{12}}{\tau'_{22}}\tau'_{12},\ \tau'_{12} - \frac{\tau'_{12}}{\tau'_{22}}\tau'_{22}\right) \\ &\quad + k_1\left(-\frac{\tau'_{12}}{\tau'_{22}}\right)'(0,\ 1,\ \tau_{12},\ \tau_{22}) \\ &\quad + \left(\frac{\sqrt{\alpha\tau'_{22}}}{\tau'_{22}}\right)'(0,\ 1,\ \tau_{12},\ \tau_{22}) + \left(\frac{\sqrt{\alpha\tau'_{22}}}{\tau'_{22}}\right)(0,\ 0,\ \tau'_{12},\ \tau'_{22}) \\ &= k_1\left(-\frac{\tau'_{12}}{\tau'_{22}}\right)'(0,\ 1,\ \tau_{12},\ \tau_{22}) \\ &\quad + \left(\frac{\sqrt{\alpha\tau'_{22}}}{\tau'_{22}}\right)'(0,\ 1,\ \tau_{12},\ \tau_{22}) + \left(\frac{\sqrt{\alpha\tau'_{22}}}{\tau'_{22}}\right)(0,\ 0,\ \tau'_{12},\ \tau'_{22}).\end{aligned}$$

So if α has no zero points, then $\dfrac{\partial C_1(k_1,z)}{\partial z}$ and $\dfrac{\partial C_1(k_1,z)}{\partial k_1}$ are linearly independent and hence the obtained ruled submanifold is regular.

In his paper [2], Bryant proved that the Lie transform of an immersed null curve of genus 1 is a holomorphic horizontal curve with a singularity. So the corresponding complex Lagrangian cone has a singularity.

If M is a branched null curve of genus 1, then there are no points satisfying $\Xi = \beta$ on the fibre at a branched point. The obtained ruled submanifold is regular, however, it is not a asymptotical conical complex Lagrangian submanifold with rate 0. This completes the proof of Theorem 1.2

By Corollary 2.1, we get the following.

Corollary 5.1. *Let M be a complex submanifold in $S^2_{\mathbb{C}}$ and f be a holomorphic function on M. If the union $\bigcup_{K\in\mathbb{C}^n} C_K(M)$ of the sets of critical points of $\tau \mapsto \mathrm{tr}\tau^t KK$ gives a complex Lagrangian cone, then $(1-t)\mathrm{tr}\tau^t KK + tf$, $0 \le t \le 1$ induces a deformation of the complex Lagrangian cone, through ruled complex Lagrangian submanifolds*

$\bigcup_{K \in \mathbf{C}^n} C_K(M; \frac{t}{1-t} f)$ *with singularity, to some complex Lagrangian subspace through* 0 *which corresponds to the critical points of* f.

Acknowledgements

The author is grateful to Professor H. Hashimoto for giving him valuable advice.

Bibliography

1. R. L. Bryant, *Surfaces in conformal geometry*, Proceedings of Symposia in Pure Mathematics 48(1988), 227–240.
2. R. L. Bryant, *Conformal and minimal immersions of compact surfaces into the* 4*-sphere*, J. Diff. Geom. 17(1982), 455–473.
3. R. L. Bryant, *Second order families of special Lagrangian* 3*-folds*, in *Perspective in Riemannian geometry*, Inv. Apostlov, A. Dancer, N. J. Hitchin, and M. Wang, eds., CRM Proceedings and Lecture Notes, vol 40, 63–98 A.M.S Providence, RI, 2006. math. DG/0007128.
4. R. L. Bryant, *A duality theorem for Willmore surfaces*, J. Diff. Geom. 20(1984), 23–53.
5. R. L. Bryant, *Surfaces of mean curvature one in hyperbolic space*, Astérisque 154–155(1987), 321–347.
6. G. Darboux, *Lecons sur la Theéorie générale des Surfaces*, Livre III, Grauthier-Villars, Paris, (1894).
7. N. Ejiri, *A generating functin of a complex Lagrangian cone in* $\mathbf{H}^n$, preprint.
8. N. Ejiri and M. Takahashi, *The Lie transform between null curves in* $SL(2,\mathbf{C})$ *and contact curves in* $PSL(2,\mathbf{C})$, in *Riemann surfaces, harmonic maps and visualization*, 265–277, OCAMI Stud., 3 Osaka Munic. Univ Press, Osaka, 2010.
9. N. Ejiri and K. Tsukada, *A remark on complex Lagrangian cones in* $\mathbf{H}^n$, in *Recent Progress in Differential Geometry and its Related Fields*, T. Adachi, H. Hashimoto and M.J. Hristov eds., World Scientific, Singapore, 2011.
10. J. A. Gálvez, A. Martínez and F. Milán, *Flat surfaces in hyperbolic* 3*-space*, Math. Ann. 316(2000), 419–435.
11. N. J. Hitchin, *The moduli space of complex Lagrangian submanifolds*, Asian J. Math. 3(1999), 77–91.
12. R. Harvey and J. B. Lawson, *Calibrated geometries*, Acta Math. 148(1982), 47–157.
13. D. D. Joyce, *Riemannian holonomy groups and calibrated geometry*, Oxford University Press, 2007.
14. D. D. Joyce, Ruled Special Lagrangian 3-Folds in $\mathbf{C}^3$, Proc. London Math. Soc. 85(2002), 233–256.
15. F. Martin, M. Umehara and K. Yamada, *Complete bounded null curves immersed in* $\mathbf{C}^3$ *and* $SL(2,\mathbf{C})$, Calc. Var. Partial Differential Equations 36(2009), 119–139.

16. G. P. Pirola, *Algebraic curves and non rigid minimal surfaces in the Euclidean space*, Pacific J. Math. 183(1998), 333–357.
17. A. Small, *On algebraic minimal surfaces in* $\mathbf{R}^3$ *deriving from charge* 2 *Monopole spectral curves*, Inter. J. Math. 16(2005), 173–180.

Received December 25, 2012
Revised May 31, 2013

POSTFACE

St. Cyril and St. Methodius University of Veliko Tarnovo in Bulgaria celebrates its 50th anniversary on May 11, 2013. Its main buildings stand on Sveta Gora Hill, which used to be the cultural and spiritual center of Southeast Europe in 12th – 14th centuries. We here recall its history briefly. On September 27, 1962, the Higher Pedagogical Institute was established in Veliko Tarnovo, and the first academic year started in the autumn of 1963. On October 13, 1971, the Cyril and Methodius Higher Pedagogical Institute was officially declared the second Bulgarian university. When we talk about University of Veliko Tarnovo, we cannot forget the names of the saintly Thessalonian brothers Constantine–Cyril and Methodius who are known as creators of Old Bulgarian alphabet, the Glagolitic alphabet, which was refined to the Cyrillic alphabet later on. To preach orthodox Christianity in Moravian mother tongue, in 863 they started for Moravia accompanied by their disciples and carried with them the first liturgical books translated from Greek into Old Bulgarian. With a historical and spiritual connection to Tarnovo Literary School established in 1371, the university has quite deep backgrounds compared with its history as a university.

On the occasion of the 50th anniversary, we would like to express our sincere congratulations. St. Cyril and St. Methodius University of Veliko Tarnovo and Nagoya Institute of Technology concluded an academic exchange program in 2008. Since then we have held the International Colloquium on Differential Geometry and its Related Fields for 3 times. This colloquium was started to provide opportunities for mathematicians from both western and eastern countries to exchange ideas and discuss mathematics quite freely. Mathematics has both the spiritual aspect in creating a mathematical world and the literal aspect as a basic language of natural sciences. In this sense we may say that our program accords with the place. We hope we would be able to give an address on the occasion of the centennial of our program in future with the hopes that it would be developed to the highest degree with the interchange of scientists between Europe and Asia and this would be a new wave not only in the area of mathematics but also in all fields of science. We would like to express our cordial wishes for a rich and flourishing future of the University of Veliko Tarnovo.

11 May, 2013

on behalf of the Japanese editors
Toshiaki ADACHI